The Pontecorvo Affair

The Pontecorvo Affair
A Cold War Defection and Nuclear Physics

Simone Turchetti

The University of Chicago Press :: Chicago and London

Translated and revised by the author from his own work, *Il caso Pontecorvo: Fisica nucleare, politica e servizi di sicurezza nella guerra fredda*, published by Sironi Editore (2007).

Simone Turchetti is an independent research fellow at the Centre for the History of Science, Technology, and Medicine at the University of Manchester.

The University of Chicago Press, Chicago 60637
The University of Chicago Press, Ltd., London
© 2012 by The University of Chicago
All rights reserved. Published 2012.
Printed in the United States of America

21 20 19 18 17 16 15 14 13 12   1 2 3 4 5

ISBN-13: 978-0-226-81664-7 (CLOTH)
ISBN-10: 0-226-81664-8 (CLOTH)

Library of Congress Cataloging-in-Publication Data

Turchetti, Simone.
 [Caso Pontecorvo. English]
 The Pontecorvo affair : a cold war defection and nuclear physics / Simone Turchetti.
   p. cm.
 Includes bibliographical references and index.
 ISBN-13: 978-0-226-81664-7 (cloth : alk. paper)
 ISBN-10: 0-226-81664-8 (cloth : alk. paper)
   1. Pontecorvo, B. (Bruno), 1913–1993. 2. Nuclear physicists—Soviet Union—Biography. 3. Nuclear physicists—Italy—Biography. 4. Spies—Soviet Union—Biography. 5. Spies—Italy—Biography.
 I. Title.
 QC774.P66T8713 2012
 530.092—dc23
 [B]

2011019135

♾ This paper meets the requirements of ANSI/NISO Z39.48-1992 (Permanence of Paper).

Contents

	Introduction: The Silent Quake	1
1	The Training of a Nuclear Physicist	13
2	Neutrons for Peace and Neutrons for War	39
3	Under Surveillance	69
4	Ten Million Reasons to Disappear	95
5	Play It Up or Down? Confronting the Pontecorvo Affair	119
6	A Political Motive	153
7	Bruno Maximovich and Professor Pontecorvo	179
8	Conclusions: The Noisy Echo of Secrecy	205
	List of Abbreviations	221
	Notes	223
	Bibliography	263
	Index	279

Introduction: The Silent Quake

The nuclear age is permeated by an ever-present fear of secret scientific information being given away. From early studies on the atom bomb to more recent investigations on "rogue" states like Iran and North Korea harnessing nuclear weapons, threats deriving from the unwanted spread of restricted scientific knowledge have been perceived as a major challenge to national and international security. And the mobility of foreign researchers has been looked upon with growing concern, because it could lead to the transfer of valuable information and expertise.

During the second half of the twentieth century, the risk of unwanted dissemination of scientific information has informed the search for instruments of national and international control. Government officials of states that possess nuclear technologies, like the United States and Britain, have counterpoised this threat by establishing security networks, whose aim is to control and prevent the transfer of technology. Their work has often been publicly praised by politicians as an astounding example of efficient policymaking against individuals, nations, and terrorist organizations that may be interested in unlawfully obtaining data, expertise, and instruments.

Yet—however incongruous it may seem—this was not the case when at the dawn of the nuclear age an Italian-born physicist defected to Russia carrying with him precious skills and knowledge in the atomic field. On 21 October 1950, British newspapers reported that the scientist Bruno Pontecorvo and his family had mysteriously disappeared in Finland during a holiday trip. The news created concern in Britain and abroad owing to the fact that the physicist was known to be a nuclear expert working for the UK Atomic Energy Research Establishment (AERE) based at Harwell. In the following months, more information was gathered about Pontecorvo's location, and it was understood that he had crossed the Iron Curtain. The scientist was now in Soviet Russia.

The defection came in the middle of a very tense period in postwar history: one year after the Soviet Union's first nuclear test; a few months after the beginning of US Senator Joseph R. McCarthy's campaign against the communist fifth column; and days before the intensification of hostilities between the US Army and the North Korean Communist troops in Asia. It also overlapped well-known cases of atomic espionage. It followed the condemnation, in 1946, of the British scientist Alan Nunn May and the arrest, in February 1950, of his German-born colleague Klaus Fuchs; both were charged with giving away classified scientific information to Soviet agents. It paralleled the FBI investigations that led to the conviction of American experts Harry Gold and David Greenglass. And it anticipated the tragic conclusion of that enquiry, in 1953, which saw Ethel (Greenglass's sister) and Julius Rosenberg executed in the electric chair.

As the defection occurred at the crossroads of important historical events intensifying the Cold War and its spy games, it also raised legitimate questions about Pontecorvo: Was he in possession of classified scientific information? Did he pass secret details on the atom bomb to Soviet Union experts? Was he a spy? Was he a communist? Was he ostracized because of the witch hunts?

Puzzled nuclear scientists considered the affair's connotations. Baffled security officers assessed it. Government officials prepared reports for their ministers. Yet none of these reviews proved satisfactory. In fact, they seemed more like attempts to close the investigation quickly rather than fully clarify its implications. If the defection promised to create an international case, especially in light of the mounting atomic espionage scandal, then it only generated a trickle of journalistic speculations. On the whole, the Pontecorvo affair was typified by the sound of its silence.

The resonance of future historical accounts was to be no louder. In the last fifty years, various accounts of Pontecorvo's flight have appeared in newspaper articles, official parliamentary enquiries, scientists' recollections, scientific biographies, books on espionage, and history monographs. However, as this academic and nonacademic literature proliferated, it hardly ever increased the amount of novel or original information. In fact, it almost always reiterated claims made earlier when the scientist had mysteriously disappeared. But what *exactly* was the impact of the Pontecorvo affair? And what place did the episode occupy in the historical trajectory of the Cold War? How does it map onto the wider history of atomic espionage? This book reveals that notwithstanding the paucity of public resonance, the affair had a knock-on effect on scientific and international relations. Its unfolding implied the mobilization of atomic scientists, science administrators, diplomats, security officers and antinuclear campaigners in a number of countries including the United States, Britain, the Soviet Union, and Italy.

Bruno Pontecorvo was the *first* and *only* scientist employed in wartime nuclear projects who in the 1950s crossed the Iron Curtain. An expert with vast experience and well-known skills in pure and applied nuclear physics, his participation in these projects made him one of the few scientists in the world aware of processes, practices, and techniques used in the production of fissile materials for atomic weapons. His disappearance marked a frightening moment for the British diplomatic and intelligence communities because of accusations about lax security measures in the management of atomic matters. US government personnel handed the case with care too, partly because Pontecorvo was a former collaborator of Enrico Fermi and Emilio Segrè, two very prominent atomic scientists in the wartime US nuclear program.

Scientific and diplomatic relations between Western states and between Cold War blocs were deeply affected by the unsettling event. From Helsinki, where Pontecorvo mysteriously disappeared, the affair created a seismic wave that could be clearly detected by those in the know. It propagated in the "corridors of power" at the British House of Commons, the Italian Parlamento, and the US Congress. It shook Scotland Yard and created unrest in the FBI headquarters. It caused havoc at the diplomatic offices and embassies of Washington and London. It reverberated in many distant places: it was clearly felt in Tulsa, Oklahoma, where former colleagues of the Italian physicist were victimized by US federal agents, and at the University of Liverpool and the AERE, where his superiors hopelessly waited for his comeback. It hit Berkeley, California, and also Chicago, where Segrè and Fermi worried about the repercussions of Pontecorvo's mysterious departure. Tremors could be registered even months later—for instance, at Lake Como in Italy, where British journalists pretended to be covert agents in order to get fresh news, or in the Kremlin, where allegedly meetings were arranged to have Pontecorvo's knowledge probed by Soviet experts. Remote places such as the Russian-Chinese border, where intelligence sources claimed that Pontecorvo was prospecting uranium, were also affected by the seismic event.

It is legitimate to question why, if the Pontecorvo case was so disruptive (catastrophic, actually) in national and international affairs, so little has surfaced until now in journalistic accounts and historical studies on its significance. The answer lies in the deceptive activities of those concerned about its impact. Diplomats, security officers, and government managers in Britain and the United States contributed to offer uniform and standardized versions of facts that could be used to accommodate vested national political and economic agendas. As the Pontecorvo case threatened these ambitions, it was quickly removed from the public eye in an attempt to silence criticism and ensure the optimization of more fictile explanations.

The priority of these political networks was to "let sleeping dogs lie" (as one British diplomat put it at the time) rather than voice concerns publicly; to quickly close the investigation rather than open inquiries; and to avoid overemphasizing the significance of Pontecorvo's research activities. Clearly, not only did these attempts limit and distort the evidence available to the journalists who covered the case; they also affected future reconstructions, thus diminishing the affair's historical relevance in future years.

Moreover, the absence of documentation regarding the defection helped journalists and commentators to fabricate new accounts of Pontecorvo's activities. When these stories appeared in the press and in literature on the subject, they confused the public further by suggesting that the scientist had been a spy for a long time. Yet they failed to substantiate the claim with factual evidence. As a consequence, Pontecorvo's flight to the Soviet Union spurred the manufacture of two distinct literary characters: the brilliant scientist and the notorious spy. These incongruous portraits have lived side by side for nearly fifty years.

This volume goes a long way in providing a better understanding of the case, reconciling these diverging portraits by using archival documentation from private and public repositories in several countries. Yet the book can also be understood as an attempt to show how much we still have to learn about the affair and its importance in the history of the Cold War. The vast amount of archival information disclosed here is accompanied by evidence that much classified data still waits to be made available.

Access to some previously classified documentation has been granted only through appeals against its current security status.[1] Other documents, such as some of Pontecorvo's papers at the Churchill College Archive, University of Cambridge, will not be available to historians before 2024. Moreover, the whole episode of Pontecorvo's defection can be read exclusively through recently declassified documents made available in the British National Archives. Nothing has been disclosed by Russian or Italian archival centers that sheds any light on the ways in which Pontecorvo's flight was planned and executed; much less why it occurred. Key documentation, such as some papers of Emilio Segrè (a key player in the affair) and the FBI file on Pontecorvo, is still inaccessible.

This inaccessibility can be understood as another indicator of the magnitude of this case in Cold War history, as its resonance continues to influence disclosure procedures even now. This is not just because of the factual knowledge contained in the archival papers, but also because of the ways in which the affair tarnished national and international affairs.

Unfortunately, this also means that this volume cannot provide a conclusive reconstruction of Pontecorvo's defection. In some important aspects, this episode is still a mystery. Nevertheless, the evidence disclosed here helps in approximating its solution and, at the same time, seeks to use the Pontecorvo affair to open to debate our present understanding of the relation between nuclear science, security, and politics in the early days of the Cold War.

Confronting Cold War History

Looking at Pontecorvo's career helps us to reconsider our understanding of important transitions in the history of nuclear science and technology. Major historical works on the history of the atom bomb project have emphasized the role of secrecy in the management of nuclear studies.[2] Thus, the completion of rival nuclear programs, such as the Soviet one, is understood as little more than a copycat affair in which the transfer or theft of scientific information was much more relevant than in-house research and expertise.[3] However, as technical aspects of the Manhattan Project and other nuclear research endeavors have been more thoroughly explored, it has emerged that key discoveries were made independently in different places.[4] By contrast, the very possibility of simply replicating any kind of scientific and technological endeavor, including nuclear experimental practices, exclusively on the basis of written information, raw formulas, or blueprints has been put into question.[5] Recent historical work has also drawn attention to the origins of the notion of "atomic secrets" in the United States, revealing that this was a journalistic myth produced in the 1950s.[6] And despite the heavy past investment in protecting these secrets, it is now clear that there is no inner logic to the use of secrecy in science. This invites us to rethink its use and efficacy.[7]

This book adds to these historical revisions. It presents fresh evidence suggesting that the role played by the unlawful transfer of restricted scientific information to Soviet Russia has perhaps been exaggerated while the defection of a nuclear scientist who played a key role in wartime nuclear projects, Bruno Pontecorvo, has been overlooked. It calls into question the empirical evidence supporting previous claims by disclosing Pontecorvo's contributions to the advancement of nuclear science and technology before and after the defection. In so doing, it also disputes the validity of the interpretative framework used in the past by suggesting that Pontecorvo's expertise was much more relevant than any written scientific documentation that could have been provided to Soviet Russia during the first decade of the Cold War.

This new interpretation is also conducive to a fresh perspective on the role played by security in the handling of atomic espionage cases. In the light of inquiries that anticipated the defection of Bruno Pontecorvo, it appears that at no time did intelligence personnel have a clear understanding of the nature of atomic knowledge and how it could be protected. This ignorance matched the more general lack of a specific intelligence program to tackle atomic espionage, and the failure to implement such a program immediately. Because of this failure, critical evidence was harnessed in circumstances that had nothing to do with investigations on suspected spying activities in the field of atomic research. And the evidence was eventually re-used after the defection even when it dealt with non-related issues. "Recycling" run-of-the-mill investigative elements helped the intelligence services to demonstrate their ability to deal with the problem of nuclear security when in fact they had not made it a strategic priority or set up suitable tools to engage with it.

Moreover, defective and biased security documentation played a part in the chain of events that led Pontecorvo to decide to leave the West, thereby further damaging nuclear security rather than strengthening it. After his defection, any future investigation was made conditional upon diplomatic and political urgencies, especially so far as political agreements on nuclear cooperation and relations between intelligence agencies were concerned. Thus, the silence that followed Pontecorvo's flight in the face of public criticism served to protect the knowledge that a secret agenda, rather than nuclear security, had played a crucial role in the management of investigations.

The new evidence about Pontecorvo's defection helps to recast the problem of atomic espionage and intelligence within a new historiographical framework. Historical accounts of intelligence organizations have far too often offered a reassuring picture of their work—one that has relied far too heavily on archival documentation, and has therefore offered the views of the protagonists uncritically. Even when these studies have been accurate, they have not examined how the agents confronted the theft or transfer of something so elusive as scientific (or even atomic) information. Scientific knowledge cannot be smuggled in plastic bags. Of course it can be written on paper or embodied in instruments and products, but even so, its real meaning is clear only to those who have the expertise to understand it.

Yet this was not seen as problematic by security agencies, nor has it been viewed in that way by historians. Atomic secrets existed; therefore, what mattered was to recount how intelligence networks in the West had attempted to protect them[8]—or, conversely, how Soviet spies had successfully got hold of the secrets.[9] Agents were portrayed either as heroes attempting to "defend the realm,"[10] or as villains whom the enemy had succeeded in corrupting.[11]

Only recently has a more nuanced understanding begun to emerge. For instance, the intelligence services' lack of preparation and knowledge on atomic matters has now been more clearly understood.[12] The fact that inquirers could be prejudiced in considering the potential suspects has also been revealed.[13] The "bipolar" setting has been considered more simplistic than exhaustive. If there is no doubt that the Cold War fueled the confrontation between spies of opposing blocs, the divergence between American and British intelligence agencies was marked as well, thus defining differences in their methods of action and strategies.[14] The importance of international diplomacy in directing the "secret hand" has also come to the fore. For instance, the atom spy cases were used by diplomats to spark political decision-making at key moments during the early Cold War years. Revealingly, some wartime international agreements allowed the allies to share information, even on scientific matters. The attempt to prosecute those scientists who favored such exchanges could thus be read as being resonant with geopolitical interests that the exploitation of the atom defined.[15]

Clearly, this is not to say that atomic espionage never existed. But we are moving forward from a somewhat distorted image of the ways in which it was dealt with, in terms of understanding its key scientific and security aspects. This volume goes in the direction of recasting the problem in comparative analysis of received wisdom and in the portrayal sketched in newly released archival records and recollections of the people involved. One element that emerges from this review is that to better understand the interplay of nuclear science, politics and security during the Cold War in general and the Pontecorvo affair more specifically, we have to look into the vested interests of the actors involved. For instance, hidden economic aspirations played a crucial role in the events that culminated in Pontecorvo's flight. These interests centered especially on the processes leading towards the exploitation of atomic energy, and ultimately exposed the many economic relations that typified nuclear physics before, during, and after World War II. Our current understanding of the origins of nuclear physics especially emphasizes the quest for knowledge about the intimate structure of nature, atoms, and particles as the catalyst of scientific change.[16] The urgency to apply nuclear knowledge to practical problems is often surmised, though it appears in connection with wartime research and the atom bomb.

Yet practical and economic concerns existed before the beginning of the war, and the economy of patents is especially revealing about how those concerns informed research directions and goals. Some historical studies have shown the role played by economic urgencies and patenting activities in the research strategies of prominent nuclear physicists.[17] Yet these have been neglected in the historical literature concerning Fermi and his group, including Pontecorvo, despite the existence of biographical material that refers to it directly.[18] We now know not only how important these patenting activities were in Pontecorvo's education as a scientist, but also how they kept Fermi and his coworkers united when they took different life and career paths by migrating abroad. One patent in particular, the "slow neutron patent," was at the center of a controversy in the postwar years between the former members of Fermi's group and the US Atomic Energy Commission (AEC). The evolution of this conflict overlapped Pontecorvo's defection in interesting and surprising ways.

Pontecorvo's flight is also revealing of a hidden political agenda. The defection was planned and executed by prominent peace campaigners in Italy who sought to use it to denounce the militaristic uses of the atom in the West. In this way they also sought to expose the dissatisfaction and disillusionment of Western scientists concerning these applications, and to convince them to be more vocal about it. These activists were thus responsible for initiating contacts with Russian leaders, using financial and logistic resources made available to them, and suggesting to Pontecorvo that he defect. This volume goes beyond our current understanding of Cold War relations between antinuclear organizations and Soviet leaders by showing the existence of a secret network that sought to control and direct the action of Western pacifists more deeply than has previously been assumed.[19]

Account of an "Affair"

This is not a biography of Bruno Pontecorvo. It does not attempt to establish a rigorous chronological setting for the events that defined his life or his career achievements. It is skewed towards a key life-changing event, using it to portray an historical period dense in scientific and political change. Unlike most scientific biographies, it does not seek to make of its subject a scientific hero or a champion of antinuclear campaigns. In fact, if it finds motives and justifications for Pontecorvo's decision to leave the West, it also highlights the disillusionment and compromises that typified his subsequent life as the Soviet citizen and scientist Bruno *Maximovich* Pontecorvo—thereby calling into question the soundness of his life-changing decision.

This work cannot be assimilated into previous biographical studies or recollections seeking to portray Pontecorvo's life and career achievements.[20] Other important biographies have recently illustrated the ethical profiles of prominent atomic scientists, especially in the effort to explain their life-making decisions.[21] My study of Pontecorvo offers a very different picture, because his decision to leave the West was based on contingent factors rather than on a solid moral or ethical foundation. This book shares much more with recent efforts to reconstruct the shaping of individual scientists in the web of their social and scientific relations; a genre recently defined as a "sociological biography."[22] However, it recasts this type of narrative in a completely different setting. Rather than explaining how the identities of illustrious atomic scientists have molded their social imagery because of the circumstances of the nuclear age, this volume shows how Pontecorvo's circumstances in the nuclear age have had the power to fragment and reshape his identity in the public arena. In particular, his defection was the event defining the emergence of two diametrically opposed characters: the notorious spy and the scientist concerned about the military uses of atomic energy.

The decision to write the history of an *affair*, however, manifests the intention to experiment with a type of historical narrative that differs from biographies. The term *affair* is etymologically neutral, indicating either an event or a connected series of events. Yet its application to historical studies has often coincided with the need to emphasize grievances attached to one's behavior or controversial circumstances that cannot be easily discerned. Political historians have used it to recount the scandal of French Captain Alfred Dreyfus, accused of treachery at the turn of the twentieth century. Historians of science have used it in connection with the seventeenth-century trials against the Italian astronomer Galileo Galilei, thereby emphasizing problematic assumptions about science and religion.[23] Controversial interpretations about the politicization of science in totalitarian regimes have also been labeled as "affairs." This is the case for the Russian agronomist Trofim Lysenko, whose plant breeding methods followed Stalin's dictates on dialectical materialism.[24] More recently, the term has been used to illustrate the hoax perpetuated in 1996 by French physicist Alan Sokal, who sought to reveal the fallacies of postmodern sociological analysis when applied to the study of science and technology.[25]

Yet, with the possible exception of some strands of literature on the Galileo affair, the merits of focusing on affairs in historical studies have very rarely been considered. The term *affair* is used here to underscore the discontinuous nature of historical change, and its condensation around catalyzing events that bring out the essence of an historical period and exemplify its tensions and conflicts. Albeit secretly administered, the Pontecorvo affair was in fact symptomatic of a malady that typified the Cold War period insofar as political and scientific circles felt less certain about the nature of security threats associated with the uses of atomic energy, and felt themselves unable to counteract them. As a consequence of this insecurity, public trust in scientific and political institutions weakened. Use of the term *affair* in connection with Pontecorvo's flight also shows the positioning of the historian in relation to the subject matter. The unfolding of an affair presents a plurality of actors with conflicting viewpoints and aspirations, which the historian seeks to represent in the context of historical inquiry.

Chapter Layout and "Version 2.0"

This study promotes a better understanding of the Pontecorvo affair through an analysis of the scientist's research interests and the circumstances that led to his departure in 1950. In particular, it looks at Pontecorvo's life and career in combination with developments in nuclear science and technology, changes in the setting up of security regulations and practices, and the existence of nuclear policies and strategies in national and international context.

It begins in chapter 1 with an examination of Pontecorvo's education and early training as a scientist, focusing especially on the role played in his formation by practical applications and patents deriving from the nascent research in nuclear physics. Chapter 2 reveals Pontecorvo's pioneering contributions to wartime research in nuclear physics, showing how they propelled industrial and military endeavors. Chapter 3 analyzes the shaping of nuclear security during and after the war, and how it affected Pontecorvo's life and career before his defection. It emphasizes the controversial nature of investigative work and how the action of intelligence services was a key factor in Pontecorvo's decision to leave the West. As chapter 4 shows, this decision also stemmed from his growing concerns about proceedings against the AEC in relation to the slow neutron patent.

Following Pontecorvo's flight, the inquiries on the defection were disappointing as nothing decisive emerged regarding his motives or accomplices. Journalists speculated wildly, looking for spies paid by the Soviets. In the meantime, British diplomats and intelligence personnel sought to play down the case to comply with their agendas. Chapter 5 reveals for the first time the vested interests behind the actions of those who were responsible for investigating the affair.

The connections between peace campaigners and Soviet leaders that made it possible for Pontecorvo to leave the West are the subject of chapter 6. The final chapter draws a comparison between Pontecorvo's activities before and after the defection, showing how his presence in Russia was far more decisive than any information that Soviet intelligence had managed to harness earlier on. The chapter also considers the proliferation of accounts about Pontecorvo, questioning their origins and reliability.

This manuscript was written following the approval of a project to translate a book in Italian on the same subject.[26] However, while completing the translation, I have reviewed the content as well. In many ways, therefore, this is a "version 2.0" of my previously published volume. I believe that I have managed to better explain some key transitions, partly thanks to recently released archival material. This is the case in the examination of Pontecorvo's political profile during his residence in Paris, his employment in Tube Alloys, the disclosure of his communist relations at Harwell, the role played by Segrè, and the attempts by MI5 to make publicly available a version of facts that was consistent with its own agenda.

Finally, the book contains quoted passages that have been translated from their original languages. The original quotation is either in the text or in the related note and, unless otherwise noted, all translations are my own.

Acknowledgments

This work was made possible thanks to the help of many people. I'd like to thank all the archivists who have been so helpful (and patient!) during my visits in repositories. In particular I'd like to acknowledge the help of archivists at the British, US, and Italian national archives; at the University of Chicago; and at the Fondazione Gramsci of Rome. In particular I'd like to thank Stephen Twigge, Margherita Martelli, and Allen Packwood. I'd also like to thank all those who have offered their recollections about the Pontecorvo case or other events described in the volume: in particular, Mary Scherbatskoy, Clara Sereni, Chapman Pincher, Bennett Boskey, and Boris Yerozolimski. I'd like to acknowledge the help of the Friends of the Center for the History of Physics at the American Institute of Physics in Maryland, who offered a travel grant for archival research in the United States.

There is a long list of colleagues who have helped me with questions, encouragement, criticism, and ideas. To all of them goes a sincere "Thank you!" The list includes my current and former colleagues at the University of Manchester: Jeff Hughes, Jon Agar, Michael Worboys, John Pickstone, Abigail Woods, Carsten Timmermann, Aya Homei and James Sumner. I'd also like to thank colleagues in the United States: John Heilbron, Angela Creager, Spencer Weart, Shawn Mullet, David Kaiser, Ronald Doel, Ian Slater, and Alex Wellerstein; in France: Dominique Pestre, Sébastien Soubiran, Soraya Boudia, and Marion Thomas; in Italy: Elena Gagliasso, Anna Guagnini, Giuliano Pancaldi, Pasquale Tucci, Paola Bertucci, Luisa Bonolis, and Lanfranco Belloni; in Spain: Nestor Hérran and Xavier Roque; in Britain: Christopher Laucht, Michael Goodman, Tim Gibbs, Greg Radick, Graeme Gooday, Martin Siegert, Simon Naylor, and Katrina Dean.

This work is the result of many revisions suggested by editors. I'd like therefore to thank Karen Merikangas Darling and Renaldo Migaldi at the University of Chicago Press and Martha Fabbri at Sironi Editore for their valuable assistance. Thanks to Sarah Rayner, Giovanni Verga, Mauro Capocci, Gianni Battimelli, and an anonymous referee for reading and commenting on the unpublished version of this work. Finally, thank you so much, for "strictly classified" reasons, to Gaia, Luna, Sonia, Pio, Sara, and Alberto.

1

The Training of a Nuclear Physicist

An Italian-born British-naturalised scientist [. . .] is missing [. . .]. Professor Pontecorvo was born in Pisa forty years ago and left Italy in the middle 1930's during Mussolini's campaign against the Jews. He went to France, and after the Nazi occupation of that country to the United States.

"Atomic Expert Missing. Gone to Prague,"
Manchester Guardian 21 October 1950

"Who the hell is he?" Imagine the journalists' bewilderment as news of the mysterious disappearance of a nuclear physicist landed on their desks. Bruno Pontecorvo, like many other protagonists of wartime research, had been by and large invisible to the wider public. Of course a few celebrated discoverers and scientific leaders of recent nuclear projects, such as J. Robert Oppenheimer, Leo Szilard, and Enrico Fermi, featured regularly in the news. But most of their colleagues' work and biographies were cloaked in secrecy.[1] Journalists thus had to browse the few newspaper articles that summarized key episodes in Pontecorvo's career, or magazine pieces popularizing major discoveries in nuclear physics, to retrieve meaningful facts on his past.

As information started to pile up, a sketchy account began to take shape emphasizing three aspects above all: Pontecorvo was an Italian Reform Jew, an émigré scientist, and allegedly a communist sympathizer. This fragmented identity constitutes in itself a remarkable illustration of the many different tensions that had typified his adolescence and early training as a scientist during the 1920s and 1930s. This chapter seeks to illustrate this fragmentation.

As a Reform Jew, Pontecorvo shared the experience of many members of his social cohort: the expectations deriving from emancipation in the post-unitary period of the Italian kingdom, as well as the tragedy of forced exile because of racial legislation during the Fascist regime. The Jews' social integration—their growing involvement in entrepreneurial, cultural, and political activities—coincided with their secularization. Only a few still maintained a strong religious identity.

As a scientist Pontecorvo learned, as did many other European intellectuals, the importance of assessing the advancement of science in the light of its applications. He was trained in considering its industrial uses as a way to build financial stability and a personal reputation. Science was discovery and invention, knowledge and wealth. Pontecorvo's encounter with one of the most promising and prolific research units in the country, Enrico Fermi's group at the Institute of Physics in Rome, sublimated these aspirations. Pontecorvo contributed to the key findings that allowed the group to propel itself beyond participation in the international debate on nuclear physics. He also envisaged practical uses for the artificial production of radioisotopes. These applications highlighted the importance of patents in ensuring a monopoly on their exploitation, thereby merging scientific research and business.

These trajectories of religious and professional identity intersect that of political participation in interesting ways. The activism of European Jews matched their wish for participation in public life as well as their scientific aspirations. The idea that science and technology could favor economic progress and modernization was transversal to political formations. In the 1930s, the Fascist regime endorsed it and sought to substantiate it through innovative policies. But it also attracted those who saw the communion between science and socialism as the true political alternative. While in Rome, the young Pontecorvo became accustomed to the fact that Fascist patronage to scientific research came on the condition that the scientists did not engage in political activities or publicly criticize the regime. He learned later on, while staying in France, about the scientists' social responsibilities and political participation.

This overlapping, contiguity, divergence, and collision of religion, science and politics created the social and cultural milieu in which Pontecorvo grew up, and it also shaped his experience as a young scientist after he left his hometown and moved to the fast-growing metropolises of Rome and Paris.

A Tuscan Clan in a Jewish Tribe

Bruno Pontecorvo was born in Marina di Pisa on 22 August 1913. He spent his childhood in nearby Pisa, the second largest city in Tuscany and a renowned port with a glorious past. His family's house was located on a street close to Piazza dei Miracoli, where Galileo Galilei had performed his famous experiments with falling bodies in the seventeenth century. Pontecorvo's family was large and wealthy, as it owned a textile company employing fifteen hundred workers.[2] His grandfather Pellegrino Pontecorvo was a shrewd entrepreneur, credited to have innovated textile production by introducing the spinning jenny, a multi-spool spinning frame, in Italy. During World War I, the Italian Air Force made use of his company's textiles to produce stouter airplane wings.

Aside from being an eminent businessman, Pellegrino was an illustrious representative of his religious community.[3] Following the establishment of Italy's kingdom in 1871, the Jews had been allowed out of ghettos, having been granted the right to actively participate in public life as a reward for their contribution to the struggle for the country's unification. Not only had the Pontecorvos actively participated in the emancipation of Italian Jewry, but they had also witnessed the pogroms of Eastern Europe and participated in international rescue operations.

This emancipation brought the Jews into public life, but it also had an impact on their religious identity. Religion continued to be important for many, yet secularism set in. For instance, Pellegrino continued officiating at traditional Jewish rituals in his family, but the younger family members' adherence was lukewarm. "We were Jewish without being aware of it," remarked Bruno's brother, the celebrated film-maker Gillo Pontecorvo.[4] "We were a typical Jewish family of those times, gentrified and liberally educated. I've never experienced religious crises," recalled Bruno years later.[5]

The circumstances of this Tuscan family illuminate the prosperity Bruno experienced in the early part of his life. He grew up as a bourgeois. A young teacher was paid to give him private lessons, and every summer the family would travel to exclusive holiday destinations such as the Dolomites and the Tyrrhenian Sea resorts. Bruno was only five years old when his grandfather Pellegrino died in 1918. The funeral was majestic, attended by political authorities and entrepreneurs who commemorated the patriarch. Following the recent Russian revolution, Italian workers were already clenching their fists at their employers. Yet no fights were recorded on the day laborers and industrialists alike mourned the late Cavaliere del Lavoro.[6]

Massimo Pontecorvo, Bruno's father, inherited and expanded the family business. Together with his wife Maria Maroni, he had five sons and three daughters.[7] As some of Massimo's sisters were married to other wealthy representatives of Jewish families, his own family became a small "clan" in a larger "tribe" that also included the Sereni and Colorni families. This tribe was a cohort that, according to the Italian journalist Miriam Mafai, became more "restless, sporty, and fashionable" in the troubled times that accompanied the establishment of a totalitarian regime in Italy.[8]

Science or Politics? The Crucible in the Wake of Fascism

At the end of World War I, Italy was in a state of turmoil. National newspapers highlighted the fact that the country's representatives had returned empty-handed from the peace negotiations in Versailles. The shortcomings of Italian diplomacy in international politics combined with a steep rise in social conflict. Farmers and workers joined forces in a string of industrial actions typifying what historians have dubbed the "two red years." Between 1920 and 1922, the recently established Communist Party of Italy (Partito Comunista d'Italia, or PCd'I) spearheaded a strong protest movement demanding better salaries and working conditions. A revolution seemed imminent. Yet the radicalization of the social conflict favored the newly established political organization led by former socialist Benito Mussolini, the Fascist National Party (Partito Nazionale Fascista, or PNF). The March on Rome of 28 October 1922 marked the beginning of a new political season in the country, as Mussolini became prime minister.

The small Jewish community greeted Fascism with mixed feelings. For instance, Massimo Pontecorvo challenged a local Fascist leader and future mayor of Pisa, Guido Buffarini Guidi, when Guidi visited his factory with the intention of taking the names of those who had participated in a demonstration, including their prominent ringleader. Not only did Massimo refuse to expose the ringleader, but he invited the brash Guidi to a duel, seeking satisfaction for what he perceived as a personal offense. Bruno's father always refused to become a Fascist party member.[9]

The advent of Fascism represented a watershed for the young members of the Jewish tribe. For some, including Bruno, it meant a renewed effort to get involved in higher education, partly to get away from political controversies. It is worth noting that since the time of their emancipation, Italian Jews had shown an aptitude for schooling and higher education that differentiated them from other sections of Italian society.[10] Not surprisingly, these scholars also represented the backbone of higher education in Italy. Bruno's elder brothers developed an interest in the sciences. Massimo's firstborn, Guido, studied agriculture at the University of Pisa,[11] while Paolo opted for engineering at the Politechnique in Turin.

Other family members were resistant to accept Fascist rule, and this opposition made them more interested in practicing radical politics. This was especially the case for Bruno's cousins, who also undertook higher education courses but were far more exposed to political interests and tensions. One of them, Eugenio Colorni, began philosophy studies at the University of Milan and another, Emilio Sereni, opted for agronomy at the University of Naples. The latter exercised an important influence on Bruno's education and choices in later life. An intellectual animated by a wide range of interests from Marxism to agricultural science, Emilio was an avid reader of all kinds of literature who engaged in animated discussions on the merits of Henry Poincaré's positivism with a young Emilio Segrè, later to be Bruno's mentor. The two Emilios, Segrè and Sereni, went on to play key roles in Bruno's life, including the chain of events leading up to the move to Russia.

In 1927 Emilio Sereni joined the PCd'I, whereas Eugenio Colorni opted for the antifascist collective Giustizia e Libertà (Justice and Liberty).[12] Emilio's brother Enzo became a Zionist and established the action group Avodah. He eventually migrated to Palestine and became a leader of the colonial kibbutz movement.[13] These choices exposed Bruno's cousins to political persecution. In 1930 Emilio Sereni was arrested by the Fascist police. His father, Samuele, who was the physician to the royal family, used his professional connections to ask for an amnesty, but was unsuccessful in obtaining one. Eugenio Colorni was also arrested in 1938 and killed by Fascist militants in 1944.

Bruno was an adolescent when these political conflicts unfolded, and although he worried about the circumstances of other members of his tribe, he did not get involved. In fact he was known as the "sporty" family member (figure 1.1), because his skills as a tennis player had gained him a national trophy.[14] Only scientific interests distracted him from mastering tennis. Following the path taken by his brother Paolo, he went to study engineering at the University of Pisa. Two years into these academic studies, he decided to go to Rome to study with a professor

FIGURE 1.1 Bruno Pontecorvo plays tennis in Canada. He won the 1948 Deep River Men's Single Trophy. Courtesy of the Chalk River Atomic Energy Laboratory.

who by then had gained reputation in the Italian scientific community: Enrico Fermi. In so doing, Bruno joined one of the most prominent research units in the country—one that in fact had managed excellently to "tune in" with the regime's agenda for the development of science and technology.

A "New Deal" for Science in Fascist Italy

Fascism did not just radicalize first, and conflate afterwards, the struggle for political power; it also represented a significant transition to a new system of scientific policy-making. At the beginning of the twentieth century, Italy was a mix of different peoples and languages—a hodgepodge of cultures and technological systems without nationally structured railways or electrical networks. In this context, the establishment of a totalitarian regime allowed its leaders to address the country's underdevelopment by propelling research in some innovative areas, directing research toward issues that concerned national industry, and promoting a synergy between publicly funded research organizations and the private sector.

In 1923 the establishment of an Italian national research council—the Consiglio Nazionale delle Ricerche, or CNR—responded to the urgent need to develop a system of new national laboratories and schools of specialization that could tie academic and industrial research together.[15] In the mid-1920s the PNF's collaboration with the liberals helped party leaders to focus on science, technology, and industrial change. From the 1930s, state funding of scientific research grew markedly to respond to the world economic crisis.[16]

In 1927 the CNR became a governmental organization directly depending on the chief of government, and Mussolini appointed the Bolognese inventor Guglielmo Marconi as its chairman. The decision was informed by a propagandist agenda, as Marconi was internationally renowned as a pioneer of radiotelegraphy—a true Italian genius. Yet Marconi was also chosen because of his dexterity in administering intellectual property rights. In the late nineteenth century he had succeeded in obtaining in London his patents on the wireless telegraph, and had established a new company, Marconi Wireless Telegraph, which accrued substantial revenues from monopolistic rights.[17] Marconi's appointment made the Italian scientific community more alert to the industrial applications of science as well as to the filing of patents.

Working in collaboration with prominent science administrators such as the chemist Nicola Parravano (chairman of the CNR chemistry section), Marconi understood that Italian inventors filed fewer patent applications than their competitors in other countries. The number of new patents had increased between 1883 and 1913 but considerably dropped between 1913 and 1929, when Italy fell behind France, Great Britain, Germany, the Netherlands, and Switzerland in its number of filed patents at home and abroad.[18] This limitation was believed to affect the national economy, as Italy depended on the import of industrial products and processes. The CNR chairman's concern translated into the laying out of new legislation on patents and the adoption of new standards for their examination. On 20 October 1932 Marconi claimed in an article published by the PNF daily newspaper that the Fascist government had gone a long way in encouraging inventors, as shown by the new intellectual property legislation.[19]

The Fascist administrators, through the CNR, were thus able to persuade large companies to invest in innovation and set up forms of collaboration with academic researchers in chemistry and metallurgy. For instance, the engineering of new ersatz materials such as alcohols, vegetable oils, and gases resulted from forms of private-public partnership.[20] This was because industrial production in Italy improved markedly between 1922 and 1929, and in some sectors (e.g., car manufacturing and chemistry) export levels nearly doubled. Yet Italy's dependence on the import of raw materials continued to hinder economic development. The regime foresaw that a better administration of new inventions could help Italian researchers to engineer new production methods and address these limitations.[21]

In the capital, the new science policy outlined by the Fascist regime overlapped the emergence of nuclear physics, thanks to the teaching and research activities of Enrico Fermi. Similarly to those of Marconi, Fermi's achievements captured the attention of the Italian press, which sought to exploit them propagandistically. Yet behind this image lay another important attempt to bind together scientific discovery, industrial innovation, and business activities through the search for novel methods to produce radioactive substances. When Bruno Pontecorvo moved to Rome to work with Fermi, he became accustomed to the importance of uniting the study of interesting and novel phenomena with the prospect of industrial advancement.

Fermi and His Group

In 1926, following postgraduate studies in Pisa (Scuola Normale Superiore, 1918–22) and Göttingen (1922–26), the young scientist Enrico Fermi returned to his hometown, where he had graduated ten years earlier, to take up the newly established professorship in theoretical physics at the Institute of Physics in Rome.[22] Despite his young age, Fermi was one of the most promising scholars in Italy, excelling in the traditional areas of physical studies as well as seeking to explore, in contrast with most of his colleagues, novel fields of research such as quantum physics. What impressed his patron, moreover, was the fact that he was equally at ease with theoretical and experimental analysis.

The Institute's director, the Sicilian Orso Mario Corbino, was equally responsible for establishing a new chair and for Fermi's appointment.

Politically a liberal, he was a staunch supporter of the Fascist new "economic deal," and one of Marconi's allies. A physicist interested in the commercial applications of electrical circuitry, in the 1920s he had given up active research to pursue a career as a science administrator and politician. Fermi's wife, Laura, claimed that Corbino divided most of his time between "political responsibilities" and "advisory offices with industries."[23] Indeed, he was a senator of the kingdom, a minister of public education (first), a minister of national economy (afterwards), and a board director at the Italian General Electric and Edison Companies. He was finally active in the Italian Bureau of Standards and the national patent office. Corbino believed that Fermi could create a new research school that, under his guidance, would be able to consider innovative teaching and research methods. For instance, Fermi worked with Enrico Persico to write physics textbooks for school teaching.[24] Yet when the opportunity came, Fermi steered the group towards exploring the applicative potential of their research.

From the onset, the prospect of interaction existed. Fermi and Franco Rasetti, his former fellow student in Pisa who also moved to Rome in 1926, attracted students from the departments of science and engineering who were fascinated by Fermi's lectures and were open to experimenting with the new applications of science. Fermi's group united various strands of the new middle classes. Ettore Majorana, Edoardo Amaldi, and Giovanni Gentile were sons of distinguished academics, whereas Emilio Segrè and Bruno Pontecorvo were sons of entrepreneurs (figure 1.2). The young Segrè was not only familiar with Poincaré's literary production, but also knew the importance of embedding novel research into industrial applications, as his father was the director of one of the hydroelectric companies that provided electricity to the capital.[25]

Corbino helped Fermi to gain positions of responsibility, and to align his work to the regime's policy of coordinating public and industrial research. In 1927 he became chairman of the CNR physics committee.[26] Later he was invited to join the boards at the national broadcasting company (Ente Italiano Audizioni Radiofoniche) and the electrotechnical firm Magneti Marelli.

Given the turmoil that typified political life in Italy at the dawn of Fascism, Fermi also tried to maintain his relationships by ensuring that none of his collaborators offered public comments on political life. In 1931, for instance, when the physicist Gian Carlo Wick, son of a well-known antifascist, was recruited for research in the institute, a worried Fermi wrote to Persico: "I would not like to have here someone who would make or has made a public statement against Fascism."[27] According to Segrè,

FIGURE 1.2 The "Via Panisperna boys." Enrico Fermi (at right) with his group (right to left): Franco Rasetti, Edoardo Amaldi, Emilio Segrè, and Oscar D'Agostino. Photograph by Bruno Pontecorvo. Courtesy of Archivio Amaldi (Amaldi Archive), Department of Physics, University "La Sapienza," Rome.

Fermi had been a lukewarm supporter of the new regime in its early days but later distanced himself, claiming to be "uninterested in politics."[28] Yet he was aware that patronage to research was made conditional on avoiding ruinous political relations. His message to the other group members was clear. Politics was to be kept out of the laboratory.

The Hunt for Slow Neutrons

Franco Rasetti was instrumental in introducing Pontecorvo to Fermi. He had known Bruno for a long time; he liked trekking in the Alps with Bruno's older brother Guido, and had first seen Bruno as a newborn baby. When Pontecorvo asked him about Fermi, Rasetti suggested he follow him in the capital.[29]

In 1934, when Pontecorvo joined Fermi's group, the young researchers had already achieved important results in the area of artificial radioactivity; the findings had brought them international recognition. From the 1920s, radioactivity was a rapidly expanding sector of physical chemistry that developed mainly thanks to the studies carried out by the New Zealander Ernest Rutherford first at the University of Manchester and then at the Cavendish Laboratory at Cambridge. It followed the groundbreaking studies on natural radioactive substances pioneered in France by Henri Becquerel and later developed by Pierre and Marie Curie at the Institut du radium beginning in 1920. Competing groups based at the Kaiser Wilhelm Institute of Berlin, the Radium Institute of Vienna, and later at the University of California, Berkeley, were also prominent in these studies. These researchers attempted to bombard chemical elements with particles of varying mass and charge. In this way they promised to provide a better understanding of atoms and their nuclei, as well as innovative ways to produce radioactive substances by particle bombardment. *Transmutation*—that is to say, the production of artificial radioactive substances, or radioisotopes—was deemed important because these substances were being introduced in medicine as diagnostic tools (as sources of X-rays) and for medical therapies (especially for leukemia and cancer). This production thus aroused commercial and industrial interests.

The group did not immediately prioritize artificial radioactivity research, but seminar activities arranged from 1931 on the recently published *Radiations from Radioactive Substances*, written by Ernest Rutherford and his coworkers, convinced them to learn more about it.[30] Reading the textbook helped the young researchers to assess the theoretical merits of Rutherford's atomic model and try to replicate the experiments. During this period they also traveled to other countries to learn what other European researchers were doing.[31] Following these exchanges, Fermi arranged a major conference on nuclear physics in Rome. Gathering together prominent researchers of the emergent discipline was a way to "denote an emergent sense of disciplinary identity" between different European scientific communities working on nuclear phenomena, and it helped Fermi's patrons to even more strongly advocate that funding be provided for nuclear research (figure 1.3).[32]

By then, transmutation had become the "alchemical dream" of this community. At the institute, Rasetti found a depleted source of natural radium that had not been used in fourteen years. From November 1933

FIGURE 1.3 International Conference on Nuclear Physics, Rome 1932. Main entrance of the Institute of Physics. Guglielmo Marconi (left) talks with Danish physicist Niels Bohr (right). Orso Maria Corbino is right behind them. Courtesy of Archivio Amaldi (Amaldi Archive), Department of Physics, University "La Sapienza," Rome.

he succeeded in extracting by chemical means some quantities of "Radium D" from this natural radium, thus making available to the group a more intense radiation source that they could use in irradiation experiments.[33] Later, the director of the Institute for Public Health, Giulio Cesare Trabacchi, made some more radium available, thereby allowing the group to design even more sophisticated experiments. The theoretician Majorana suggested that Fermi consider the potential of neutrons as projectiles in nuclear reactions following recent developments at other research centers.

Two years earlier Marie Curie's daughter Irène, together with her husband, the chemist Frédéric Joliot-Curie, had succeeded in activating some chemical substances at the Curies' Institute of Radium using alpha particles as projectiles. In the same year Chadwick had highlighted the existence of a new type of particle, the *neutrons*, with mass comparable to that of alpha particles and no charge. These findings made Majorana aware that repeating the Joliot-Curie experiment with these projectiles would have caused stronger activations, mainly because neutrons would not be repelled by electrical forces in the nuclei.[34]

Following Majorana's suggestion, Fermi, Amaldi, Segrè, and Oscar

D'Agostino (a chemist who had recently worked with Frédéric Joliot-Curie) decided to bombard fourteen elements with neutrons from a tube filled with beryllium and radium, obtaining important radioactivations. After publishing their results in the CNR journal *La Ricerca Scientifica*, they sought international publicity and recognition by sending a report to the prestigious journal *Nature*. Later on, Segrè and Amaldi went to Cambridge and brought to Rutherford a comprehensive account of the group's experiments. At his suggestion, the account was eventually published in the *Proceedings of the Royal Society*.[35]

Enter Pontecorvo

Bruno Pontecorvo joined the group following these achievements. Earlier on, he had simply been one of many students gravitating around the group. On 10 November 1933 he graduated in physics and mathematics with the highest marks.[36] Even so, the opinion on Pontecorvo's experimental skills was initially mixed. According to Rasetti, he and Eugenio Fubini, another one of Fermi's pupils, were "extremely clumsy in experimental work."[37] In one instance the two used some cement to case glass windows at the ends of tubes used for spectroscopic analysis, but the cement dropped onto their hands and clothes, and dripped onto the floor. When Amaldi saw them dirty he said, "My gosh, you work like pigs here."[38]

Yet Pontecorvo learned fast, focusing on experiments that would prove pioneering in enhancing the radioactivation technique. In the summer of 1934, while carrying out new tests with neutron sources, he and Amaldi observed that similar nuclear reactions seemed to have greater yields if the apparatus was placed on a wood desk rather than on a marble one, thus suggesting that different media might alter nuclear reactions.[39] This is how Amaldi would recall the experiment thirty-five years later, when answering the question of what had informed the group's decision to focus on slow neutrons:

> I don't know, because when the discovery of slow neutrons was made, it was also a purely experimental discovery. [. . .] After Segrè and I came back from Cambridge, I was asked with the help of Pontecorvo [. . .] to try to prepare a kind of absolute scale of the activations of the different elements. That means to find a standard condition of irradiation and to see how the different elements become active when irradiated all in the same conditions. This was the idea, because in the first paper that was

> published in the Proceedings of the Royal Society, we had just given a qualitative scale, by saying "very strong activity," "weak activity," and so on. We felt it was unsatisfactory [. . .]. Pontecorvo and myself started to do this in a very precise way, but we found irregularities. We were not able to obtain all the same results irradiating under the same conditions [...] .[40]

As by then Pontecorvo's clumsiness was known, Rasetti thought that it was "virus-like"—it had affected Amaldi as well. The two were messing up quantitative analysis. Amaldi claimed his innocence: the results would be inexplicable even if the experiments had been accurate throughout. Fermi, who was abroad for conferences, returned to the laboratory in October and suggested that they use a hydrogenous compound, paraffin, in new experiments. The suggestion was put forward because Pontecorvo had noticed that the lead casing used to shield the experimenter from radiation seemed to substantially weaken the reaction. Fermi's intuition was thus to use a substance with a molecular structure completely different from lead to see how it worked.

The effect was much more noticeable with paraffin. Later, Fermi understood that substances rich in hydrogen, such as paraffin, could slow down neutrons, thereby increasing their efficiency as projectiles in nuclear reactions (in contrast with substances containing heavy elements such as lead). The breakthrough paved the way to the definition of a new process for the production of radioactive elements. Pontecorvo played a key role in what was ultimately the result of a mixture of hard work and accident.

This time, however, Fermi was more circumspect about publishing research results than he had been on previous occasions—that is to say, when the role of neutrons in nuclear reactions had yet to be fully understood. He sent no communication to *Nature*, nor were meetings sought with Rutherford. The only publication was a sketchy account sent on 22 October 1934 to the CNR journal. Until then, the group had been very eager to publish. What made them now so reluctant to disclose information on their studies?

Exploiting the Discovery

Their unwillingness certainly derived from Corbino's directives. The old physicist and policy maker insisted that the group file a patent application before publishing further, foreseeing that the discovery represented an opportunity for synergy between nuclear physics and industry—exactly what the regime had wished for. According to Laura Fermi,

> One morning [. . .] Corbino came into the laboratory [. . .]. They were preparing to write a more extensive report on their experiments [following the report prepared for the CNR journal]. Corbino became incensed. "What? Do you want to publish more than you have already? [. . .] Are you crazy? Can't you see that your discovery may have industrial applications? You should take a patent before you give out more details on how to make artificial radioactive substances!"[41]

On 26 October 1934 a patent application prepared by Fermi was filed by the patent agent Letterio Laboccetta. It included the names of six inventors—Fermi, Rasetti, Segrè, Pontecorvo, Amaldi, and D'Agostino—along with that of Giulio Cesare Trabacchi, who had provided the neutron source for the experiments.[42] Shortly afterwards, a similar letter patent, this time written in English, was sent to the British patent office. It claimed intellectual property rights on instruments and processes for producing radioactive substances by placing a neutron source and a target together with a moderating material, such as paraffin or another hydrogen-containing substance.[43] The "slow neutron patent" played a very important part in the future lives and careers of all members of the group—especially, as we shall see in the next chapters, in that of Pontecorvo, including the episode of his mysterious disappearance.[44]

Between 1935 and 1937, commercial activities centered on the invented process, and the patent became as important for Fermi and the group members as the development of further research on the subject. Although they had filed the patent in Italy first, they thought that its merits could be appreciated by industrialists in other European countries and in the United States. The inventors could now profit from their relationship with Gabriello M. Giannini, another one of Fermi's students. Giannini had left Italy in 1930, a few months after graduating in the same session as Amaldi and Majorana.[45] Exploiting his father's network of relations, he had been able to establish an agency specializing in trading patents and inventions in New York City. His office was located in Rockefeller Plaza, in the heart of Manhattan.[46]

Giannini was instrumental in administering the slow neutron patent in the United States, as well as contacting industrial companies interested in the invention. Filing the patent in the United States proved troublesome.

Following a year-long diplomatic crisis, Italy and Ethiopia entered in a conflict over colonial territories in Eritrea in October 1935. The League of Nations imposed sanctions on the Italian kingdom, and the decision envisaged the possibility of the League's member states requisitioning Italian properties abroad, including patents. Fearing that the Fermi group's patent could be seized by US authorities, Giannini proposed to establish a new American company that would figure as assignor for the patent filed by the Italians. Soon after, G. M. Giannini & Company Incorporated was established, and on 3 October 1935 application no. 43,462, describing the process for slowing down neutrons, was submitted.[47]

Giannini also contacted patent managers at General Electric and Westinghouse to promote the industrial development of artificial radioactivity. As early as July 1935 Harvey Rentschler, director of research at the Westinghouse Lamp Division in Bloomfield, New Jersey, recommended the acquisition of Fermi's patent and set aside funding to start research on the neutron process. The company was willing to acquire a license for twenty-five thousand dollars. However, Rentschler eventually concluded that that Giannini's assignment was incomplete, and that his agreement with the inventors did not protect the company in the case of litigation between the two parties.[48]

Meanwhile, in Europe, Segrè contacted the Dutch physicist Cornelius Bakker, whom he had known in 1931 while working for Pieter Zeeman in Amsterdam. Bakker was now employed by the Dutch firm Philips Gloeilampenfabriken.[49] The company immediately showed some interest in the slow neutrons process, partly because it produced isotope-filled vacuum tubes for X-rays, such as the newly designed Metalix, which was used in radiology and skin carcinoma therapy.[50] In October 1935 its patent manager visited Rome, and eventually Philips became responsible for administering the process in Europe in exchange for 5 percent of the deriving royalties on profits from instruments and radioisotopes.[51] While finalizing these agreements, Fermi and his coworkers were made aware, in a manner not dissimilar from that used by Corbino the previous year, of the importance of controlling the publication of findings. Philips provided the Italians with guidelines indicating that research findings should be "promptly" communicated to the company, but also that "in no case should publications of any sort be made before a patent was filed."[52] But was the envisaged business of slow neutrons really promising?

Fermi was known to be a very cautious individual who would hardly ever let enthusiasm prevail over analytical thinking. Yet, during a meeting of the Italian Society for the Advancement of Science (Società Italiana per il Progresso delle Scienze, 12–18 October 1935), he claimed that in a few years' time natural radioactive substances would be replaced by artificial radioisotopes. The difficulties that made it impossible to replace them now were, he stressed, purely technical.[53] Moreover, using radioisotopes in the medical industry was only one of many industrial prospects. For instance, the Hungarian physicist Leo Szilard, another radioactivity pioneer, was adamant that transmutation could also be used for energy production and that "the production of energy and its use for power production would be possible on such a large scale and probably with so little cost that a sort of industrial revolution could be expected."[54] Corbino shared this viewpoint. In 1934 he claimed that nuclear physics was a kind of new "super-chemistry" allowing the release of more energy than traditional chemical methods such as combustion. Corbino thus used the comparison to make even blunter claims about how nuclear physics could circumvent Italy's traditional lack of natural resources. If only one could extract the amount of nuclear energy contained in fifty grams of hydrogen, this would suffice to replenish the kingdom's yearly electricity consumption.[55] Most certainly the claim, albeit unrealistic, was aimed at enticing Corbino's colleagues in Palazzo Venezia, Mussolini's headquarters in Rome (figure 1.4).

These expectations motivated attempts to trade the slow neutron patents and became an important aspect of Bruno Pontecorvo's training as a scientist. Even when his contribution to these business activities was limited—mainly because he was too young—he learned to appreciate that understanding the link between discovery and industrial applications was just as important as knowing laboratory techniques. In turn, he understood how to manage intellectual property rights. An important reason for Pontecorvo's limited contribution to patent trading was that he sought to carry out some research abroad in order to complete his training. Already in 1934 he had applied for a CNR scholarship to study in Berlin.[56] In 1935 he was appointed as lecturer (*assistente incaricato*), but at the same time he succeeded in getting a scholarship from the Italian Ministry of National Education to study nuclear physics abroad. He thus moved to Paris to work at the newly established Laboratoire de chimie nucléaire under the supervision of Frédéric Joliot-Curie.[57]

Science as Militancy sur les Boulevards

When Pontecorvo moved to Paris in 1936, the city was bursting with political activities. In June the Parisian Léon Blum became prime

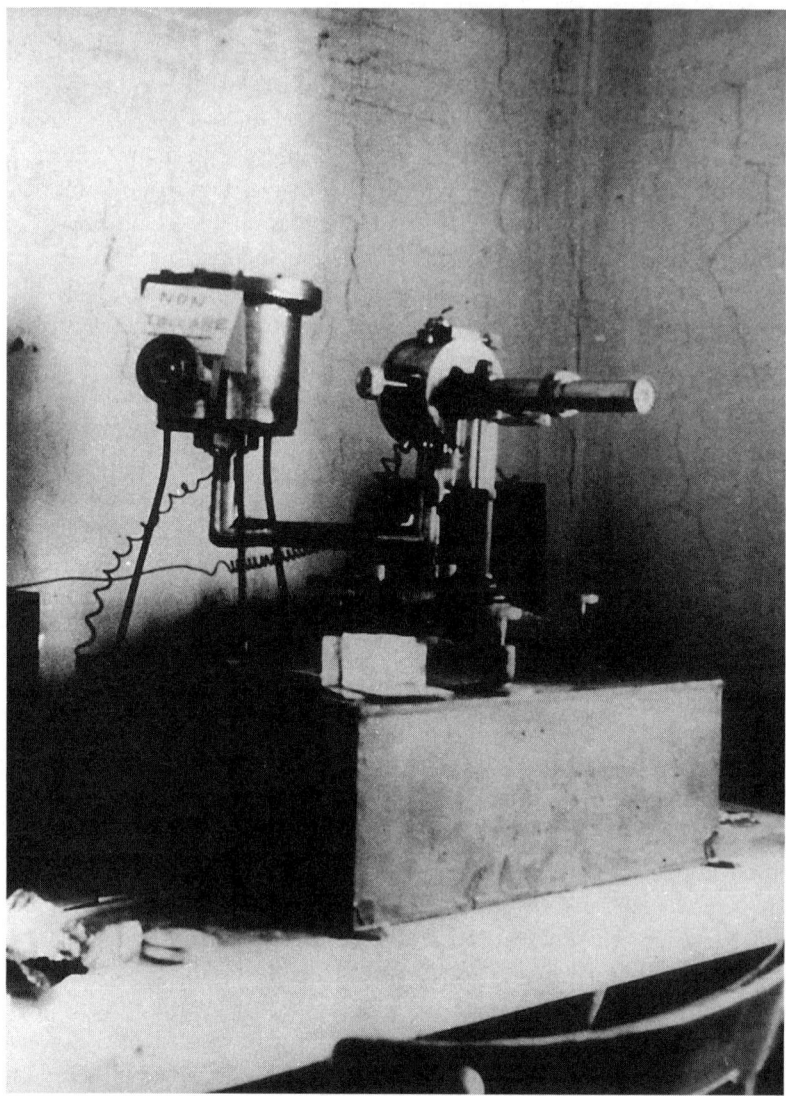

FIGURE 1.4 A "Fascist" design? The ionization chamber used by Fermi and his coworkers to carry out radioactivity measurements in summer 1934 was named "Roman Sign," thus mocking the gesture featuring in the Fascist symbolic repertoire. Courtesy of Archivio Amaldi (Amaldi Archive), Department of Physics, University "La Sapienza," Rome.

minister; the first socialist to lead the country in France's history. The advent of the Popular Front (Front Populaire, a coalition of socialists, communists, and radicals) meant that, in contrast with the rampaging Nazi fascism typifying other European countries, France was administered through a democratically elected left-wing government. Blum's experiment was short-lived, as it lasted only from 1936 to 1938. But in that period, France became the favorite destination for political refugees from Nazi-Fascist countries. Giorgio Amendola, the PCd'I leader and a Parisian resident, referred to the city as a sheltering island in the sea of totalitarianism that had by then engulfed Europe.[58]

Science continued to be the key element of Pontecorvo's life, but now it became an integral part of his activism, thus marking a discontinuity with the period of his life spent in Rome. Now Pontecorvo could see science not only in terms of practical and business benefits, but also in connection with the political goals of the radical left. He thus considered his engagement with scientific activities not only in terms of personal gain, but also in terms of social advancement. These two inner motives, individual and social enrichment, continued to propel Pontecorvo's scientific career, thereby demonstrating themselves as foundational to his professional identity.

Living in Paris helped Pontecorvo to be less restrained in approaching politically active individuals than he had been in Rome. He could now see relatives and friends who had migrated abroad when Mussolini had risen to power in Italy. This freedom had an impact on his private life, too, when he met Marianne Nordblum, a Swedish woman four years his junior who was also studying in Paris. The two met at a hostel in the popular Place du Panthéon, just outside the Cité Universitaire, where most students and researchers had their accommodations. In 1937 they decided to live together without getting married, and the first of their three children, Gil, was born the following year.[59]

In 1936 an amnesty allowed Pontecorvo's cousin, Emilio Sereni, to leave prison. He fled to Paris, where together with his partner Xenia Silverberg he campaigned for the clandestine PCd'I. In the French capital his relationship with Pontecorvo flourished as Sereni's intellectual acumen left a lasting impression on the young scientist. The perceptive cousin had by then become a high-ranking official in his organization.[60] Most certainly Sereni introduced Pontecorvo to party officials. Two of those officials, Ambrogio Donini and Giuseppe Berti, were leading figures in the clandestine organization coordinating the activities of young European militants at the Leninist school of Moscow.[61] Aside from political activists, Pontecorvo continued meeting with researchers who had also left Italy, such as the radiobiologist Salvatore Luria, a graduate of the University of Turin who had also studied with Fermi, and the cosmic ray physicist Sergio De Benedetti. From 1937, the beginning of the Spanish civil war mobilized the small refugee community in Paris, and some left to fight Francisco Franco's nationalist militants.

The available documentation does not help to more precisely ascertain the relationship between Pontecorvo and these communist militants. At no time does Pontecorvo seem to have been directly involved in actions, political propaganda, or any other sort of political activism with the PCd'I. Even more astonishing seems to be the naivete of Pontecorvo's adherence to communism in this period as, according to some recollections, he decided to join the party on 22 August 1939, the day when the Soviet and German foreign ministers, Vyacheslav Molotov and Joachim Von Ribbentrop, signed a pact of nonaggression between their countries—an event that famously left many militants shaken and dubious about their loyalty.[62]

Yet an analysis of Pontecorvo's acquaintances and readings allows us to shed some light on the nature of his political interests, and consider the practice of scientific research as a form of militancy in itself. He shared with both Sereni and Frédéric Joliot-Curie the idea that the very process of getting involved in science and technology was already a form of political participation insofar as science and technology would contribute to social change through industrial applications and progressive values.

Bruno's brother Gillo claimed that it was mainly because of Bruno that he became a communist during his short stay in Paris, as allegedly Bruno suggested that he read a popular, albeit tediously doctrinaire, book entitled *Précis du Marxisme*, written by one René Vallon.[63] The anecdote is puzzling, as there is no René Vallon who has ever published such a work. It was instead the French socialist *Louis* Vallon who published works on socialism that became very popular during this period. A prominent member of the economic group X-Crise and a former student of the prestigious École polytechnique, Vallon advocated economic planning and technocracy before joining the Popular Front. His "experimental" version of socialism was especially trendy among young *universitaires* and, if anything, it confirms even more the influence exercised by Frédéric Joliot-Curie on the young Pontecorvo.[64]

The Parisian chemist and physicist who had married Irène Joliot-Curie was, in contrast with Fermi, a politically active scientist. He took part in Popular Front campaigns through the local committees of his hometown. Irène played a key role in the coalition, as she became the undersecretary for scientific research.[65] Similarly to Vallon, Frédéric Joliot-Curie and other French scientists considered the merits of socialism in its relationship with science. Following the Russian Revolution they considered the interplay of socialism and science in terms of social progress. Frédéric Joliot-Curie and his colleagues Paul Langevin and Francis Perrin (who took over Irène Joliot-Curie's governmental duties) conceived their commitment to socialism in terms of contributing to scientific research as well as science policy. Indeed, in the light of these activities, socialism *was* science, as it was the only political doctrine of which the soundness could be scientifically demonstrated. As Perrin put it: "Je suis socialiste comme je suis physicien. J'assaie d'utiliser dans l'un et l'autre domaines les mêmes règles logiques, les mêmes notions de lois, et surtout les mêmes méthodes de pensée libre."[66] The scientists' participation in Blum's coalition defined the creation of a government committee on scientific research that included Marie Curie, Perrin, and Langevin. The ideas developed by the socialist scientists were now shared by many among the French scientific community, including Pontecorvo.[67]

This novel approach to scientific research, which united it with political interests rather than casting them aside as in Rome, fascinated Pontecorvo to the point of causing him to reconsider his plan of a short stay in Paris before returning to Rome to take up a lectureship. He was enthusiastic about becoming part of an ebullient political and scientific context.[68] If for Pontecorvo going to Paris meant a discontinuity in terms of experiencing political participation, he stayed involved in politics mainly by further developing the experimental studies of nuclear physics already started in Rome, with an eye on their application. Thus, the most significant "political" contribution of Bruno Pontecorvo during his stay in France was exactly this one.

From Isotopes to Isomers

In January 1937 Frédéric Joliot-Curie joined the Collège de France as professor of nuclear chemistry. There he set up a new research laboratory devoted to nuclear phenomena. It was staffed with researchers from many different countries, including the Austrian Hans Von Halban and the Russian Lew Kowarski, who eventually went on to become Joliot-Curie's closest collaborators. When Pontecorvo came to Paris from Rome he joined the laboratory and worked together with them, as well as with other promising scientists such as Pierre Auger and Bertrand Goldschmidt.[69]

Nuclear physics was by then becoming more dependent upon the provision of highly sophisticated and expensive machinery able to produce highly energetic particles. The new laboratory was equipped with an electrostatic generator of the Van de Graaff type. Between 1936 and

1939 a new cyclotron was also installed there, partly thanks to funding from the Rockefeller Foundation.[70] These machines were used to accelerate nuclear particles used in reactions, and they helped Pontecorvo to develop experiments that were unthinkable in Rome.

Together with the French physicist André Lazard, who had previously been responsible for designing a Van de Graaff–type generator with Irène Joliot-Curie, Pontecorvo looked into nuclear isomerism, then a poorly understood subject of nuclear physics.[71] Revealed in 1917 by Rutherford's assistant, Frederick Soddy, and then studied more extensively by the German physicist Otto Hahn four years later, nuclear isomers were chemical species of identical atomic number and mass that decayed accordingly to different processes. Pontecorvo focused on the isomeric forms of radioactive rhodium and cadmium, concluding that isomerism appeared to be much more common than previously assumed. His publication in *Nature* proved his ability to master the subject and caused excitement in the Parisian scientific community. Joliot-Curie renamed the process as *nuclear phosphorescence* because isomers produced highly energetic X-rays.[72]

But looking into Pontecorvo's research in this period also reveals its inner motives and ties them to his career and political trajectories. In November 1937, Szilard visited Paris and bombarded a foil of indium with fast electrons. In this way he found out that isomers do not change nuclear structure once they stop emitting radiation. This eliminated one of the main problems associated with the use of radioisotopes: their further transformation into other elements.[73] Once again Pontecorvo looked at his own research with an eye to future applications. Nuclear isomers in a sense could be considered as "recyclable" radioisotopes, thereby contributing further to lowering the costs associated with producing them. Their properties could thus be extremely useful, especially in new medical treatments—findings Pontecorvo reported to the French newspaper *L'Oeuvre*.[74]

This research undoubtedly followed the path of Rome's research efforts to find cheaper ways to produce radioisotopes, but also embedded itself into the political meanings attributed to scientific advancement in the Parisian intellectual milieu. The research was so promising that it gained Pontecorvo a longer stay in Paris. In 1938 he won the Curie-Carnegie prize for his research on nuclear phosphorescence, and in the following year his research was funded by the French National Research Council (CNRS).[75] These scholarships kept him away from the problems experienced by his research group in Rome, which was now struggling for survival.

Troubles in Rome

From 1937 onward, Fermi and his coworkers sought to use an accelerating machine that would enable them to compete with other groups in the United States and Europe that were experimenting with more energetic particles. Yet the funding available to the group was greatly reduced, while the slow neutron process—now being patented by Philips—needed further development to prove commercially viable. These shortcomings led to the group's collapse.

The accredited inventor of the cyclotron was the director of the Radiation Laboratory of Berkeley, Ernest Orlando Lawrence. In the 1930s he pioneered transmutation research by trying to obtain radio-sodium by bombarding sodium with deuterons in cyclotrons. In 1935 Rasetti visited the Radiation Laboratory and recognized that the production of radio-sodium with the cyclotron was far more efficient than the slow neutron process. He thus understood that to compete with other nuclear physicists working in the field, his group needed to build a machine capable of producing more energetic projectiles.[76]

The setback worried Fermi. With Pontecorvo now in Paris, Segrè teaching at the University of Palermo, and D'Agostino permanently employed at the CNR, he felt less certain about future prospects. Fermi and Giannini tried to regain leadership in the design of new methods for the production of radioisotopes by establishing some form of collaboration with prominent groups in other countries. In 1937 Giannini went to Britain to meet with Thomas Allibone of the company Metropolitan-Vickers, which owned the rights to a high-tension generator designed by the Cavendish researchers John Cockcroft and Ernest Walton. He suggested an agreement that would have allowed the British company to use Fermi's process and the Italians to build a similar generator without paying royalty fees. Allibone had his own plans to buy a license on Fermi's process, but Giannini's agreement with Philips had made these plans impossible to pursue.[77] A meeting with Szilard, who was the owner of a patent on a high-energy generator, also proved inconclusive. Giannini left Britain empty-handed.[78] In the meantime, Lawrence started producing more isotopes such as radio-phosphorous and radio-iron, while the medical therapies based on radiation emitters gained a considerable share in the American market and generated profits between two and five hundred thousand dollars per annum.[79]

Whether or not Fermi knew about these advancements and commercial successes, he now had to face far more compelling issues. In 1937 Marconi and Corbino died, leaving the group without political protection. Soon Fermi began feeling vulnerable, especially when, in the same year, his plan for the establishment of a national radioactivity center housing a newly built high-tension generator was rejected by the CNR executive committee. Moreover, the organization offered him research funding in an amount much reduced from what he had requested. Fermi thought about leaving the country and working abroad, and new racial legislation introduced by the regime convinced him of the soundness of such a proposition.

Racial Laws and Leaving Europe

While Fermi was busy with research and administration tasks, key government authorities started an all-out campaign against the Jews. Anti-Semitism was not an indigenous ideology in Italy, but it allowed the regime to strengthen its relations with Nazi Germany. In August 1938 a special census with racial intent was introduced, and in the following month two decrees allowed the expulsion of all Jews who had arrived in Italy after 1 January 1919 and the forced resignation of all academic staff of Jewish origin. Ninety-six university professors and lecturers, 133 assistants and research fellows, and 279 high school teachers had to leave their institutions of learning.[80]

The academic diaspora that followed the new legislation deprived the country of its most talented scholars, including Fermi, whose wife was of Jewish heritage. When in 1938 he was awarded the Nobel Prize for physics because of his group's findings on the action of hydrogen-containing substances on nuclear reactions, he decided to migrate to the United States. Although the move was permanent, he disguised it from regime officials by claiming that he intended to travel abroad for conferences.

The racial laws also had a disruptive effect for other members of Fermi's group. Segrè decided to stay in Berkeley, where he was a visiting fellow, rather than return to Italy. Although Bruno Pontecorvo had plans to stay in Paris, he was on the list set out in the decree, which made it impossible for him to find employment in Italian universities. Amaldi took over the responsibility for what remained of the group's activities, together with some young recruits. The legislation was even more disruptive for Pontecorvo's family, whose members moved abroad. Paolo found employment in the United States with the electronic firm Raytheon. Guido succeeded in obtaining a grant from the United Kingdom's Society for Protection of Science and Learning, which was busy rescuing academic refugees from Nazi-Fascist countries, and which helped Guido to continue genetics studies at the University of Edinburgh. Another brother, Giovanni David, and a sister, Anna, found refuge in Britain thanks to a "Miss Clayton," a family friend.[81] In November 1938 new anti-Semitic laws forbade the employment of Jews in public offices and limited their property rights. Massimo sold the family business and the house in Pisa before moving to a small apartment in Milan.[82]

Pontecorvo's haven in Paris was short-lived. When, on 10 June 1940, German troops invaded France, it became clear that the émigrés had to leave at once. Bruno and Marianne married in a rush on 9 January 1940 with plans to move to the United States, where Bruno had already secured a job. Marianne obtained a US visa following a recommendation by the Swedish consulate.[83] Bruno then went to Bordeaux seeking to validate his passport. The two briefly regrouped in Toulouse, where they met with Bruno's sister Giuliana. The city by then had become a center of resistance activities against the Germans, and the physicists Goldschmidt and Langevin, as well as some militants such as Sereni, had also found refuge there.[84] In late July Pontecorvo and his family, together with Giuliana and his fellow researchers Luria and De Benedetti, left for Lisbon; from there, on 9 August 1940, they boarded the passenger liner SS *Oranza*, bound for New York.[85]

An Émigré Scientist

Analyzing Pontecorvo's childhood, adolescence, and training as a scientist provides us with interpretative elements that can enrich our understanding of the Pontecorvo case; these elements will feature prominently in the narrative that follows this chapter.

Pontecorvo's tragic circumstances were undoubtedly common to many Jewish migrants who in the wake of World War II were also seeking a haven. Yet, in contrast with the circumstances of others, they did not combine with a strong religious identity or propel life-changing decisions. If anything, they favored the emergence of his pressing need to accommodate his career interests in choosing new destinations. Pontecorvo was an émigré but he hardly ever considered himself an exile, as his working experience led him into prioritizing career options even when it entailed moving to new countries. He already had a job lined up when he moved to New York. Although the occupation of France motivated him to speed up his plans to move to the United States, it is likely that the move would have happened anyway.

He inherited from his secularized Jewish family, and especially from his grandfather Pellegrino, an aptitude for privilege and for finding a place in society through his professional activities. He gathered from his older brothers that scientific research could be a way to do this. This foundational element propelled most of his future career decisions, including that of going to Soviet Union. Traveling and moving abroad was indeed part and parcel of his professional experience.

This aspect of Pontecorvo's formation maps onto his understanding of science in general, and nuclear physics more specifically, especially as it was practiced in Italy. This was a promising field that united important discoveries with equally important applications that opened up new industrial uses. There is no doubt that financial constraints and racial legislation halted the scientific project set out by Fermi and his colleagues in Italy. But they went on exploring the practical applications of nuclear physics abroad, uncovering its utility in fields as diverse as the search for raw materials (an application pursued especially by Pontecorvo) and the development of warfare technologies. They also continued protecting their intellectual property rights, as they had plans for profiting from their inventions.

This emphasis on science as contiguous to "business" and "social emancipation" is also important in terms of capturing the essence of Pontecorvo's political allegiance. Pontecorvo's stay in France helped him to understand that science was a means of political participation because of its social function. It is worth noticing that after Pontecorvo's defection this participation was flagged in journalistic accounts as shedding light on his membership in communist organizations. Yet our study offers a different portrayal, emphasizing the nature of his militancy as participation in innovative and socially minded scientific research. In fact, although Pontecorvo maintained his connections with communist militants, his ideas were far more lenient towards socialism as he understood it from his reading of Vallon's works and from Frédéric Joliot-Curie's teaching. Furthermore, political participation never led Pontecorvo to relinquish his perception of science as a "business." His very nuanced but possibly naïve adherence to socialist ideas inevitably presented contradictions that would be decisive in the events and decisions of 1950.

2

Neutrons for Peace and Neutrons for War

> The Minister, Mr. George Strauss, who is responsible for atomic energy research in Britain, said: "Though Dr. Pontecorvo has not had direct access, except in a very limited way, to secret subjects for some time, it would be quite impossible to say that he has not been able to gather information, while residing at Harwell or in Canada, which might be of value to an enemy." Earlier, in reviewing the professor's record as a scientist, Mr. Strauss stressed that for several years Pontecorvo's "contacts with secret work have been very limited."
>
> **"Minister to MPs: Missing A-Man Could Help Enemy,"** *Daily Mirror*, 24 October 1950.

When Pontecorvo's defection was announced, the reaction of the British government to his disappearance was extraordinarily understated. In two parliamentary briefings, the British minister of supply claimed that Pontecorvo's contact with secret work had been negligible. Although George Strauss could not foresee what use could be made of Pontecorvo's knowledge in Russia, a security threat existed only because he might have gathered restricted information, not because of his research. Even when the British press drew attention to the case, implying that more should be said by government officials on its implications, only a few statements came from the higher echelons.

The German émigré physicist Klaus Fuchs had just been pronounced guilty of espionage when Pontecorvo flew to Russia. Fuchs's case created a sensation. Vital secrets regarding the production of atomic weaponry were now deemed to be in the hands of the Soviet Union. So was it really the case that the notorious Fuchs had given away critical details, whereas the narrowly accredited Pontecorvo had little to offer the Soviet atomic program? Could the loss of restricted scientific information really be as threatening for scientific security as one scientist in "flesh, bones and brain?"

The analysis of previously classified archival evidence casts serious doubt on this interpretation. Not only had Pontecorvo been among the few scientists who actively participated in wartime nuclear projects, but he had gained access to the most confidential aspects of wartime and postwar atomic programs. His almost unique knowledge of technical steps that anticipated the production of atomic weapons, including the search for uranium and the production of atomic explosives in nuclear reactors, made him a resourceful and knowledgable expert capable of contributing to important research.

What is seen here is how Pontecorvo's expertise in nuclear physics paved the way to a variety of applications before, during, and after World War II. In particular, the new narrative offers a novel picture of how, through Pontecorvo's work, nuclear physics "colonized" research domains as far afield as oil prospecting and pile physics, the search for strategic minerals, and nuclear detection technologies. In this way Pontecorvo had become one of the most prominent scientists in the Western nuclear research establishment. His presence in the Soviet Union was thus far more problematic than what a few government officials led the public believe.

Caught between Rope Chokers and Doodlebuggers

Before the war, the city of Tulsa, Oklahoma, was a stronghold for applied science and technology; a Silicon Valley before its time. This was mainly because of the search for petroleum; the advent of prospecting and exploration was attracting scientists and engineers who were eager to escape Europe. Prominent oil companies and small research organizations alike recruited these refugees, as well as American nationals, to develop new technologies for the extraction of precious natural resources. Following his lucky escape from Paris, Pontecorvo joined them, with plans to innovate the field.

Since the early 1920s, new geophysical studies had reversed a trend that had prevailed at the beginning of the twentieth century, when drillers or "rope chokers" were considered key experts in finding oil deposits. Drilling revealed little about the potential capacity of a well, so when the demand for oil grew, new methods were sought. Oil companies first recruited geologists who could analyze different types of strata, limestone and sandstone especially. Wells can be found in cavities or "domes" that lie between these geological formations. Thus, an analysis of the strata made it easier to identify oil traps.[1]

But even the crafty geologist was eventually replaced by the geophysicist, as various techniques based on the detection of physical phenomena such as gravity, electricity, magnetism, and seismology contributed to better map the oil-bearing ground. Dynamite proved immediately handy, as detonating small charges of explosives in boreholes and recording the resulting waves enabled geophysicists to find out about structures likely to be trapping oil; as a result, less rough-and-ready techniques were developed. For instance, the Société de prospection electrique, owned by Conrad and Marcel Schlumberger, pioneered electrical coring methods.[2] Following its resounding success in Europe, in 1940 the company moved its headquarters to Houston, Texas.[3]

The shrewd French were not alone. Owing to the considerable amount of oil in the area, new prospecting companies were established in the South. In 1920 the first US prospecting firm, the Geological Engineering Company, was set up, shortly to be followed by its main competitor, the Geophysical Research Corporation.[4] Soon, crews of prospectors were seen in oil fields carrying bulky apparatus on their shoulders. Locals labeled them "doodlebuggers," comparing them to water diviners carrying newly engineered dowsing rods (i.e., doodlebugs). Alternatively, customized automobiles would carry sensing devices that were attached to long ropes and used in boreholes (figure 2.1).

Oil prospecting was no conventional academic science, and not just because its laboratory was a van crammed with equipment. Most of its practitioners made only limited use of academic publications, being more interested in patents. The fierce competition led practitioners to avoid disclosing details of their methods to rivals, in order to ensure profitable contracts with oil companies such as Texaco and Standard Oil. Petroleum was indeed the strategic resource of the US economy, which was responsible for 63 percent of worldwide oil production, followed by Mexico and Canada, which produced 14 each, and the Soviet Union, responsible for 10 percent.[5]

In the 1930s the search for innovative geophysical techniques led the doodlebuggers into exploring nuclear physics, which in turn brought them into contact with European experts. In 1929 William G. Green, a graduate of the University of Oklahoma, established a new company, the Seismic Services Corporation (SSC), using funding from the Mid-

FIGURE 2.1 When the laboratory is a truck, 1941. The van used for prospecting by Scherbatskoy, Pontecorvo, and their coworkers was equipped with electronic apparatus. It was made available by the "U.S. Government" (see license plate). Courtesy of Serge Scherbatskoy's family.

Continent Petroleum Company. In the fall of 1933 Green joined Gerald W. Westby in a partnership that also attracted local practitioners such as William Russell (from Texas A&M University), Robert Fearon, and Gilbert Swift.[6] Green employed a "young Turk," Serge Alexander Scherbatskoy, who had just landed in the United States without a passport. Born in 1908 at Buyukdere, Turkey, he was the son of a Czarist diplomat based at the Russian embassy in Constantinople. After engineering studies at the prestigious Technische Hohschule of Charlottenburg, Germany, and at the Sorbonne in France, Scherbatskoy had found temporary employment at the American firm Philco. His wife had then convinced him to move to Tulsa, Oklahoma, her hometown.[7] The Polish nuclear physicist Jacob ("Jake") Neufeld followed Scherbatskoy to Tulsa shortly after. Born in Lodz in 1906 with the name of Jakov ben Itzhak, Neufeld had trained at the University of Liege in Belgium before moving to the United States, where he worked at Cornell University and then became one of the SSC consultants (figure 2.2).[8]

Green's doodlebuggers had far-reaching plans. During the 1930s they started considering alternative logging methods to Schlumberger's electrical coring, as by then the focus of attention in oil prospecting had shifted from the practice of coring to the more refined one of logging: recording on paper the structure of geological formations through the use of remote sensing devices. The doodlebuggers considered the natural radioactivity of rocks possibly an excellent indicator of petroleum-bearing geological structures.[9] Neufeld and Scherbatskoy began to design and use gamma ray detectors, including the ionization chambers and Geiger-Muller counters now routinely used in nuclear physics.[10]

Green, Scherbatskoy, Neufeld, and Fearon soon understood that gamma rays functioned as shale indicators. Natural radioactivity is very weak in pure sands, salt, quartz, coal, and limestone. It reaches a maximum in shale that contains radioactive materials (chiefly uranium and thorium). Thus natural radioactivity could reveal the "shaliness" of strata, making the technique ideal for prospecting shale-rich oil fields—

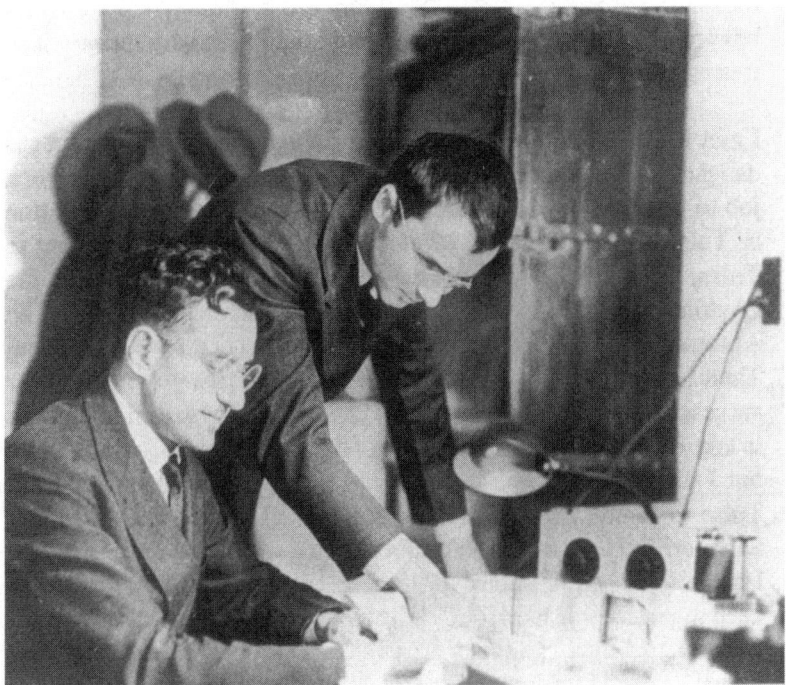

FIGURE 2.2 Jacob Neufeld (seated) and Serge Scherbatskoy working at the Well Surveys, ca. 1940. Courtesy of Serge Scherbatskoy's family.

exactly those in which electrical coring was of no avail. Eventually the gamma ray log made it again possible to use recently abandoned wells in the states of Oklahoma, Kansas, Texas, and Louisiana. Between two hundred and one thousand barrels of oil per day were extracted using the new technique.[11]

Following this success, money from Standard Oil and another firm, Socony Vacuum, poured into the cash-needy SSC, allowing Green to establish Well Surveys Inc. (WSI), of which Scherbatskoy became research director.[12] WSI's chief goal was to pioneer nuclear methods of finding oil, and its research team begun looking at neutrons as a possibility. A skilled nuclear physicist was needed for the task ahead. In 1939 Scherbatskoy and Neufeld first tried to arouse the enthusiasm of Emilio Segrè, who was in California. They had met Segrè for the first time at a gathering for European émigrés, and in May 1940 he spent a week in Tulsa with plans to work there. Yet despite the precariousness of his job in Berkeley, Segrè showed little interest in well logging, and instead recommended Pontecorvo:

> When I turned down the Tulsa job, Scherbatskoy asked me about Bruno Pontecorvo, who then was in Paris with Joliot-Curie, and I warmly recommended him. The oil-exploration firm, Wells Surveys, then decided to offer him a job and cabled him. Thus Pontecorvo, escaping on a bicycle from Paris, about to fall to the Nazis, and in imminent peril of his life, suddenly found himself with an assured job in America. A true miracle![13]

Neutron Well Logging

Miraculous or not, Pontecorvo's employment could take place because he had by then already envisaged in a research note that neutron-induced reactions had potential for the development of prospecting apparatus that would exploit radioactivity. In fact it was just such a proposition that had prompted the attempt of Henri-George Doll, Schlumberger's research director in Houston, to hire him. Yet it was Schlumberger's much smaller competitor WSI that actually acquired him.[14]

Pontecorvo worked for WSI from the second half of 1940 to the first half of 1943, producing forty company reports, two papers for specialized journals and four patents.[15] Following a line of inquiry already suggested by Segrè, he first tackled the problem of using neutrons in logging apparatus from the theoretical point of view. He went back to the definition of neutron density curves as expounded five years earlier by Amaldi and Fermi, and considered whether the "AF curve" could be successfully applied to prospecting problems.[16] He then went on to build an artificial well that could be used to simulate what happens in the strata as an effect of neutron bombardment (figure 2.3). The well continued to be a valuable research tool throughout the period in which the neutron log technique was developed.

By then the WSI team had already considered how neutron-induced reactions could function as strata indicators. Fearon and another researcher working for Shell had filed patents from 1938 that were not dissimilar from those of Fermi's coworkers: they had made use of a neutron (radium/beryllium) source. A patent was issued in November 1940 that gave Shell a priority towards the licensing of uses of neutron rays in geophysical prospecting. In turn this forced the WSI to find, with Pontecorvo's help, alternative ways to use neutrons in well-exploring devices. Fearon's patent was issued only in 1942.[17]

Pontecorvo realized that neutrons work in a completely different manner from gamma rays. Neutrons would exclusively give an indication of the efficiency of the materials adjacent to the hole to diffuse the neutron radiation. These considerations followed the analysis carried out years earlier on the slowing of neutrons in the presence of hydrogen-containing substances. Indeed, these materials were more efficient than any other in absorbing neutrons, and thus the neutron log provided valuable information on the porosity of rocks and their chemical composition.[18] A geological examination of the strata would eventually provide more detail on their oil-bearing capacity.

The design of instrumentation for the neutron log kept Pontecorvo busy as he looked into issues such as the best neutron source to be adopted in trials, and the type of detector to be used. Although these subjects were not so dissimilar from those that had typified his earlier career, their application to prospecting problems presented important elements of novelty. For instance, in order for the neutron log to be sufficiently accurate, Pontecorvo deemed it necessary to have a source that produced little gamma radiation. He thus sought rarer sources of neutrons (such as polonium, Radium-D, and actinium) and initiated contacts with companies producing them.[19]

The detector could register the neutrons either *captured* or *scattered* by hydrogen nuclei. Since the process of neutron capture yields gamma rays, a detector such an ionization chamber was a suitable instrument; but what was needed to register neutron scattering was a more sophisticated detector, such as a counter that could indirectly detect neutrons. These alternatives paved the way to the definition of two different methods

FIGURE 2.3 Apparatus for neutron well logging, ca. 1941. The detector was introduced into an artificial well (at left) and wired to electronic apparatus and a logging device. Courtesy of Serge Scherbatskoy's family.

of prospecting: the *neutron-gamma log* and the *neutron-neutron log*.[20] These considerations about detectors instigated laboratory work and prompted innovative designs and academic papers.[21] On 8 April 1941, Fearon and Pontecorvo discussed details of their ionization chambers in relation to a US patent office objection to issuing a new patent on the subject.[22] A few days later they held private talks with researchers from one of their competitors, the Texas Company, about innovative types of counters. By then, Pontecorvo also realized the potential of boron trifluoride (BF_3) counters in the application of the neutron-neutron log.[23]

By 1942, a WSI team that included Pontecorvo had tested the neutron log—mainly with ionization chambers—at twelve locations in Oklahoma, Texas, and Louisiana. As they produced more accurate logs than their electrical and gamma ray equivalents, the neutron logging trials were considered "highly encouraging," also revealing the transitions between limestone and sandstone.[24] Deflections to the left identified transitions from sandstone to shale, whereas deflections to the right indicated transitions from limestone to sandstone. Other variations could be analyzed through gamma ray logs, which appeared to be an ideal complement to neutron logs.[25] Yet neutron well logging could work in areas where electrical logging could not, such as the salt beds in west Texas and western Kansas.[26]

In April 1942 Pontecorvo visited some East Coast laboratories in an effort to further develop neutron logging. But in fact this was his first move towards a career change. The trip made him aware of wartime nuclear projects associated with neutron research. At the same time, the scientists who met him wondered whether he could represent a valuable addition to these projects.

From Tulsa to Montreal

The research trip led Pontecorvo into visiting the metropolises of Chicago, Philadelphia, and New York. In New York he met Fermi, the Czech-born scientist Georg Placzek, and Hans Von Halban, whom Pontecorvo knew from Paris. He informed them about his recent research on neutron well logging, and discussed the neutron absorption coefficient of elements such as calcium and potassium, whhich are very important in the identification of limestone structures. Pontecorvo learned that these data also had great importance for Fermi and Von Halban, but at first he could not understand why, since they were not engaged in oil prospecting. He thus discovered that there was more than oil at stake in neutron research. His report on the visit is telling: "The data I obtained in New York have not been published and cannot be published for a long period to come, because of their confidential character." [27]

This knowledge further strengthened Pontecorvo's suspicions about the drying up of sources of neutrons in America, which suggested that someone somewhere was harnessing impressive amounts of radioactive minerals. These doubts had been another important reason for his trip to the East Coast, as since 1941 his search for other sources of neutrons had produced no result. Although sufficient radium was available, rarer by-products were more difficult to purchase. In several circumstances the WSI put pressure on its suppliers, the Canadian Uranium and Radium Corporation and the Radiochemical Company. The shortage even prompted Pontecorvo to set up a new research program, as he was asked to consider the merits of different types of detectors in the search of radioactive minerals.[28]

The shortage of neutron sources presented a major obstacle to the development of neutron well logging, and eventually made Pontecorvo restless. In November 1942 he was about to accept a job offer from the Russian-born entrepreneur Boris Pregel to work in his New York laboratory.[29] Bernard Goldschmidt, who already had worked there, considered the offer unworthy of Pontecorvo's talent. He thus suggested to Hans Von Halban that he hire him. Goldshcmidt and Von Halban had been setting up a research team in the context of a secret military project that was codenamed Tube Alloys and based in Canada.[30] As Von Halban had met Pontecorvo a few months earlier, it was not difficult for Goldschmidt to convince him. Following a customary job interview, which also took place in November 1942, Pontecorvo and his family moved to Montreal in January 1943, where he would pioneer another important branch of neutron research known as "pile physics." Actually there was no such thing as "pile physics," properly speaking, because there were no piles. Reactors became the major technological features in the military application of nuclear physics, especially thanks to Fermi's pioneering studies at the University of Chicago.

Neutrons for War

After receiving the Nobel Prize in December 1938, Fermi fled to New York, where he worked at Columbia University in the physics department headed by George B. Pegram.[31] There his understanding of neutron-induced disintegrations greatly improved. Between 1939 and 1941, several studies conducted by researchers in Europe and the United States revealed that uranium atoms could fission if they were bombarded with neutrons, which would thus liberate more neutrons and energy that in turn could kick-start a chain reaction. The theory behind these studies was further developed by the Danish physicist Niels Bohr and his American colleague John Wheeler. Bohr and Wheeler understood that only one uranium isotope—U-235, which represents a negligible constituent of natural uranium (0.72 percent)—undergoes fission. On 26 January 1939, Fermi conferred with Bohr during a conference in Washington, DC.[32]

Fermi sought to explore the subject further. Funding for neutron disintegration analysis was made available by a "Uranium Committee" headed by Lyman Briggs and established by the US National Defense Research Committee (NDRC). Assisted by the American Herbert Anderson and the Canadian Walter Zinn, Fermi and Szilard (now a refugee scholar in the United States) developed nuclear fission studies, also becoming more alert to their military implications. Szilard was especially concerned that German scientists might have developed similar work, and he decided to discuss these matters with Albert Einstein, who by then was already in the United States. In a letter delivered to President Franklin D. Roosevelt by the economist Alexander Sachs, Einstein mentioned Fermi and Szilard's research on uranium and Germany's attempt to seek control of uranium deposits in Czechoslovakia, as well as ongoing research at the Kaiser Wilhelm Institute where "American work [...] is being repeated." The letter clearly referred to the possibility of using such a reaction for an explosive device.[33]

In the meantime, American science administrators such as the Carnegie Institution director, Vannevar Bush, the NRDC chairman, James B. Conant, and Arthur H. Compton, head of a National Academy of Sciences committee on the military uses of atomic energy, guided a major restructuring that made it possible for the project to develop. The 1941 Uranium Project was now administered by the US Office of Scientific Research and Development (OSRD), of which Bush was made director. Together with Compton, he took responsibility for the research program. The Uranium Committee was now renamed OSRD Section S-1, and on 19 August Fermi took responsibility for one of its four subsections, dealing with fission theory.[34]

While Pontecorvo was busy designing oil prospecting devices, Fermi had started working for the US government, and it was *exactly* in relation to this work that he wanted to exchange information with Pontecorvo about neutrons in geophysical systems. In this period, Fermi was on the brink of discovering the conditions critical to ensuring the feasibility of a self-sustaining chain reaction deriving from neutron bombardment. In the meantime, Fermi and his coworkers were also busy designing an experimental device that could produce such a chain reaction. In their discussions at Columbia University, they frequently used the word "pile" to refer to a device in which various chemical compounds were stacked together to generate a critical reaction.[35]

Fermi reasoned that in such a device, a neutron source irradiates pellets of uranium oxide so as to obtain a reaction in which more neutrons are emitted in fission processes than absorbed. He understood that such a reaction could occur only in the presence of some hydrogen-bearing substance that could slow neutrons down to thermal energies in order to enable them to split the uranium nuclei. Together with Szilard, he thus went on to investigate which substance would be best suited for the task. After considering several possibilities, including boron, water, and heavy water (containing deuterium), they opted for graphite, a by-product of coal production.[36] Several experimental "heaps" were thus assembled at Columbia with graphite and natural uranium; but none went critical.

Towards the end of 1941, Compton made plans for a shake-up in the Uranium Project to concentrate experimental work in key areas. In research that from 1940 had involved Segrè and US chemist (of Swedish ancestry) Glenn Seaborg, it appeared hat another isotope, plutonium (Pu-239), could undergo fission and be considered an explosive compound like U-235. Plutonium could be artificially synthesized from fission reaction products of natural uranium, and piles could be used to manufacture the isotope. Thus Compton recommended that pile work be moved to the University of Chicago, where the research on plutonium synthesis was ongoing at the Metallurgical Laboratory. Fermi reluctantly accepted Compton's request, and in April 1942 he moved to the university along with Szilard. There, experiments started again with a new experimental device that was built underneath the west stands of the university's now unused football stadium, Stagg Field. Blocks of graphite were now filled with baseball-sized pellets of uranium oxide (figure 2.4).[37]

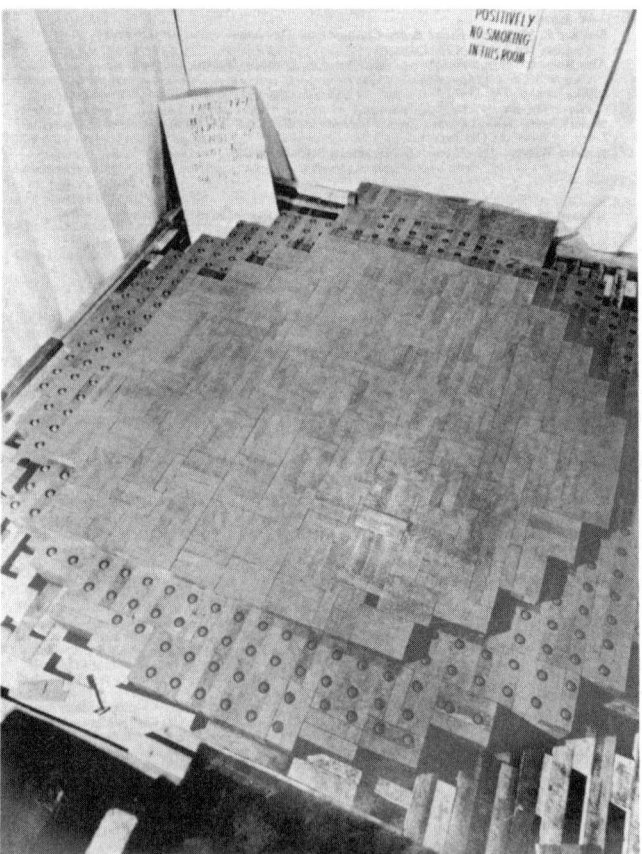

FIGURE 2.4 "Positively no smoking in this room." The Chicago Pile 1, assembled beneath the grounds of Stagg Field Stadium in 1942. Courtesy of Argonne National Laboratory.

The project reached completion on 2 December 1942 when the Chicago Pile (CP-1) produced the first self-sustaining nuclear fission reaction. It was an event of a few seconds, as reaction-controlling cadmium rods were inserted into the device for the amount of time necessary to demonstrate that the reaction was indeed critical. A bottle of Fermi's favorite wine, Tuscan red Chianti, was opened to celebrate the event. Compton cabled Conant at the OSRD: "The Italian Navigator has reached the New World."[38]

The event changed the history of nuclear physics, proving that a uranium-based chain reaction was indeed feasible. Fermi's experiment marked the conclusion of the US-government-led Uranium Project, and the beginning of the Manhattan Project. This is because the US Army Corps of Engineers headquartered in Manhattan took control of future work on uranium. General Leslie Groves was nominated as the project's head. The experimental confirmation led toward the solution of other problems, such as the establishment of supply and production facilities and the setting up of a secret research laboratory at Los Alamos. More nuclear reactors were planned just outside Chicago (at a new facility in the Argonne National Forest), at Hanford, Washington, and at Oak Ridge, Tennessee. Yet, as pile work had just started, more research was needed on nuclear reactors.

Tubes, Alloys, and Pontecorvo

It wasn't only the Americans who were mobilizing to research fission. British scientists had considered the military potential of atomic energy much earlier; in April 1940 the government committee MAUD considered the feasibility of a fission bomb.[39] The establishment of this committee followed a report by German émigré scientists Rudolf Peierls and Otto Frisch that theoretically demonstrated its feasibility. Experimental work began in British universities and was boosted by the contribution of prominent French scientists. Frédéric Joliot-Curie's coworkers Von Halban and Kowarski secretly fled to Britain when Paris was occupied following the initiative of Charles H. G. Howard, the twentieth Earl of Suffolk (the scientific liaison officer with the French government). A precious stock of heavy water, another hydrogen-bearing substance used in fission experiments, was also rescued.[40]

From July 1941 the deliberately deceptively-named Directorate of Tube Alloys (DTA) was established to replace the MAUD. The DTA's newly appointed head, Wallace Akers (formerly of the British firm Imperial Chemical Industries), decided to transfer the fission project to

Canada, fearing that precious knowledge and materials could fall into German hands if Britain was invaded.[41] The Canadian authorities enthusiastically approved the plans for a joint Anglo-Canadian project that retained the code name Tube Alloys. A new laboratory was set up in Montreal in collaboration with the Canadian National Research Council. Von Halban became the project's research director, taking responsibility for assembling a team that included Britons, Canadians, émigrés from Nazi-fascist countries, and French researchers who joined in as a contingent of the Free French Forces (figure 2.5). The presence of such an international environment created concerns among administrators. Akers initially opposed Pontecorvo's employment on the grounds that the number of foreign scientists was already too high. Eventually, however, Von Halban convinced him to accept Pontecorvo, pointing out that no specialists in neutron physics of the same caliber could be found in America.[42] Von Halban realized that the project needed someone who could develop experimental research and theoretical calculations similar to what Fermi was developing in Chicago. Pontecorvo was one of Fermi's pupils, had worked with him before, and could consult him if necessary. Moreover, his recent work on neutron well logging had led him to consider the interaction of neutrons and hydrogen-baring or fissionable substances.

Pontecorvo was the ideal candidate for the job in pile physics. Von Halban, not unlike Fermi and Szilard, had already considered alternative pile designs, distinguishing between *homogeneous* piles, in which all chemical components were in the same state of matter, and *heterogeneous* piles, in which the components were in different states. He believed that the best option was to use liquid heavy water as a moderator and uranyl nitrate, a water-soluble salt, as a fissionable material. But Pontecorvo's study on neutron fluxes in heavy water ruled out this design, having shown that it would not allow a divergent chain reaction.[43] The option of a heterogeneous system with uranium in the form of metal bars placed inside a heavy water tank, called the calandria, thus prevailed.[44]

In the first half of 1943, delays and misunderstandings brought the whole project to a stalemate. The laboratory lacked sufficient experimental equipment, and the Anglo-Canadian project received no support from the United States despite the fruitful collaboration established on radar research after the mission of British war research planner Henry Tizard in 1940. In August 1943 Roosevelt met with the Canadian and British prime ministers, William Lyon Mackenzie King and Winston Churchill, in Quebec City to establish some collaboration between

FIGURE 2.5 From left to right: Pierre Auger, Hans Von Halban, Jules Guéron, Bernard Goldschmidt, Bruno Pontecorvo, and Henry Seligman at the Montreal Laboratory. AIP Emilio Segrè Visual Archives.

atomic projects. Later a new organization, the British Central Scientific Office (BCSO), was established under the directorship of Charles Galton Darwin in Washington, DC.[45] The setting up of more coordinating organizations, such as the Combined Policy Committee (CPC) and the Combined Development Trust (CDT), dealing with the provision of raw materials) shortly followed.[46] The managers of the atomic projects also agreed to diversify the Tube Alloys research program, which in turn created further support for the idea of building a heavy water pile in Canada. This was a design that the Americans had not consistently pursued.[47]

In 1943 John Cockcroft replaced Von Halban as Tube Alloys research director.[48] Cockcroft's experience in dealing with large research teams at the Cavendish Laboratory and in wartime radar research made him well suited to take the project forward to the stage of operational planning. In the same year, the site of the nuclear facility housing the Nuclear Reactor X (NRX) was also chosen. The location was near a small village called Chalk River, about 130 miles west of Ottawa.[49]

Pontecorvo played a vital role in designing the reactor, and he also benefited from Fermi's advice. In March 1944 he was one in a group of theoretical scientists who met in Montreal to define the reactor plant's dimensions through calculations and experiments on neutron fluxes in the heavy water/uranium assembly. This so-called Lattice Group was responsible for making decisions about the pile's size and geometrical shape (the lattice) on the basis of the required energy output. The group also included four other scientists, of which the Russian-born Canadian mathematician George M. Volkoff was the theoretician. Volkoff's estimations eventually led Pontecorvo to analyze the coefficients of neutron absorption in heavy water and uranium in laboratory experiments, and to discuss them with the other team members.[50]

Fermi's support in this effort was very important. Pontecorvo traveled to Chicago six times in 1944 to exchange physical data and opinions with Fermi and his coworkers.[51] These data referred to the absorption of neutrons in fissionable materials such as U-235 and Pu-239; discrepancies between the results obtained in the two research laboratories had been noticed and discussed by Fermi and his former pupil.[52] Shortly after Pontecorvo's first trip to Chicago, the Lattice Group agreed on the reactor's size. The available documentation does not tell us, however, whether these exchanges were aimed exclusively at helping Pontecorvo. An enquiry by the US Joint Committee on Atomic Espionage (JCAE) revealed in 1953 that in those visits Fermi and Pontecorvo compared the characteristics of different nuclear reactors, and Fermi shared data on the Hanford pile, which was the main US plutonium production plant.[53]

Pontecorvo was also responsible for other important design features. In July 1944 he considered shielding problems in connection with heat and neutron diffusion.[54] As heavy water is a moderator much more efficient than graphite, it made the erosion of a reactor's structural components much more likely—including the shielding system, which consisted of concentric shields of graphite, water-cooled cast iron, and concrete.[55] Pontecorvo conducted experiments to determine whether the concentration of fission products could be problematic. The NRX was aimed at producing plutonium from natural uranium and the uranium isotope U-233 from thorium, thus insigating analysis of their decay products and fission properties.[56] In August 1944, Cockcroft reported that "the lattice and rod dimensions together with the tank reflector and shield dimensions have now been frozen." Design work was completed.[57] In July 1946 the new Chalk River laboratory, adjacent to the nuclear installation, was ready. Another nuclear reactor, the Zero Energy Experimental Pile (ZEEP), was built under Lew Kowarski's direction, and on 5 September 1945 it went critical.[58]

In these years the study of nuclear science and technology reached a turning point as atomic weaponry revealed its destructive power. A plutonium nuclear device was tested in the Alamogordo desert in July 1945 in the context of the Los Alamos laboratory activities. Two more nuclear devices, one containing U-235 and the other Pu-239, were exploded on the Japanese cities of Hiroshima and Nagasaki to end World War II, with Nazi Germany already defeated. Nuclear physicists divided over the political implications of nuclear studies. The famous remark later made by Los Alamos research director J. Robert Oppenheimer, "The physicists have known sin; and this is a knowledge which they cannot lose," remarkably synthesizes those tensions.[59]

Scientists like Pontecorvo who had provided important, even if not fully acknowledged, contributions to this effort were now called upon to make decisions about their future activities. Tempted by the remunerative jobs that flourished after the power of nuclear physics was finally fulfilled, Pontecorvo worried about the political overtones of nuclear physics. Leaving Canada seemed a viable option. For instance he was offered a job by Seaborg at the University of California, Berkeley, to which that physicist had returned from Chicago, and the only reason why Pontecorvo refused was that he wanted to see his brainchild, the NRX, finally working. He thus replied to Seaborg: "The fact is that although your offer was extremely tempting, I feel that my coming to Berkeley before the end of this year would make it impossible or extremely difficult for me to use the Canadian pile for research. [. . .] Since I have been with the Canadian project for several years I feel that it would be a pity to leave just when research, as distinct from engineering starts."[60]

Pontecorvo had to wait for quite some time, as the NRX did not work. Deadlines were missed, and not until the night of 21/22 July 1947 did the pile reach criticality. Pontecorvo was present in the NRX control room, but this time no Italian navigator was noted to have landed in a new world. The customary cables announced matter-of-factly that the NRX had gone critical, claiming only that the "big day for which we have waited for so long" had finally arrived (figure 2.6).[61]

Prospecting Once Again

No nuclear reactor could work or atom bomb be built without precious supplies of radioactive minerals. Most accounts of the history of the atom bomb have focused mainly on the few scientists who dealt with nuclear reactions rather than those who sought to design new methods for prospecting radioactive minerals to cope with growing demand. This may be the reason for another important contribution of Bruno

FIGURE 2.6 The Nuclear Reactor X at Chalk River. Courtesy of the Chalk River Atomic Energy Laboratory.

Pontecorvo to the advancement of nuclear science and technology: his work as a prospector for "strategic" minerals (i.e., uranium and other radioactive raw materials), which has yet to be fully appreciated.

Prospecting for strategic minerals had occupied the WSI team since 1940. Although the team prioritized its work on neutron well logging, it also tried to expand the company's activities beyond oil prospecting.[62] Pontecorvo had considered one method to distinguish different types of strata by measuring the hardness of the gamma rays they emitted. If the method found application in oil prospecting (as the uranium/ thorium ratio is high in shales, but significantly lower in sandstone and limestone), it could also be used in the search for radioactive minerals (specimens of the radium/uranium family emit lighter gamma rays than those of the thorium family).[63]

Pontecorvo's expertise had turned out to be very useful in Canada, as it brought him closer to the Canadian entrepreneur Gilbert Labine, the director of Eldorado, Canada's largest uranium mining company. In 1926 Labine had established the company in the belief that gold deposits could be found in the Northwest Territories of Canada. He later realized that the area lacked gold, but did contain uranium- and radium-rich pitchblende.[64] In 1940 the Russian Boris Pregel, who in 1942 would try to hire Pontecorvo, rescued Labine's company from bankruptcy. Owner of the world's largest uranium deposits (in Katanga, Belgian Congo) as well as an OSRD supplier of natural uranium, Pregel had provided the uranium for the CP-1 project. As demand for strategic minerals shot up, the Canadian government forced Pregel out of the company. Labine was demoted to managing director, while the Canadian administrator Lesslie Thompson took responsibility for developing uranium prospecting; the two went on to coordinate their activities with the Montreal laboratory.[65]

When Pontecorvo informed Labine and Thompson about the innovative technique designed at the WSI, Labine went to Tulsa to discuss with Scherbatskoy the design of new portable ionization chambers that could be used in field exploration. In the summer of 1944 Scherbatskoy flew to Ottawa and met Pontecorvo and Thompson with plans to carry out a survey in Port Radium, near Great Bear Lake (figure 2.7), using the newly designed apparatus.[66] Scherbatskoy's detector was used across a vast territory that had no roads and was covered by barren rocks, tundra, and muskeg. Fieldwork with the apparatus took place between 9 and 14 September 1944. A secret report was filed.[67]

The joint WSI/Tube Alloys team carried three instruments: a Geiger-Muller counter; a single ionization chamber, and a double ionization chamber. Although counters and ionization chambers were equally efficient in detecting outcroppings of pitchblende, ionization chambers were better suited for detecting pitchblende at lower depths.[68] This is also the period in which prospectors started considering how to detect uranium deposits of greater concentration rather than those offering, as one expert put it, many "nibbles" but few "bites."[69]

After the war, the Tulsa-based journalist Roger Devlin claimed that Scherbatskoy's laboratory had been responsible for finding the uranium

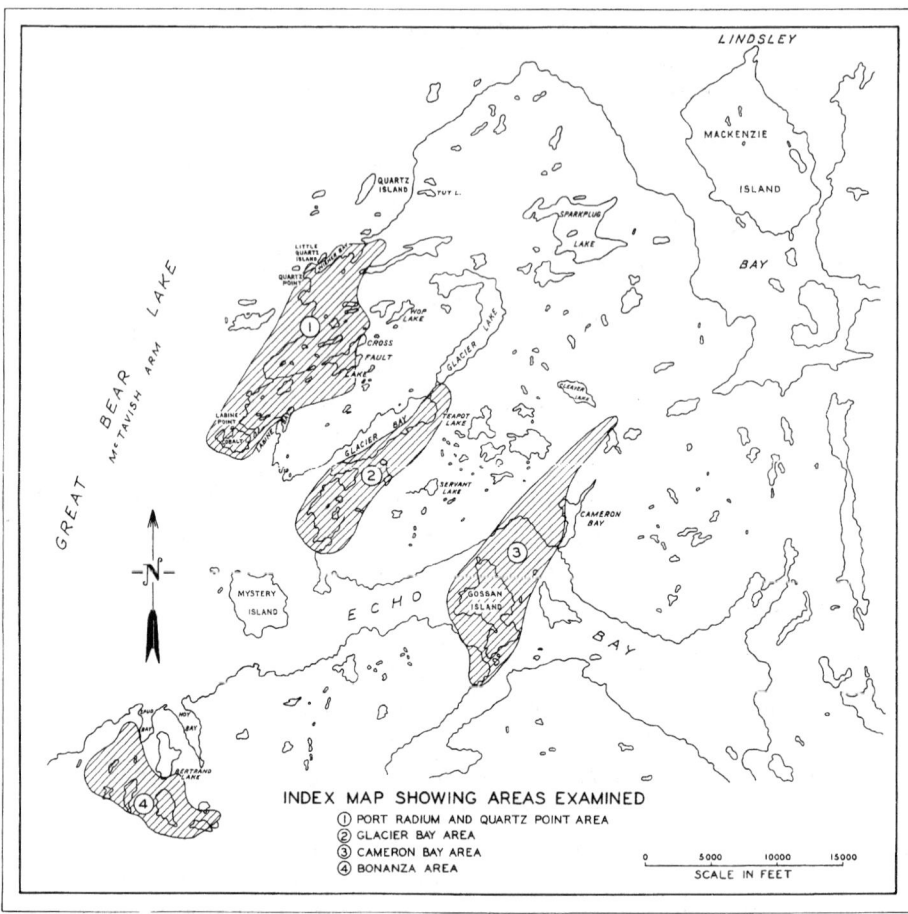

FIGURE 2.7 Map of Great Bear Lake, 1944. Deposits of uranium were found in the running of an exploration arranged by Pontecorvo and his former coworkers at the Well Surveys. From file AB 1/648, "Report on Trip to Port Radium, September 1944," National Archives, Kew Gardens, London.

needed for the Manhattan Project.[70] The tale contained an important truth, as in the prospected area two important veins were found. These finds enabled Eldorado to cope better with the growing demand for radioactive minerals, after initially having been unable to satisfactorily supply both the US and Anglo-Canadian projects.

As Pontecorvo played a key role in the exploration, Thomson and Cockcroft recommended him as the Montreal laboratory envoy for the 1944 trilateral meeting on prospecting problems that took place in Washington, DC. Two days after the Quebec Agreement was signed,

Groves recommended that the United States "allow nothing to stand in the way of achieving as complete control as possible of world uranium supplies."[71] Eventually the CDT was assigned the task of conducting the worldwide survey, which was codenamed "Murray Hill Area."[72] US Army Colonel Paul Guarin, one of Groves's assistants and a geologist who had worked for Shell, signed a contract with Union Mines Development Corporation (UMDC) to carry field explorations in thirty-seven countries. On 31 October 1944 he arranged a meeting in the Federal Works Agency Building of Washington, DC, to discuss prospecting problems and examine what advanced instrumentation existed.[73] Sixteen scientists representing the US Geological Survey, UMDC, the Canadian National Research Council, Eldorado, and the Chicago Metallurgical Laboratory attended the meeting. All the groups brought their own instrumentation. Guarin emphasized the security aspects of the meeting and, more generally, of the work connected with uranium prospecting. As the whole edifice of nuclear research was now dependent upon uranium provision, it was extremely important that information on search methods was kept secret.[74] The fact that Pontecorvo attended this meeting tells us a great deal about his expertise on prospecting problems and how it informed his participation in wartime nuclear research, especially in restricted missions that dealt with designing innovative prospecting instrumentation and mapping deposits of strategic minerals.

Don Quixote Charging at Neutrinos

The portrayal of Pontecorvo's wartime activities in this chapter suggests that his work was exclusively in the applied field. In fact, towards the end of World War II his studies began to blend pure and applied problems in innovative ways. In the postwar years the development of very sensitive instrumentation for the detection and measurement of radiation played a key role in the advancement of nuclear physics. Pontecorvo sought to contribute to this advancement by pioneering the study of neutrinos.

Nuclear reactors are sources of energy; but more significantly they produce radiation that can be analyzed to make assumptions about atoms and nuclear particles. In 1946, Pontecorvo thought that the NRX could be used to analyze particles of no charge and negligible mass, the neutrinos (v), whose existence had been postulated for the first time in 1930 by the German physicist Wolfgang Pauli. Only the existence of neutrinos, according to Pauli, would have demonstrated that the beta decay (the decay of a neutron into a proton and an electron) followed the law of energy conservation. In 1934 Fermi wrote a paper on the subject that was sent to the prestigious British scientific journal *Nature* but was cast off because the journal's editor believed that it "contained speculations that were too remote from physical reality."[75] Pontecorvo believed that by using reactors, he could vindicate his master's work.

If nuclear piles produced neutrinos, then either "the pile itself, *during operation*," "hot uranium metal extracted from the pile," or "fission fragments" could be used as neutrino emitters. Neutrinos are undetectable in radiation measurements, but Pontecorvo believed it possible to analyze their products in an *inverse* beta process, when intensive neutrino bombardment allowed the production of new isotopes.[76] The proposed experiment presented substantial difficulties, as the amount of isotopic substances produced through such a process would be extremely low. Thus, Pontecorvo needed a pool of several cubic meters filled with the irradiated chemicals in a liquid state if he wanted to detect the neutrinos. But the NRX was not yet available when he thought of detecting neutrinos in this way, and it would not be so for two years.[77]

The search for neutrinos thus informed Pontecorvo's decision to set up a small cosmic ray unit at Chalk River together with the British scientist E. P. "Ted" Hincks, whom he had met while studying U-233 fission products.[78] Cosmic radiation is produced in the form of high-energy nuclear particles outside the earth's atmosphere. The impact of these particles with the atmosphere produces nuclear reactions that physicists use in making hypotheses about their characteristics.[79] It was especially the meson decay that, in Pontecorvo's opinion, lent support to a novel understanding of the beta process, and in turn led him to infer something about neutrinos. Mesons are charged particles with a mass one to two hundred times greater than that of electrons. Their decay was analyzed in the 1930s by the Japanese scientist Hideki Yukawa, but from 1945 new experimental evidence—such as that presented by Marcello Conversi, Ettore Pancini, and Oreste Piccioni—challenged Yukawa's conclusions.[80]

In 1947, these contradictions led the American physicist Robert Marshak to postulate the existence of two types of mesons: a heavy one that disintegrates in the upper atmosphere and a lighter one that results from this disintegration. Later that same year, research carried out at the University of Bristol by Cecil Powell, Giuseppe Occhialini, and Cesar Lattes confirmed Marshak's assumption with the use of a novel technique based on photographic emulsions. The lighter muon (μ-meson) was produced in the decay of the heavier pion (π-mesons).[81] Pontecorvo later recalled that the debate on mesons fuelled "severe antidogmatism."[82] What made him eager to follow it even before the Bristol group published its groundbreaking results was an alleged analogy between the beta processes taking place in reactors and in cosmic rays. He assumed that muons were comparable to electrons in the reactor's beta decay and that their own cosmic ray beta decay should have produced an electron and two neutral particles—of which one would presumably be a neutrino and not, as suggested by Yukawa, a photon.[83] The assumptions about these interactions prompted Pontecorvo's attempt to examine muon decay products. In turn, this led him to produce more sophisticated detectors, mainly electrical counters, to use in these experiments.[84]

The conclusions reached by Pontecorvo and his coworkers confirmed the findings of Powell's group, even if they attracted less interest in the scientific press. The technique used in Bristol had been much more fashionable, as it produced photographs that looked like artworks whereas electrical counting produced only clicking noises registered in headphones. Furthermore, if the results found by Pontecorvo's group lent support to the claim that two types of mesons existed, they did not shed light on the existence of neutrinos. In 1949 Hincks and Pontecorvo concluded that none of the existing hypotheses about the decay could be justified with experimental evidence. Their experiments could be understood to mean that the muon decay consisted of the emission of an electron and two neutrinos only because all other solutions had been ruled out. It was a typical case of falsification of alternative hypotheses—which in itself, of course, is not a validation.[85]

If the experiments on the meson decay did not produce the needed "piece of magic," then the chlorine-argon method failed as well. Pontecorvo confessed his difficulties in a letter to Amaldi in which he defined his attempt at detecting neutrinos as "incredibly acrobatic."[86] Fermi, too, was skeptical about Pontecorvo's research, as Pontecorvo recalled years later:

> I discussed the chlorine-argon method [. . .] with Fermi in Chicago (I think in 1948) and later in 1949 at a conference in Basel-Como. Fermi was quite unenthusiastic concerning the application of this method to neutrinos [. . .]. Looking back, I can very well understand Fermi's reaction. I think it was Segrè who said that Don Quixote was not one of Fermi's heros [sic]. He could not regard with sympathy an experiment which, although it did have a brilliant conclusion due to the heroic efforts of R. Davis, but only many, many years after it was first conceived.[87]

FIGURE 2.8 Enrico Fermi (second from right) and Bruno Pontecorvo (at left) during a visit to the plant of the Italian manufacturer Olivetti, 1949. Courtesy of Archivio Amaldi (Amaldi Archive), Department of Physics, University "La Sapienza," Rome.

Fermi was far more convinced about the validity of another important method Pontecorvo pioneered during this period: the proportional counting technique (figure 2.8).

At the end of the war, the British physicist Wilfrid B. Lewis had replaced Cockcroft as head of the Chalk River facility. Lewis was a world-class expert in the development of counting techniques, having written the first textbook on the subject in 1942,[88] and he sought their advancement at Chalk River. Together with the British Geoff C. Hanna (Chalk River director from 1972) and the Canadian David H. Kirkwood, Pontecorvo considered the merits of using the technique to investigate the spectra of beta-emissive isotopes such as tritium and argon.[89] The three investigators concluded that the proportional counting technique allowed them to achieve greater accuracy in analyzing electron capture, beta-ray problems, and measurements of fluorescence yields.[90]

The Implications of Pontecorvo's Physics Program

We shall see that the postwar transition of Pontecorvo's research into areas of nuclear physics such as cosmic rays was used by British officials in the wake of his defection to demonstrate that his studies were of little relevance in the development of the Soviet nuclear program. But the development of innovative technologies such as those described above emerged in conjunction with the demand for instrumentation for NRX monitoring purposes as well as, more generally, with prospecting problems. This reveals that even the most theoretical aspects of Pontecorvo's physics program were influential in the solution of applied tasks.

For instance, in 1946 Pontecorvo designed a large proportional counter (a BF_3 neutron counter) that was used to detect NRX fuel failures through the detection of delayed neutron emissions from fission products. He also manufactured sensitive neutron monitors for the initial NRX start-up. This is the reason why he was one of the four physicists allowed in the control room when the pile went critical for the first time.[91] Another reason for his development of cosmic ray experiments at Chalk River was to test instruments used in measurements of the neutron production in power reactor systems.[92]

The development of innovative counters also marked a progress in the neutron well logging program. Pontecorvo continued advising Scherbatskoy and Neufeld about the potential of new counters, as well as on new logging techniques more generally. In the postwar years the WSI merged with the Los Angeles–based company Lane Wells, becoming one of the leading prospecting firms in gamma ray logging.[93] In 1948, Scherbatskoy left the WSI and established his own firm, the Tulsa-based Geophysical Measurement Corporation (GMCO).[94] In the same year, Neufeld was employed as a member of staff of the Health Physics Division at the AEC Oak Ridge National Laboratory.[95] From that point on, the three researchers sought to collaborate on a consultancy basis.

Before leaving Tulsa, Pontecorvo had considered that the improvement of any logging instruments exploiting the detection of radioactivity relied upon the combination of gamma ray and neutron logging. Natural gamma ray logging makes it possible to distinguish between radioactive shale and limestone or sandstone, and the neutron log provides details on the concentration of hydrogen in the strata. In this way the combinatory logic of detector signals and electric circuitry (also used in cosmic ray research) was used to understand the structure of oil-bearing geological formations.

One promising technological solution that allowed this combination was the scintillation counter, in which crystals such as anthracite, cadmium tungstate, or calcium tungstate convert incoming radiation into light impulses. The accuracy of such counters was higher than that of traditional neutron counters and, instigated by Schebratskoy to find novel techniques, Pontecorvo sought to better understand their uses at Chalk River.[96] For instance, in 1947 Scherbatskoy sought Pontecorvo's advice on radioactive tracers that could be used in the mud from oil well drilling, and he received information on the potential of indium, which had been the subject of a recent investigation by Pontecorvo and the British physicist James Dunworth.[97]

Although it is mainly thanks to recently released archival documentation that important details about the implications of Pontecorvo's research have now been revealed, it is worth noticing that Pontecorvo's expertise in nuclear geophysics was well-known in its own time, as it prompted requests for collaboration from colleagues and industrialists alike. In May 1947, Amaldi wrote to Pontecorvo that Italian oil prospectors wanted to share information on his recently developed techniques.[98] Similar attempts were made by personnel at the Institute for Applied Geophysics based at the Polytechnics of Milan, at the Azienda Generale Italiana Petroli, and at the Anglo-Iranian Oil Company. The Italian physicist Italo Federico Quercia sought to develop a program on neutron well logging in the context of the newly created Centro di Studio per la Fisica Nucleare (Center for Nuclear Physics, or CNR). In 1949 Goldschmidt and Frédéric Joliot-Curie asked for information on the nuclear log.[99] The merits of Pontecorvo's pioneering technique also did not go unnoticed in the Soviet Union, as we will see in the following chapters.

Nuclear geophysics was experiencing unexpected commercial success. In 1949 Scherbatskoy's patents had ensured him an important advantage in field development. The GMCO began selling prospecting instrumentation to American and Canadian companies, and it produced eight commercial counters, of which seven were deployed in subsurface radioactive techniques for oil exploration and one was used in uranium prospecting. By then, Scherbatskoy and Russell had also become AEC consultants, and they designed new counters used to prospect uranium in the Colorado Plateau.[100]

Commercial interests overlapped governmental interests too, as by then uranium had become a strategic commodity. In 1946, 97 percent of uranium resources worldwide were in CDT hands. Low-grade ores were also found in Sweden and South Africa, and the CDT promptly ensured their output. However, the forecast was bleak. In ten years' time the Soviet Union would catch up by acquiring at least 65 percent of newly found deposits.[101] Nuclear technologies were closely monitored exactly because of this. Regular checks were made on existing instrumentation, and declassification was allowed only after it was definitely ascertained that no useful knowledge could fall into the wrong hands. Russell and

Scherbatskoy did not publish details on their work until 1951.[102] Archival documentation suggests that Pontecorvo was one of the few world experts who had access to this type of instrumentation and knew its secret details.

Uranium deposits worldwide were also closely monitored by British research agencies, and Pontecorvo knew about this work because in 1948 the disclosure of information relating to uranium prospecting instrumentation had caused a major outburst. In 1945, the UK Geological Survey set up a uranium committee on behalf of the Ministry of Supply. Intelligence gathering provided the committee with a list of uranium deposits in several countries. Most of the prospecting work was done with instruments designed in Canada.[103] In 1947, following another meeting in Washington, DC, on prospecting instrumentation, the British Telecommunication Research Establishment (TRE) was put in control of the design and provision of new counters. The reports written by Pontecorvo on Great Bear Lake geophysical exploration were passed on to TRE experts, together with details of Pontecorvo's neutron counters.[104]

"Uranium intelligence" dealt also with the study of whether or not the Russians knew about their own deposits, and also provided a general idea about their detectors' sensitivity. Following Pontecorvo's written recommendations on uranium prospecting, the TRE developed two research programs: portable devices for field exploration and instrumentation for aerial surveys. This work was supervised by a governmental organization, the Strategic Minerals Committee, that worked for the Ministry of Supply.[105]

In April 1948, following his publication in a TRE newsletter of details of a new type of gamma ray detector, the physicist Dennis Taylor was reprimanded by his superiors—and just two months later, Taylor met with Pontecorvo to carry out radioactive element analysis with proportional counters.[106] The instrumentation used for determining the proportions of uranium and thorium in ore samples was considered secret. In one piece of correspondence between managers of the Geological Survey, it was also stressed that "to publish reports in technical journals is surely to make a quite unnecessary free gift of information to certain countries particularly interested in what we are doing in this field. Perhaps it is salutary to remember that release of *any* information concerning prospecting of non-ferrous ores within the USSR is a capital offence."[107]

By this time it was clear to British officers that the Cold War loomed large over matters of prospecting. Instrument designers like Pontecorvo knew that the makeup of counters could make the difference in speeding up nuclear programs.

Harwell

In 1948 the British journal *Nature* announced that Pontecorvo, "who was responsible for pile development in Canada and is now conducting fundamental atomic energy research in nuclear physics," had been appointed senior principal scientific officer at the AERE.[108] By 1949, when he settled in Britain, Pontecorvo had in many ways already contributed to the creation of the new research facility.

When Cockcroft left Canada to take up the AERE directorship, he considered the importance of having Pontecorvo working at Harwell. Initial plans indicated Pontecorvo as AERE's pile physics expert. But as the physicist stayed in Canada for two more years, the plans were never implemented, and James Dunworth was appointed instead for the research position.[109] Yet even when an immediate transfer of Pontecorvo proved unfeasible, Cockcroft sought to find ways to have his research team at Harwell exchange ideas with him and with the other researchers at Chalk River.[110] The Harwell plant was designed along lines similar to those that typified the Chalk River facility. Thus, a new zero-energy pile (GLEEP) was built at Harwell together with the main research reactor, the British Experimental Pile. Collaboration between the research groups at Chalk River and Harwell was strengthened through a "pile discussion group" that included Pontecorvo, and Dunworth stressed that pile designing at Harwell ought to be managed by a group similar to the NRX Lattice Group established in 1943.[111]

When Pontecorvo moved to England in March 1949, he joined the AERE Power Steering Committee (PSC), which was designed to consider recent developments in reactor technology, fissile materials, and pile design. Cockcroft stressed that Pontecorvo's presence was required exactly because he had played "a considerable part in pile development in the past."[112] In this period the committee was involved in the planning of new power-generating reactors to be deployed both for energy supply on land and to power submarines. Joint meetings with representatives of the British admiralty had also taken place in collaboration with the British firm Metropolitan-Vickers. Classified papers for both projects were circulated among the participants. Pontecorvo took an active role in these discussions, stressing that, as far as these new programs were concerned, heavy water reactors offered better prospects owing to their higher production rate.[113]

Pontecorvo also sought to further develop his nuclear physics studies, even if he was puzzled about the directions of research. Called to participate in a project of cooperation between several European countries on cosmic ray research, he indicated his intention to use the laboratory in the Alps that had just been set up by the CNR.[114] However, Pontecorvo's participation in pure and applied nuclear physics programs was abruptly interrupted following an internal security enquiry. In March 1950 the PSC chairman Klaus Fuchs was arrested and charged for espionage activities. In the same month, fresh evidence emerged that Pontecorvo's relatives were communists; Pontecorvo was invited to leave Harwell. By this time Pontecorvo had achieved a remarkable position as a scientist involved in key projects typifying the nascent nuclear science and technology—a position that British government officials avoided publicizing when he fled to Russia. Indeed, they used Pontecorvo's expertise in cosmic rays and nuclear physics to claim that it was his only research interest. They were successful in that effort because Pontecorvo's scientific reports outlining the building of nuclear reactors and the prospecting of uranium in Canada and Britain were covered by secrecy. Our treatment restores some consistency to the deliberately incoherent and deceptive account given at the time of Pontecorvo's defection by showing that studies of cosmic rays and nuclear physics represented just two important aspects of his many different interests, which also included the application of neutron physics to oil and uranium prospecting, the study of fissile materials, and the design of nuclear reactors. More significantly, the portrait that emerges here is one of an eclectic scientist who was able to transfer knowledge between apparently distant research fields. This dispels one of the many myths that followed his defection. What made Pontecorvo's defection compelling wasn't just the threat of his revealing important details, but also his ability to use his accumulated expertise to solve new problems that might emerge in different research programs. Pontecorvo's expertise in nuclear geophysics helped him in shaping trajectories of nuclear reactor research, and in turn this allowed the beginning of a research program on neutrinos that eventually fed into both neutron well logging and pile physics. It was this flexibility in the application of theories and experimental techniques that was most dangerous.

3

Under Surveillance

Bruno Pontecorvo, my Italian colleague, had characterized the situation well: "You see," he said to me, "This business [of secrecy] will never cease to be both unpleasant and fascinating at the same time." **Goldschmidt, *Atomic Rivals*, 188.**

Pontecorvo's decision to flee Britain and cross the Iron Curtain followed a security inquiry at Harwell whose content was never fully divulged. Following his defection, George Strauss informed MPs in the House of Commons that Pontecorvo's vetting operations in the seven years preceding his flight had been rigorous. It was only in April 1950, when new information on Pontecorvo's communist acquaintances prompted the recommendation that he be released from a job exposing him to sensitive information, that MI5 decided to hasten his transfer to another position that presented no security risks. Strauss's words provided a reassuring picture. The conduct of the security services had been sound insofar as they had reviewed Pontecorvo's case frequently and thoroughly, and action was taken as soon as the threat to nuclear security had emerged. The management of British science administrators was equally satisfactory inasmuch as they had quickly intervened by finding new employment for the Italian-born scientist. The chain of events leading to Pontecorvo's departure was unpredictable and unpreventable.

If Strauss tried to restore confidence in the nuclear security system, the inconsistencies, contradictions, and flaws that had typified the security services' action before Pontecorvo's flight were enshrined in secrecy. Nothing was said about the divergence between interpretations by US and British intelligence agencies of the nature and significance of security information that the agents had put together on Pontecorvo from as early as 1943. Nothing was said about how these conflicts in security management had continued throughout the war years, also affecting counterespionage operations in connection with atomic matters. Not a word was proffered on the fact that the services' action may have caused, rather than prevented, Pontecorvo's departure. Nothing was said that could shed light on the connection—if it existed—between the Pontecorvo affair and the atomic espionage cases.

The reasons for Strauss's omissions will be fully explored later in this volume. This chapter, however, seeks to promote an analysis of how investigative work was carried out, and security provisions applied, in the years that anticipated Pontecorvo's defection. In looking at these activities, some important aspects of security administration in relation to nuclear matters emerge. The harnessing of intelligence did not respond to the urgency of protecting nuclear security and tackling atomic espionage. It emerged in the context of intelligence activities that had nothing to do with it, and which were marred by bias that in fact revealed the agencies' ambitions. In particular, Pontecorvo's status as a refugee from an enemy country immediately attracted unwanted attention from the secret services, as did the presence in his family of communist militants.

However, when security information was harnessed, it hardly ever helped to assess whether Pontecorvo was a risk to nuclear security, because it was too vague and generic to clearly recommend action. In fact, in different stages it paved the way to contradictory decisions in relation to Pontecorvo's recruitment, access to restricted information, and travel abroad. In contrast, critical evidence showing the theft or illegal transfer of scientific information was never produced—in fact, it seems that it was never even sought.

Nevertheless, the harnessed evidence helped the intelligence agencies to avoid blame and criticism when Pontecorvo defected. In fact, recycling nonspecific details on Pontecorvo helped to decrease public pressure on the security services when they were accused of laxity in protecting nuclear secrets.

MI5 and the Enemy Aliens

The atom bomb was not yet a prospect, nor was atomic espionage a threat, when security personnel at Scotland Yard (MI5 headquarters in London) assessed Pontecorvo's circumstances for the first time. In the spring of 1940 Pontecorvo's name was in the list of colleagues of Frédéric Joliot-Curie to be rescued and transferred to Britain in the context of the secret mission organized by the Earl of Suffolk. This was the mission that, as we have seen in the previous chapter, led Hans Von Halban and Lew Kowarski to Cambridge. Pontecorvo's rescue never occurred, however, as it met with opposition from MI5. A file of its Special Branch indicated that Pontecorvo was not welcome in Britain: "One Dr. PONTECORVO, a collaborator of Professor JOLIOT, is regarded as 'mildly' undesirable: might possibly be allowed to work if vital to the war effort, but even if working, should be watched."[1] At the time, no incriminating evidence was available against him. Thus, the allegation that he could potentially represent a security threat did not derive from restricted knowledge about his life or acquaintances in Paris. In fact, it was the result of a broader analysis of the refugees' circumstances, drawn on the basis of those exiles now living in the Britain. Scotland Yard agents had always feared that refugees from enemy countries could be saboteurs secretly working on behalf of their governments. MI5 historian Nigel West has claimed that the secret services had been efficient in carrying out surveillance and vetting operations. But it is also true that their agents' approach to the problem seemed to be marred by prejudice.[2]

The agency known as MI5 had been established in 1916 to replace the Secret Service Bureau. The Directorate of Military Intelligence, Section 5, was born in the midst of an anti-alien hysteria that had typified Britain during World War I. Its main duty was shadowing and vetting thousands of refugees now in Britain, some suspected of secretly working for enemy governments.[3] During World War II, Scotland Yard dealt with the refugees on the basis of the lesson learned during World War I. It thereby emphasized the need to limit refugees' admission to the country or curtail their enlistment for war work. So by 1940 the rescue or recruitment of enemy aliens was frowned upon, even in cases where the exile would be fighting against the regime that, for political or religious reasons, had been responsible for his forced emigration from his homeland.[4]

The way in which war developments were portrayed in Britain seemed to confirm the need for stricter measures. In turn, this allowed

MI5 personnel to tighten security further and handle the problem of enemy aliens without restraint. In May 1940, the British press went on a rampage against an invisible fifth column (including "fascists, communists, pacifists and refugees") that had allowed the fall of Norway and Denmark into German hands.[5] On 10 June 1940, when Italy declared war on Britain, all male Italians between fourteen and sixty years of age on UK soil were rounded up and arrested. Most were sent to internment camps, and at no time was a decision taken to divide those who had openly supported the Fascist regime in the past from those who had nothing to do with it, or even opposed it. After the war ended, former internees even claimed that MI5 filled lists with the names of antifascists secretly handed over by the OVRA (i.e., the Fascist secret services).[6] During these security operations Bruno's older brother, the biologist Guido Pontecorvo, was arrested in Edinburgh, where he lived, and was interned on the Isle of Man.[7] One of Bruno's cousins, Enzo Sereni, was arrested in the kibbutz he had helped to establish in Palestine—at the time still a British colony—and sent to an internment camp.[8]

The security measures implemented by the British secret services did not always meet with approval. Eventually, criticism surfaced as it emerged that they had been of no avail against imaginary fifth columnists while, by contrast, causing avoidable suffering to innocent people. Whatever the case, an MI5 intervention, right in the middle of some of the world conflict's darkest days, hampered Pontecorvo's transfer to Britain.

Political Refugees in the United States

When Pontecorvo reached the United States, he found a situation better than the one in Europe so far as refugees were concerned, but one that did not last for long. On 7 December 1941 the Japanese attack on the naval base at Pearl Harbor, Hawaii marked the US entrance to the conflict and the beginning of warfare with Japan's allies: Italy and Germany. Italian refugees now living in the United States paid the price for these decisions as they, like their compatriots in Britain, were now considered enemy aliens. Panic spread within Italian-American communities as new regulations prohibiting travel and communication were put in place; there were also rumors that soon Italian citizens in the United States would be sent to internment camps.[9] However, the situation in the United States was somewhat different from that in Britain, as the US Federal Bureau of Investigation (FBI) did not directly intervene in the political debate on the refugees. In the "land of the free," national security management depended on more than one agency devoted to internal affairs; and each one of those agencies possessed a certain degree of autonomy. Both the US Army and Navy, for instance, had intelligence organizations whose duties included the management of countersabotage and antisubversion operations.

The FBI had been born in 1908, and for some time its main duty had been the control of organized crime and the Mafia. Its second director, J. Edgar Hoover, had widened the agency's ambitions to include the surveillance of political organizations. In 1936, US President Franklin Delano Roosevelt allowed the agency to take care of investigations outside criminal activities, such as those of fascist and communist groups. Moreover, the US House Un-American Activities Committee (HUAC), established in 1938, sought to match the FBI investigations with its congressional inquiries into political activism.[10] From 1940 the FBI could put under surveillance anyone suspected of belonging to fascist or communist groups, and could even indicate the need to dismiss any such individuals who worked for the Federal government. New directives also allowed the imprisonment of foreigners suspected of belonging to such organizations.[11] By the end of 1941, investigations against refugees from Germany, Italy, and Japan intensified. The alliance between the United States and the Soviet Union did not change these circumstances. Communists continued to be put under surveillance, even though they were not openly prosecuted.[12]

In the meantime, other US security agencies had begun investigating scientists who had been recruited for military research projects. These surveillance operations drew in the Italian physicist Enrico Fermi because, as has been shown in previous chapters, at the beginning of the 1930s he had been a lukewarm supporter of the Fascist regime, especially in accepting its new science policy. By the end of the 1930s, when Fermi moved to the United States, it had become obvious that he was no longer associated with the regime or its leaders. His move had, in fact, become the subject of backbiting in Mussolini's secretary office. When Fermi asked to meet the Fascist leader Luigi Federzoni—a prominent member of the Grand Council of Fascism and the chairman of the Royal Academy of Italy—before leaving for Stockholm, someone had penciled on his request: "His wife is a Jew, and so is his daughter."[13] It is true that once he was in the United States, the Nobel-prize winning physicist tried to avoid a scandal in his own country by keeping quiet about the real reasons for his emigration. Yet the Fascist press had taken his departure as clear evidence of his enmity to the regime. In January 1939 Federzoni wrote to Fermi asking him to explain why he had moved to the United States. Fermi answered: "My visit to America has no connection whatsoever with racial matters. It is exclusively due to scientific interests, as were my previous five visits. I'd like you to dismiss on my behalf any other misleading interpretation."[14]

Fermi's real intentions could not be easily disguised. Before leaving Italy, he had left to Amaldi all administrative matters regarding the Physics Institute in Rome. Two years later, the Ministry of Foreign Affairs in Rome received fresh news from the Italian embassy in Washington that "even though he does not directly campaign against fascism, he has all his acquaintances [. . .] with intellectuals who are fierce enemies of the regime."[15] Even so, once in the United States, Fermi was put under surveillance by security agencies. According to intelligence documentation produced by the US Army, there was no doubt that he was a Fascist. Thus it was deemed extremely important that he not be involved in research of a classified nature.[16]

Despite his position as an enemy alien under surveillance, Fermi became one of the key figures of the Uranium Project and, later, one of the chief scientists involved in nuclear research. We now know that the FBI was against his recruitment, but the OSRD director, Vannevar Bush, personally guaranteed his integrity and loyalty. While working in military research programs, Fermi complained to his superiors about the actions of security personnel, being aware of these surveillance activities.[17] By then he was resigned to the restrictions on enemy aliens traveling across the country, even if they forced him to seek a permit every time he visited New York from Chicago. Yet he was very annoyed when he found out that his correspondence was being inspected by intelligence personnel.[18]

Emilio Segrè, now living in Berkeley, California, was also annoyed about these inconveniences, and he feared that soon he would be sent to an internment camp. The FBI, like MI5, had filled lists of Italian refugees in the United States to whom restrictive measures on travel and communication applied. These measures were eventually revoked only on 12 October 1942, when Roosevelt decided against them. This was Roosevelt's personal contribution to the 450th anniversary of Columbus landing in America.

Framing (and Recruiting) Pontecorvo

Pontecorvo's position also became the subject of secret inquiry in those tense times. Just before the Roosevelt administration ended enemy alien status for Italian refugees, FBI agents began investigating those on

American soil to verify their adherence to the provisions. In September 1942, two FBI officials from the Oklahoma City command center inspected Pontecorvo's house in Tulsa. We don't know exactly what they found, or what they reported to their superiors.[19] Yet in the wake of the Pontecorvo affair, the information contained in the FBI report produced by the two agents was forwarded to MI5 through its B2a (counterespionage) division, which was headed by James C. Robertson:

> On September 25, 1942 the subject's residence at 219 South Florence Street, Tulsa, Oklahoma, was searched in view of his status as an enemy alien. During the search it was noticed that he possessed 25 or 30 books or pamphlets containing Communist literature. Subject's wife was reported to have been interviewed and asked whether or not her husband was a Communist, to which she replied that she did not understand and did not know what a communist was, although she spoke English well and appeared to understand all questions which had previously been asked. Mrs. Pontecorvo stated that her husband studied under Madam Curie while in France.[20]

From that moment, Pontecorvo was aware that the federal agency had framed him as a communist sympathizer, and therefore as a potential security risk. It is likely that the episode played a part in his decision to leave Tulsa and seek employment elsewhere; we know that in November 1942 he was desperately looking for a new job. It is also possible to speculate that the visit by FBI agents ruled out Pontecorvo's employment in sensitive posts in the United States. The agency continued collecting information on Pontecorvo's relations with other Italian émigrés known to be militants of the Italian Communist Party, and it knew that Pontecorvo's sister Giuliana, her husband, Duccio Tabet, and the militants Giuseppe Berti and Ambrogio Donini had all moved to the Bronx in New York, one of the largest settlement of Italian emigrants in America.

Bizarrely, the FBI enquiry did not hinder the scientist's recruitment in the Anglo-Canadian project or his employment in Tube Alloys. There is sufficient evidence showing that Pontecorvo's appointment followed an accurate analysis of his circumstances through refereeing, vetting, interviews, and even coordination with American science administrators from both scientific and intelligence viewpoints. The case of his employment was first handled by the British chemist John H. Wolfenden at the BCSO in Washington, DC. When Wolfenden was appointed as the

BCSO's principal scientific officer in 1942, he was assigned the task of dealing with the request to employ Pontecorvo and Auger in Montreal.[21] After Von Halban had first considered recruiting the Italian physicist, Wolfenden sought to obtain more information about him.[22] In turn, two prominent scientists were asked to provide information on Pontecorvo's circumstances. The first, the American nuclear physicist Hugh C. Paxton of the Department of Physics at Columbia University, was most certainly aware of Pontecorvo's scientific credentials. One of Lawrence's "cyclotroneers" at the Radiation Laboratory, Paxton had been dispatched to Frédéric Joliot-Curie's laboratory at the Collège de France in 1937 to help in completing and testing the cyclotron installed there.[23] In the meantime, he had also studied phenomena of nuclear resonance with Von Halban.[24] Paxton left Paris in September 1938 to help start up another cyclotron at the Columbia University, becoming a member of John Ray Dunning's group. It is worthy of notice that Dunning and his coworkers collaborated with Fermi after his employment at Columbia on problems associated with nuclear fission. Thus Fermi, through Paxton, might well have been aware of Pontecorvo's appointment—and may even have contributed to the refereeing process. Paxton went on to work in the Manhattan Project, and in 1948 he became one of the most prominent scientists at Los Alamos, where he worked until 1976.

The second referee, the French biologist Louis Rapkine, was also a former Collège de France scholar, but most importantly he was a prominent figure at the Rockefeller Foundation, where he advised on the rescue of European émigrés. He had been responsible for making it possible for many French scientists to find refuge in the United States and establish forms of cooperation with the Americans. Actually, Rapkine had "used adversity" to establish a French scientific mission in America and contribute to its participation in the Allied war effort.[25] Following an agreement between French and Americans, in December 1941 the Bureau Scientifique was established in New York, and under Rapkine's directorship it became the scientific arm of Charles de Gaulle's Free French Movement.[26] The agreement did not fully remove the suspicion of US science administrators about French activities, but it did allow Rapkine to independently manage a number of issues, including arranging the employment of French and other European nationals in America.

British security also checked Pontecorvo's circumstances thoroughly before agreeing on recruiting him. Once references were obtained, Wolfenden passed the documentation on to E. K. Balls, an official of the British Security Coordination (BSC). This was an organization based in

New York and directed by William Stephenson; it was concerned with British security matters in the Western Hemisphere.[27] On 9 December 1942, less than a month after the FBI agents visited Pontecorvo's house in Tulsa, Wallace Akers and Michael Perrin, Tube Alloys' two leading administrators, concluded that "British Security people here give PONTECORVO an *unusually* enthusiastic report" [my emphasis].[28] Balls gave his final approval to the recruitment on 30 November, after having acquired a "rather detailed report."

Balls made the same comment in correspondence with Wolfenden in March of the following year, when the BSC received information from the FBI regarding the visit to Pontecorvo's house.[29] The FBI had cabled the information to the British organization on 3 February 1943 from their Tulsa office. They received a request for clarification from the BSC a week later, when a Mr. Stott, the BSC officer who had taken notice of the FBI report, relayed information about the recent refereeing process. The FBI's reply to Stott reiterated the content of the recent investigation, but was noncommittal about whether the findings should preclude Pontecorvo's employment in Canada.[30] It is unclear whether Balls saw the report and considered it too flimsy to rule out the Pontecorvo's employment—or, as seems less likely, whether Balls and Stott had not communicated about the circumstances of Pontecorvo's vetting.

We shall see that it was exactly the discrepancy between the FBI report and the BSC evaluation that sparked a major trans-Atlantic row in the wake of the Pontecorvo affair. When the episode of Pontecorvo's vetting was reviewed, the MI5 deputy director-general, Guy Liddell, alleged that presumably the FBI report had never been passed on to the BSC. Even if it had, Liddell remarked, it was unlikely to arouse suspicion.[31] It is also important to notice that the BSC vetting procedures had been approved at the highest level and, after the 1943 Quebec agreement, had also received approval by the Manhattan Project executives.[32]

Liddell's attempt to justify the discrepancy in judgment between the FBI and MI5 is telling. During the war the significance of evidence gathered on refugees' security circumstances varied considerably with regards to national context, period concerned, and agency involved. Toward the end of 1942, the strategy adopted by some security agencies with regard to refugees had radically changed. In Britain, for instance, refugees interned at the beginning of the war had been not only liberated but recruited into special units of the Foreign Office and the MI6 to carry out war work, notwithstanding the MI5 fear of sabotage. Not dissimilarly, the FBI's approach to involvement of refugees and left-wing sympathizers in war activities opposed that of other federal agencies—

for instance, the Office of Strategic Services, later to become the Central Intelligence Agency (CIA). William Donovan, the OSS head, remarked in one circumstance: "I know they're communists; that's why I hired them [. . . .] I'd put Stalin on the OSS payroll if I thought that it would help defeat Hitler."[33] Finally, the same individuals who were considered a security risk in the United States could be considered precious allies in Britain. For instance, the administrators of the Manhattan Project were deeply concerned about Frédéric Joliot-Curie, whereas in Britain he was considered an important collaborator. One document by Churchill's minister John Anderson toned down Joliot-Curie's communist profile by noting that: "He has been known to describe himself as a Communist but his communism is of the intellectual variety and he is first and foremost a patriotic Frenchman."[34]

This throws a different light on the vetting cases in the context of Tube Alloys, and on the checks of Pontecorvo more specifically. The Tube Alloys project was the result of a collaborative effort by British, French, and Canadians. The most intimate collaborators of the communist scientist Frédéric Joliot-Curie, Hans Von Halban and Lew Kowarski, had now become leading figures within the project. Others such as Placzek and Auger also had positions of responsibility. If Pontecorvo was considered a potential threat, then many others—even key line managers—could be thought of in the same way.

Even just looking at security circumstances from the British, rather than the French, perspective reveals that the knowledge in possession of the FBI was unlikely to impress British science administrators. Since the attack on Pearl Harbor, the Soviet Union had been a British and US ally. Following the experience of fruitful scientific collaboration that typified Anglo-American relations after 1940, Henry Tizard was eager to establish a similar partnership with the Russians. In 1943 the British War Cabinet approved a mission to Moscow aiming at exchanging scientific information with the Soviet Union. It was scrapped in the summer of 1943 only because of opposition from the Americans. Even so, it is clear why Liddell believed that the FBI's information was unlikely to arouse suspicion in the political climate of 1942. By then, leading figures within the British scientific organizations saw collaboration with the Soviet Union as being as compelling as collaboration with the United States.[35]

Once recruited, Pontecorvo was no longer the subject of investigations. In two circumstances in 1943 and 1944, British agents sent their counterparts at the FBI a copy of lists of scientific personnel employed as part of the Anglo-Canadian project, highlighting that they had been vetted. Despite the existence of the 1942 report, they received no objection or criticism from their American colleagues. In one circumstance Von Halban claimed, possibly because of Paxton's reference, that "there was no doubt in his mind that Pontecorvo was cleared by the American authorities and that he, himself, was informed of the clearance, which he stressed was satisfactory."[36]

French Lambs and Russian Wolves

The management of security in the context of the atomic project was not just typified by paradoxical situations. It was also hampered by tensions of both a political and a personal nature that led to fear of security flaws where there were none, and which didn't help in understanding of where the real danger could actually come from. The case in point was the continued suspicion of US science administrators concerning the French contingent in Montreal. This made it impossible for security personnel and science administrators to clearly see the threat coming from individuals of different nationalities.

Towards the end of the war, the administrators of the Manhattan Project became obsessed with security. General Groves's main concern had now become avoiding cases of information leakage or sabotage that could get in the way of the much-awaited moment when nuclear weaponry would be tested for the first time. Security intensified especially in proximity to the first atomic trial (Trinity, 16 July 1945, Alamogordo Desert, New Mexico). The regime of collaboration between Allied projects worried Groves even more, especially because Hoover and his assistants at the FBI kept reminding him about security defects, especially with regard to the presence of French scientists in the Anglo-Canadian project. This added more fuel to an already inflammable situation.[37] The American general was especially anxious about the request put forward by members of the French contingent to return to their country in order to participate in its liberation after the campaign of Normandy in the summer of 1944. Groves perceived these requests as unacceptable because traveling could represent a security breach.

It was mainly thanks to James Chadwick, the British scientific officer responsible for coordinating collaboration between the American and Anglo-Canadian projects, that the French requests were kept at bay for some time.[38] In September 1944 Winston Churchill's chief scientific adviser, Frederick Lindemann (Lord Cherwell), visited Canada and requested that all visas in the possession of non-British scientists in Tube Alloys ought to be withdrawn. The following month, Chadwick discussed their positions with Cockcroft, and explicitly mentioned the case of Goldschmidt, as well as those of the Italian Pontecorvo and the Czech Placzek; all of these scientists wanted to visit their home countries as soon as possible.[39]

The application of these restrictions proved difficult. In November 1944, Von Halban (whom Groves already considered an unreliable individual) left for France, where, ignoring the imposed prohibitions, he met with Joliot-Curie. Von Halban informed him about recent developments in fission research, but did not approach him with the idea of leaking information in mind. As the two had important contractual agreements on fission patents, Von Halban felt compelled to update Joliot-Curie, as a patent owner, about the circumstances of their intellectual property rights. But clearly this did not prevent Groves from becoming furious with Von Halban when he returned to Canada. This time Chadwick could do little to avoid repercussions.

In January 1945 the CPC decided to keep the French scientists in Canada, and prohibited them from traveling to Europe for the rest of the year. Other nationals recruited in the project, such as Pontecorvo, were also prohibited from traveling until the following summer.[40] The only exception was that traveling to the United States was allowed if an authorization was provided by Chadwick. Von Halban was hit hard by the new restrictions, as he was confined in US institutions until September 1947. Eventually the measures were revised so as to allow Goldschmidt, Kowarski, and Guéron to visit France; but only under surveillance. Not only did these measures cause enmity among the scientists', but they prevented the spotting of an important case of information leakage. After so much "crying wolf," when the wolf really came into sight, nobody noticed it.

On 5 September 1945 Igor Gouzenko, who had moved to Ottawa in June 1943 to work as a clerk in the Soviet embassy's encryption office, informed Canadian authorities about the existence of a spy network. Canadian authorities initially considered him mentally ill, and his request to meet with security personnel was rejected. Eventually the case was jointly handled by the Royal Canadian Mounted Police (RCMP) and the British services. BSC director William Stephenson, Roger Hollis of MI5 Section "F" (Overseas Control), and MI6 Peter Dwyer (counterespionage officer in Washington, DC) all featured prominently in these investigations. Gouzenko was transferred to a secret location together with his family. His revelations led to the uncovering of a spy ring that included Canadian and British nationals. Fourteen individuals belonging to the researchers' trade union, the Canadian Association of Scientific Workers, were incriminated for espionage.[41] Among them there was the British scientist Alan Nunn May, who was arrested on 4 March 1946. Nunn May eventually confessed to passing secret information to Russian nationals and was sentenced to ten years in a forced-labor detention center. Nunn May had formerly been a researcher at the universities of Liverpool and Bristol in Britain, where he had worked before joining the Montreal team. Codenamed "Alek," he was tried for passing several pieces of information to Soviet agents, including a report on plutonium separation and a minute quantity (162 micrograms) of U–233. Its disappearance had been noticed by the French Guéron, who had immediately alerted Cockcroft. Being already aware of the investigations on Nunn May, Cockcroft held the loss as another piece of evidence in the investigation against him.[42]

The legal proceedings however, confirmed that there were important vested interests behind the portrayal of Nunn May's actions as espionage. Nunn May's defendants contested the allegation that he was a spy, stressing that in fact the information had been passed to the Soviet Union, which was a US and British ally, in an effort to help defeating their common enemy. The strategy designed by Nunn May's solicitors did not pay off, however, as he was convicted for having passed restricted information without the necessary authorization. In his closing speech, prosecutor Hartley Shawcross juggled the evidence so as to justify conviction but dispel the impression that the act implied enmity towards a wartime ally: "My Lord, I think I ought to make it abundantly clear that there is no kind of suggestion that the Russians are enemies or potential enemies. [. . .] The court has decided that the offence consists in the communication of information to unauthorized persons—it might be to Your Lordship, it might be to me, it might be to anyone."[43]

The case against Nunn May engendered contradictions that could hardly be solved exclusively at security and judicial levels, because they were greatly affected by political controversies. For instance, Gouzenko's line manager, the GRU Colonel Nikolai Zabotkin, was in charge of an organization that had a long-standing agreement with the Canadians. It entailed exchange of military information and had been approved by the Canadian War Cabinet. For some time the British war planners also continued to suggest sharing scientific information with the Soviets; a proposal to begin "regular disclosures" was put forward at the second Quebec conference of 1944. In these circumstances Nunn May's crime can be seen more as proving the frustration and confusion of the scientists recruited in the war effort, rather than as a deliberate attempt to deceive their employers.[44]

Notwithstanding these contradictions, the cases of Gouzenko and Nunn May had considerable repercussions for the scientists working in Tube Alloys. For some months, the small research community was secretly watched by security personnel in order to find out whether Nunn May's colleagues were complicit in what the court ruled as an espionage crime. Just before the Gouzenko affair came out, Groves had complained to Chadwick that Nunn May and Dunworth had been too "nosy" during their visit to Chicago.[45] A further visit by Pontecorvo was not allowed.[46] If the case of Von Halban had prolonged Pontecorvo's forced stay in Canada until summer 1945, the Nunn May case made that stay even longer—which, not surprisingly, annoyed the Italian.

These circumstances made Pontecorvo's future employment precarious. When in the summer of 1945 Chadwick rejected once again his request to visit Italy, Pontecorvo made it clear that he would continue to work at Chalk River only if allowed to travel. He felt that the restrictions on traveling could no longer be accepted. By then he had already applied for positions at the universities of Berkeley, Rochester, and Michigan. With Christmas approaching, he asked to meet with Chadwick at his BCSO office in Washington, DC, and—since the investigations on Nunn May were nearly complete—only then were his requests to travel abroad finally approved.[47]

The release of information on the Nunn May case gave impetus to the action of the security services. Uninterested in political tinkering or the implications of wartime alliances, intelligence administrators sought to use the case exclusively to extend their clout. When the agents concluded that nobody else at Chalk River was implicated in the Nunn May case, they allowed the scientists to know about his circumstances.[48] In the meantime, restricted information on the espionage cases was made available to the FBI, which used it to explore the ramifications of the espionage network in the United States.[49] By then Pontecorvo had been interviewed by security personnel for a second time, and nothing against him was recorded. On 25 September 1946, following another security review at Chalk River, a RCMP senior officer reported that "no member of the British team was reported upon in a directly adverse manner," with the exceptions of Von Halban, Nunn May, and Norman Veall (who was also in the list of suspects implicated in the Gouzenko case).[50]

The British secret services also completed their investigations, and recorded nothing against Pontecorvo. Documentation produced at this time by the British Deputy Chiefs of Staff (DCOS) highlighted that "[Pontecorvo] is one of the younger nuclear physicists and is acknowledged to be an expert on slow neutron physics—a subject on which he was doing research before the At.[omic] En.[ergy] Pr.[ogramme] was started."[51] The point was reiterated by the UK director general of scientific research (defense), stressing that as many as possible of the scientists comprising the British and Canadian teams employed on atomic energy research should be recruited in the British program. Pontecorvo's name was explicitly mentioned in the director's memorandum.[52] By the end of 1946 it was firmly established that Pontecorvo's security status was satisfactory for both Canadian and British security services.

Nevertheless, as these investigations ended, it also became clear to British military administrators that a different treatment of UK nationals and foreigners was not beneficial to nuclear security. The Nunn May blunder had made them realize it for the first time. The DCOS provisions had required the periodical review of foreigners' security status at quarterly intervals, but these checks were now deemed an unnecessary burden on those scientists and their employers. In April 1946 this regulation was reviewed, and the employment of a small number of aliens in secret defense work was authorized in view of their being "men of very exceptional ability who cannot be replaced by British scientists of equal experience and ability." Furthermore, the secret memorandum discussing the position of foreign workers in defense establishments contended that their contribution was considered to outweigh the potential risk to national security deriving from their employment.[53]

On the other hand, science administrators like Groves continued to judge the suitability of foreign scientists for work of national importance mainly in relation to the future development of nuclear programs in their home countries. When on 1 January 1946 Goldschmidt's contract was terminated, Groves discussed his circumstances with him, concluding that he was recruited "on assignment from the French government," which in turn hindered the continuation of his employment. Conversely, Groves "had just agreed to let [Pontecorvo] stay on at Chalk River" because of lack of ties with the Italian government.[54] Was nuclear security *really* about national security? Or had it more to do with the international control of atomic energy in the postwar years?

Witch Hunts: Another FBI Report

From March 1947 the doctrine of containment advocated by the new US president, Harry S. Truman, deeply changed the international political landscape, also having important repercussions in terms of foreign alliances and home security. The Marshall Plan provided financial support to the reconstruction of European nations and, at the same time, strengthened US influence in the internal affairs of Western European states. The Berlin Crisis of 1948 divided war-stricken Germany into two countries affiliated with the two conflicting alliances: NATO (from 1949) and the Soviet bloc. Churchill's 1946 warning about an Iron Curtain descending across the European continent was now proving more compelling than ever. The division that marked the beginning of the Cold War implied a renewed attention to the diplomatic and military significance of atomic weaponry. On 29 August 1949 the Soviet Union exploded its first nuclear device. Less than a month later, Truman announced the beginning of a new research program on thermonuclear devices of destructive yield far superior to that of any atom bomb yet used.

These events deeply affected the management of US home security, as national newspapers trumpeted that the Soviet Union could have completed its nuclear program so quickly mainly thanks to the existence of a fifth column in the United States that had helped the Soviets gather the needed scientific information. These fears of espionage and subversive activities instigated a revision of security provisions, and hit American political dissenters and radicals. The effort to contain domestic communism paved the way to the witch hunts that reached their peak in the summer of 1950, when Pontecorvo defected.[55]

The "red scare" played a crucial role in reinforcing the activities of HUAC, which for some years had effectively been a dormant congressional board but was now in full revival. In 1947, under the chairmanship of Representative John Parnell Thomas, Republican of New Jersey, the HUAC focused especially on Hollywood actors, directors, and screenwriters, as it feared that the cinema industry was now infiltrated by subversive leftists secretly propagandizing communist values. But its activities also encompassed inquiring about those scientists who were active against nuclear proliferation and who campaigned for the international control of atomic energy. Their activism came to be perceived within the HUAC as unpatriotic, thus being evidence of their antimilitarist beliefs as well as proving their affiliation to suspect organizations. Noticeably, HUAC was fed confidential information about individual scientists by the FBI, which reinforced its auditing.

FBI surveillance of high-ranking American scientists was also implemented from March 1947 through Truman's federal loyalty program. This increased federal agencies' power of investigation, and considered the vague standard of sympathetic association with any organization or group of persons deemed by the attorney general to be totalitarian, fascist, communist, or subversive as sufficient grounds for dismissal. "Reasonable doubt" was the new yardstick to judge the scientists' loyalty.⁵⁶ This put US scientists and government officers in a state of perpetual jeopardy, as "old charges were never settled definitely" and "once cleared, an individual could confront the same accusation in subsequent loyalty-security investigations."⁵⁷

Victims of this new regime of control, vetting, security, and persecution included physicists such as Eugene Rabinowitch, Harlow Shapley, and Edward U. Condon. The parallel work of HUAC and FBI was influential in undermining the political activities of organizations such as the Federation of Atomic Scientists (FAS), which had been campaigning for the international control of atomic energy since the end of World War II, and whose political activism was now perceived as a Soviet-led propagandistic effort. By 1949 the FAS had lost one-third of its members owing to open and secret blows delivered by the HUAC and the FBI.⁵⁸

Tightened security measures had been adopted by the US government partly in an attempt to shift public attention away from the campaigning of Senator Joseph R. McCarthy, Republican of Wisconsin, as he tried to show that the Democratic administration was too soft on security matters. McCarthy's tactics fueled fears about "fellow travelers" that catalyzed political debate in the early years of the Cold War and continued to be prominent up until 1953. American universities were deeply transformed by this renewed attention towards loyalty and security. From 1949, a loyalty oath became a necessary requirement for employment at the University of California and in another nine universities across the United States.⁵⁹ Many academics refused to adhere with the new regulations, thus having to resign.

Among those academics was Pontecorvo's friend and colleague Gian Carlo Wick, professor of physics at Berkeley. Wick joined the executive committee of the local Group for Academic Freedom, which was established on 6 July 1950 and included thirty-one nonsigners of the new oath, providing them with financial and legal assistance.⁶⁰ Segrè, also employed at the University of California, did not oppose the regulations, but feared that America had plunged into a collective hysteria that could lead the nation towards totalitarianism, as university oaths had also typified Italy's transition to fascism.⁶¹ In 1950 the microbiologist Salvatore Luria had to move to the University of Illinois in Urbana owing to his role as president of the local Teachers' Union, which made him an easy target in the running of these purges.⁶²

Domestic anticommunism allowed the FBI to intensify its activities. These consisted of new means to tap potential suspects as well as expanding the network of FBI informants.⁶³ It is in this context that fresh evidence on Pontecorvo was put together; the circumstances in which that evidence was acquired will be explored in the chapter 4. The report *Information Regarding Possible Communist or Pro-Communist Tendencies of Two Nuclear Physicists and One Biologist* contained evidence on Joliot-Curie's group in the 1930s. However, Pontecorvo's acquaintances when he was in Paris were now revealed thus:

> Informant "A" of proven reliability vouched the reliability of Informant "B" who advised that he was acquainted with three individuals in Paris under Prof. Langevin of the Institute of France who were exposed to the virus of Communism. He identified them as Bruno Pontecorvo, Sergio De Benedetti, and Salvatore Luria. Luria met Pontecorvo on the ship Quanza [sic] the last time they saw them was in 1944. Informant B reported that all 3 individuals were friendly with Mr. Sereni, a Communist, Mr. Ambrogio Donnini [sic] in addition he reported they were friendly with a Mr. Giuseppe Berti.[64]

This is how James C. Robertson, head of MI5 counterespionage section B2a, summarized it in 1949. Robertson claimed that the evidence was transferred to the FBI, and acquired not directly by it, but by another "government agency."

Although the information referred to a much earlier period, it was instrumental in instigating security operations for the individuals concerned. Sergio De Benedetti, who was now employed at the Carnegie Institute of Technology at Pittsburgh, Pennsylvania, became the subject of an FBI investigation. Luria was investigated too, but the circumstances of his political activism did not lend support to any claim of communism—not even radicalism. Aside from his role as trade unionist in the local university association, he was by then associated with the ill-fated Progressive Party, which attempted to "capture the imagination of the moderate Americans" against the "social retrenchment" of the Truman years.[65] Surprisingly, the only one who did not appear as the subject of a security review was Pontecorvo—mainly because he was by then employed by a scientific agency of a foreign country. Nor, as we shall see, was the FBI information forwarded to British security services. Yet it did eventually reach Britain.

The Fuchs Case and Pontecorvo

Europe had been only mildly affected by the anticommunist hysteria. It had most certainly entered into European countries' political life, but had not provoked major security operations. Everything changed, however, with the case of the German émigré physicist Klaus Fuchs, who was arrested in Britain on 9 February 1950 following secret investigations begun in the United States. In the summer of 1949, American cryptographer Meredith Gardner and FBI agent Robert Lamphere succeeded in cracking a wartime Soviet cipher system that led them to believe Fuchs had been passing secret information to covert Soviet agents during his stay in Los Alamos. The revelations prompted Wing Commander Henry Arnold, who was responsible for security at Harwell, to investigate the allegations. But as Arnold had known Fuchs for quite some time, MI5 managers felt that the task of interviewing Fuchs should be left to William James Skardon, of the MI5 counterespionage division. The hearings took place between November 1949 and January 1950, when Skardon obtained a confession. He found out that Fuchs had been passing restricted information to Soviet agents since 1942.

The Fuchs case opened a new phase in security management on both sides of the Atlantic. In Britain the case led towards the prosecution and trial of the German émigré scientist, who on 1 March 1950 was convicted (the prosecutor was once again Hartley Shawcross) and sentenced to fourteen years' imprisonment.[66] In the United States the case was shrewdly used by McCarthy as evidence that there was indeed a spy network that had infiltrated even the most security-tight research units in the country. In October 1949, eleven leaders of the American Communist Party were convicted for conspiracy. In early 1950, the Soviet United Nations attaché was convicted of involvement in spying activities for the Soviet Union.[67] Six days after Fuchs's arrest, McCarthy delivered a speech at Wheeling, West Virginia, in which he claimed to have a list of 205 communists who worked in the US State Department. In an attempt to capitalize on evidence of spying activities, on 20 February 1950 McCarthy claimed that the Democratic administration was infiltrated by Soviet moles. By then he had become a symbol of Republican extremism and "a political force of major proportions."[68]

In the same period, J. Edgar Hoover's FBI started an all-out campaign of investigation of the spy network that had allowed Fuchs to get in touch with Soviet spies. In May 1950, the identification of Fuchs's contact in the United States, Harry Gold (codenamed "Raymond") led to the arrest of US Army officer David Greenglass. Shortly afterwards, Greenglass's sister-in-law Ethel Rosenberg and her husband Julius were arrested. Accused of espionage in 1951, they were sentenced to death in 1953.[69] At the beginning of October 1950, ninety foreign nationals were retained by US security authorities for their political ideals. More researchers and government workers went on to lose their jobs or be convicted on flimsy evidence. Others left the United States, seeking refuge in Europe or elsewhere.[70]

After Fuchs's arrest, Cockcroft informed Arnold that he had heard "very much at third hand" that Pontecorvo and his family were communists. Cockcroft received information about Pontecorvo's alleged communist association in February 1950 during the fourth declassification conference that took place at Harwell and included delegates from Canada and the United States. MI5 records show that Cockcroft was informed by an "American delegate" about Pontecorvo's alleged political sympathies. He also asked Arnold to consider whether or not the matter should be better investigated by questioning the Italian-born scientist. However, the allegation followed the FBI's acquisition of restricted information that pointed in the same direction. The information referred to Pontecorvo's acquaintances in Paris, and therefore alleged not that Pontecorvo was a communist, but that he had been one ten years earlier. In any case, it set off the "chain reaction" of events that eventually let Pontecorvo out of Harwell (figure 3.1).[71]

Cockcroft, who had known and worked with Pontecorvo for nearly five years, was surprised by the revelation, as in no circumstance had Pontecorvo manifested political interests. Arnold was surprised too, as he had been responsible for vetting operations and had reached similar a conclusion. When on 26 November 1947 Pontecorvo became a British citizen, Arnold had filed the final report indicating that there was nothing detrimental that could halt his change of nationality. On 6 January 1948 he had consulted other scientists working at Harwell, concluding that Pontecorvo was "a straightforward fellow with no political leanings."[72]

As the Fuchs case kept Arnold busy for the rest of the month, he was only able to discuss these matters with Pontecorvo by the end of February, when the scientist "disclosed that he had a brother [. . .] who was an active Communist." The next day, a report from a Swedish source described Pontecorvo and his wife as "avowed Communists." But when information about the reliability of this source was requested by British intelligence officers, it was not provided.[73] A second interrogation took place on 6 April 1950, when Pontecorvo disclosed that his brother Gilberto ("Gillo") and his wife were communists. Two other sisters were a communist and a communist sympathizer respectively.[74] Pontecorvo denied that his wife was a communist, but confirmed that his politics were, if anything, "labor" (as in agreement with the Labour Party in the UK), but that he was much more interested in scientific research than in

FIGURE 3.1 Alexander, "Chain Reaction," *Philadelphia Evening Bulletin*, 22 May 1948. This cartoon—which appeared after a loyalty inquiry at the Oak Ridge Atomic Laboratory, resulting in two scientists bring suspended after a security hearing—highlights one key aspect of the Pontecorvo affair. The information is passed from one investigator to the next in a typical game of "telephone" to make the case about a security risk. Photo in the public domain.

politics. Arnold concluded once more that Pontecorvo had no clear political leaning, even if his views were definitely to the left and he seemed unhappy about describing the communist associations of his relatives.[75] One could speculate that Pontecorvo's unhappiness stemmed from the knowledge that, given the climate of hysteria concerning security matters, he could lose his job because of these relations even if they were neither influential to his professional activity nor subject to his control. In any case, Pontecorvo was invited to resign from Harwell.

These relations had featured so prominently in the interrogations partly because of the Fuchs affair and the pressure from the United States to tighten nuclear security. Just before the FBI cabled to MI5 the revelations contained in the cracked Soviet code, Fuchs had asked Arnold for advice about his father. Emil Fuchs had recently joined the University of Leipzig in East Germany, as professor of theology. Fuchs had asked whether he should resign from the AERE as a consequence of his father's new appointment. Arnold replied that he should do so only if he felt that he might release secret information if his father were threatened.[76] On these grounds we can deduce that Arnold may have seen the relations between Gillo and Bruno Pontecorvo as a threat to Harwell for the same reason—that is to say, the possibility that Soviet agents might threaten Gillo was a liability for atomic security.

For instance, Arnold gave some weight to the knowledge that Pontecorvo had met his brother Gillo in Menaggio, near Lake Como, during the international conference on nuclear physics that took place in the summer of 1949. With hindsight, however, we can see that Arnold's worry was misplaced. During the war Gillo was involved in anti-Fascist activities as an eavesdropper on behalf of senior members of the clandestine Italian Communist party (formerly PCd'I, now PCI). After the armistice, Gillo continued clandestine work for its section in Milan. But at the end of the war he was dissatisfied with political activism. He moved back to Paris, where he became more interested in cultural activities, getting closer to Jean-Paul Sartre, Raymond Queneau, and Pablo Picasso. He eventually became a filmmaker, and in that year he was busy contributing to Yves Allégret's latest production.[77] Exactly when the closeness between Bruno Pontecorvo and his brother had been creating concerns at Harwell, the distance between Gillo and active politics was increasing significantly.

An Easy Way Out

In the summer of 1950, Pontecorvo accepted an invitation to join the Department of Physics at the University of Liverpool. His decision followed the recent investigations, and was a convenient solution to the recent security concerns—which, not surprisingly, pleased MI5. It made it possible for them to keep Arnold's findings secret, thereby not generating the same sensation that the Fuchs case had created. It also let Pontecorvo work on nuclear physics in a new environment that offered fewer security restrictions.

The possibility of Pontecorvo transferring to Liverpool had already emerged at the beginning of that year after the appointment of the Austrian-born physicist Herbert Wakefield Banks Skinner as the physics department's director. Another Cavendish Laboratory "pundit," Skinner had previously been Harwell's acting chief when Cockcroft was still in Canada.[78] Like Pontecorvo, he had been the subject of security investigations, as he and his wife Erna were among Fuchs's closest friends. From November 1949, Scotland Yard agents shadowed them and inspected their correspondence.[79] Arnold trusted Skinner, even if he was suspicious of his wife.

In Liverpool Skinner replaced Chadwick, who had directed the department for the previous ten years, and whose wartime activities as science manager had worn him out.[80] With Skinner the physics department underwent an important expansion that saw the construction of a new laboratory to house the biggest particle accelerator (a synchrocyclotron) in Europe, a project that the British government had substantially sponsored.[81] In January 1950 Skinner asked Pontecorvo to join him in Liverpool, believing that he was the ideal candidate to lead experimental activities. When in 1949 the Polish émigré scientist Joseph Rotblat, later to become famous as secretary general of the Pugwash Conferences on Science and World Affairs, resigned from his post as department reader, a position became available. In trying to interest Pontecorvo, Skinner even dictated that the Faculty of Sciences agree to establish a professorship rather than assign Rotblat's readership to someone else. This was because Pontecorvo had made it clear that he would leave Harwell only if a better academic position was available. Yet when Pontecorvo visited Liverpool, he was not particularly impressed. The vice chancellor James Mountford was not so happy about Pontecorvo either, because of his English.[82]

From the spring of 1950, Skinner used the recent security investigations to put pressure on his colleague to accept the new position. He also convinced the university's administrators of Pontecorvo's suitability without making them aware of the ongoing inquiry. By then the security services and the Foreign Office had also warmly recommended that the Ministry of Supply hasten Pontecorvo's transfer to Liverpool.[83] The university administrators who formed the selection committee continued to raise issues about the appointment, partly because of the lack of enthusiasm shown by one of the external referees. One stressed that Pontecorvo be given a junior chair because of "his possible present lack of knowledge of the teaching and organization methods."[84] Cockcroft and Chadwick, the other two referees, were more positive, which is not that surprising given their previous knowledge of the applicant (as well as the "confidential" transfer requests put forward by the MI5).[85]

Pontecorvo's prospective placement put Chadwick in an awkward position. As he knew Mountford well, he was puzzled about whether to inform him of the ongoing security investigations. On 30 April 1950 he wrote Mountford that he was going to meet Cockcroft to discuss Pontecorvo's appointment. But he never did.[86] Pontecorvo's interview did not change Mountford's negative feelings, even if an official offer was finally made.[87]

Skinner had to fight hard to have Pontecorvo employed at Liverpool. In July 1950 Pontecorvo was still undecided, and he wrote to the vice chancellor that he wanted to wait another month "before making a final decision."[88] Skinner now stressed that a definitive decision had to be taken soon. On 24 July 1950, Pontecorvo replied to Mountford: "I am very glad to say that I have definitely made up my mind and I accept your offer of a Physics chair at the Liverpool University. [. . .] I will leave Harwell definitely at Christmas, and I might come to Liverpool only on a part-time basis from October to Christmas [. . .] in order to facilitate the completion of my work at Harwell. [. . .] I will be absent from Harwell from July 25th to September 3rd for my vacation."[89]

Pontecorvo's appointment at Liverpool University had been engineered by the security services in connection with the Ministry of Supply in order to avoid the unpleasant consequences that had typified the Fuchs case. Skinner had outmaneuvered Mountford and his colleagues to make them accept Pontecorvo's appointment, despite their very obvious resistance. Skinner's action was meant to avoid an explicit call for a dismissal from Harwell, which ultimately could be justified only in the light of secret evidence that nobody wanted disclosed. This was partly because it would have put the existence of investigations about security matters beyond the Fuchs case in the public eye. But at the same time it was also evidence that decisions on nuclear security matters were not taken exclusively on the grounds of the existence of security threats, but on the basis of a wider analysis of the employee's circumstances.

In this respect, Pontecorvo's "warmly recommended" appointment fits well in the portrait this chapter has offered of inconsistencies and contradictions typifying nuclear security. The information available to the US and British security services tells us more about their bias than about Pontecorvo's alleged espionage activities. In 1940 Pontecorvo was judged as mildly undesirable by the MI5 only because he was an Italian enemy alien. In 1942 the FBI found compromising evidence in

Pontecorvo's house in Tulsa, only because of its anti-alien investigations. Between 1943 and 1947, the effort to tighten atomic security seemed to originate more from the urgency to reassess atomic diplomacy at an international level than from genuine security concerns. As a consequence, Pontecorvo and others who worked in secret research tasks, were subjected to severe security restrictions. Meanwhile, the discovery of an espionage ring centered at the Soviet embassy in Canada came as a lucky consequence of Gouzenko's defection. After 1947 it was domestic anticommunism, and not spy-hunting, that put fresh evidence on the desks of FBI officials. And it was because of anticommunism that Gillo Pontecorvo's unrelated activities could be perceived by the security officer at Harwell as being problematic to his employment.

The investigations that raised reasonable doubt about the continuation of Pontecorvo's employment alienated him, and set the conditions for him to consider a career change. In August 1950, another important issue made him feel even more concerned about his life and career prospects.

4

Ten Million Reasons to Disappear

> Some sources have even suggested if the United States had been more prompt and generous in its payment for the process of which he was one of the inventors, perhaps his desertion to Communists may have been prevented.
>
> **Lawrence J. Bernard to Ross G. Gray,**
> **12 November 1954, in Carte Nuovo Amaldi, AMF.**

Nobody could imagine that the day Pontecorvo and his family left their house in Abingdon, near Harwell, to begin their holidays could produce startling events. It was a very ordinary day indeed. On 25 July 1950 their car was packed with holiday gear, a brand-new tent, and one son's bicycle laid down on the roof rack. Pontecorvo's sister Anna, a teacher living in London, joined the five holiday-makers on their route to continental Europe. The travelers took the ferry at Dover to land in Dunkirk, France. Then they headed south to the French Riviera, stopping at Saint Tropez. They decided to visit the Alps, resting in Switzerland. On 31 July 1950 they camped at Menaggio, near Lake Como. Between 6 and 8 August they spent some time in the Dolomites before moving south once again to visit Rome. Bruno's sister, Giuliana Tabet, owned a holiday home in the sea town of Ladispoli, a few miles north of the capital. Then they went to another sea resort, Circeo, two hours south of Rome. There they met with

Bruno's brother Gillo and spent the following week scuba diving and fishing.

Pontecorvo's "surf and turf" approach to holidaymaking might seem surprising. But, as we have seen, it had also typified his childhood and adolescence. What happened from 22 August 1950 was far too whimsical and extravagant to be explained in the light of his holiday habits. In less than a week, several unexpected adjustments to the agreed-upon plans were made, and the movements of the carefree holiday brigade suddenly became erratic. Several events—including a road accident with a cyclist and a few trips back and forth between Rome, Ladispoli, and Circeo—show that Pontecorvo was by then somewhat in distress. Heated discussions with his wife, Marianne, ensued, and excuses were made for visits to his parents in Milan, his colleagues in a laboratory in the Alps, and others—none of which ever occurred. This was the wavering prelude to Pontecorvo's mysterious disappearance and his flight of 1 September 1950 to the Soviet Union. What *exactly* happened after 22 August that caused Pontecorvo to vacillate? What worried him so much, and convinced him to make such drastic holiday and life changes?

There is a mystery within the mystery of Pontecorvo's disappearance; one that drew scientists and nuclear research managers into patents and intellectual property rights matters. It unfolded in parallel with the events of Pontecorvo's career and his relationship with American and British intelligence agencies. At times these important business affairs surfaced in newspaper articles, but most of the time they were kept confidential. This is the reason why an important episode that had an impact on Pontecorvo's decision to flee the West is yet to be fully understood. This chapter sheds new light on the importance of patent proceedings that on 22 August 1950 precipitated a court case that was widely publicized in the United States and worldwide, and which promised to have major consequences for Pontecorvo's already precarious security situation. In turn, new evidence suggests that Pontecorvo felt the looming danger like a cloud hanging over his head, and it led him to risk ruining his already shaky career and private life. These preoccupations made him behave erratically during his holiday in Italy, and ultimately propelled him toward life-changing decisions.

The Slow Neutron Patent Turns "Atomic"

We have seen that from the late 1930s, not only did the study of the slow neutron process pave the way to new avenues of research in nuclear physics, but it also convinced Fermi and his coworkers of the importance of filing patent applications on the new method. Two of these applications were filed in the United States, and they concerned the technique itself as well as the radioisotopes it was used to produce. On 2 July 1940 the first of these patents was finally issued, whereas the other was never approved, and was issued only in Canada.[1]

Before and during the war, Fermi's pupils sought to expand these patenting activities, also profiting from the experience that negotiations with the representatives of leading industrial firms had provided them with. Between 1941 and 1943, Pontecorvo went on to further exploit this experience in the field of nuclear geophysics by filing four patents, some of which concerned well logging methods based on the slow neutrons process.[2] These expectations concerning the commercial merits of fruitful inventions overlapped the problems the refugee scientists experienced in the United States as enemy aliens. For instance, in September 1942 Fermi was informed that because of his status as migrant from a country at war with the United States, he ought to file a statement on the slow neutron patent with the Alien Property Custodian in order to avoid seizure of the patent.[3]

During the war, none of Fermi's former collaborators was fully aware of the real value of the patent pool originating in the slow neutron process, even if previous commercial exploitation in Europe had shown encouraging results. Pontecorvo succeeded in selling part of his future royalties, thus demonstrating the slow neutron patent commercially profitable. In the early months of 1942 he alienated 5 percent of royalties to Scherbatskoy in exchange for fifty dollars.[4] In October of the same year he renounced a further 50 percent to the Italian émigré scientist Eugenio Fubini (who had worked with Pontecorvo in Rome and was now employed at Columbia University) for five hundred dollars.[5]

But Pontecorvo greatly underestimated the patent's value at the very moment when it increased exponentially. The CP-1 experiment performed in December 1942 at the University of Chicago was a turning point in the history of the slow neutron patent, as it exhibited the invention's importance in the design of nuclear reactors. From that moment, the principle of using hydrogen-containing materials to slow down neutrons was renamed *moderation*, with the materials deployed for this purpose (e.g., graphite, heavy water) known as *moderators*. As the Uranium and Manhattan Projects went on, moderators became a critical feature of reactors, and therefore an essential ingredient in the industrial production of plutonium. Fermi and Segrè realized that their patent was now worth much more than ever before.

However, this realization came with the knowledge that obtaining royalties for the patented invention could be difficult, as its applications were in a military field. By then a new policy existed in the United States that enforced governmental control over patents on military inventions. In 1942 the OSRD, under the guidance of Vannevar Bush, outlined new patent provisions that gave the government "the power to determine the disposition of all rights in discoveries and inventions" for national defense purposes. The decision was taken in an attempt to keep scientists from accusing companies of unlawfully exploiting their innovative technologies.[6] Yet it immediately became clear that this would be difficult, as some of the scientists involved in the Uranium Project became more vocal in defending their rights. For example, on 4 December 1942, only two days after Fermi's CP-1 experiment, Szilard was asked by a military administrator to file a patent relating to the chain reaction. The Hungarian objected, wanting to know the policy adopted on patents protecting "inventions made and disclosed before we had the benefit of the financial support of the government."[7] Arthur H. Compton, the Manhattan Project's science administrator, referred the Szilard case to Robert A. Lavender, the Navy captain and attorney who had been appointed by Bush to administer the OSRD patent office in Washington, DC. From 1943 Lavender tried to avoid similar conflicts by designing a more comprehensive system that included four different patent procedures. These measures reaffirmed absolute governmental monopoly on "atomic" patents, but also envisaged some alternatives such as licensing, and even exclusive licensing (only for uses outside the atomic energy field), to companies that showed an interest in them. One procedure allowed dealing with patents already issued and whose purchase was "off-the-shelf"; in such cases, the government would assume liability for infringement.[8]

The development of the nuclear program entailed strict control of patent matters. When the Los Alamos laboratory was set up, it included a patent office as part of its administration. Headed by Captain Ralph Carlisle Smith, one of Lavender's assistants, it began operating in July 1943. The office controlled the scientists' work so as to make sure that new inventions could be protected through patents. Patent applications on sensitive inventions, such as those relating to atomic weaponry, were placed under a "secrecy order."[9] The patent program designed by Bush and Lavender entailed the filing of 1,250 new patents with the US patent office, of which around one hundred were handled as secret items. Eighteen described Fermi's inventions on nuclear reactor technologies and were assigned to the US government for the symbolic royalty fee of one dollar.[10]

In the meantime, discussions ensued between science administrators on the possibility of trilateral collaboration on patent administration. In 1942, Akers and Perrin visited Bush in an attempt to negotiate some agreement on exclusive reciprocal licensing. At that stage, however, "Bush had blown cold over the whole question of patents."[11] By then Joliot-Curie, Von Halban, and Kowarski had struck an agreement with the British administrators of DTA on the exploitation of their nuclear fission patents (referred to as the "goldfish pool"). But as the processes described in these patents overlapped claims that were also contained in patents obtained by Fermi and Szilard, Bush felt that no gain might come from this tripartite collaboration.[12]

Scientists working at Los Alamos also wanted an agreement on off-the-shelf patents. In 1943 Fermi and Segrè (figure 4.1) approached Smith and Lavender; on 14 July 1944 the four met at Los Alamos but reached no agreement, and even considered alerting Groves about the controversy. Segrè believed it necessary to lodge a formal protest in order to avoid losing the related intellectual property rights. Fermi foresaw that an outright sale of the slow neutron patent for a sum of approximately

FIGURE 4.1 Enrico Fermi and Emilio Segrè at Valle Grande, Los Alamos, 1945. Courtesy of AIP Emilio Segre Visual Archives, Segre Collection.

$450,000 would satisfy the inventors.[13] These negotiations should have been administered by the patent assignor, Gabriello Giannini. Yet Fermi and Segrè could not relay information on the negotiations without contravening the Espionage Act.[14]

When they left Los Alamos and the potential of atomic energy was finally revealed, Giannini found out about the patent's changed circumstances. In October 1945 he wrote directly to Bush, who promised him that compensation would be effected and agreed on a sum of about $900,000. Yet we shall now see that further negotiations involving Bush, Lavender, and Giannini proved unproductive.[15]

A Soviet Ruling in the Atomic Energy Act

Bush did not keep his promise. After the war, the management of atomic energy matters became the subject of a heated political debate in the United States. Following the explosions at Hiroshima and Nagasaki, the US Congress established the Joint Committee on Atomic Energy (JCAE) to deal with nuclear matters. Chaired by Senator Brien McMahon, Democrat of Connecticut, the committee went on to make recommendations about the country's future atomic policy, and singled out the issue of whether atomic energy affairs should be administered by a military or a new civilian agency. McMahon advocated a civilian agency, whereas other Democrats, such as Edwin C. Johnson of Colorado, proposed legislation favoring the military. McMahon's approach prevailed because many Democrats felt that industrial and peaceful uses of the atom could flourish under a new civilian regime.[16] Public Law 585 (S.1717), known as the Atomic Energy Act, was prepared in the fall of 1945, amended by the Congress, and approved by the Senate in April 1946. On 1 August 1946 the US Atomic Energy Commission (AEC) was created.[17] It included a General Advisory Committee (GAC), with Fermi as one of its members.

Patent legislation featured as one of the most controversial items in the new law, which recognized the imperative need to safeguard atomic secrets by prohibiting the granting of patents on inventions that could be used in the production of fissionable materials. It also allowed the AEC to purchase (or requisition if necessary) all patents relating to such production. Although these measures had already been in place in the wartime patent system, McMahon wanted to avow them.[18] Yet the call for transparency in the administration of atomic patents met with dissent in congressional talks, as it contravened key tenets in patent law, namely the right to freely dispose of newly designed processes and instruments. The new provisions were attacked as a "Soviet" edict. The chairman of the House Patents Committee denounced them as "the end of the patent system," while a former commissioner on patents defined it "dangerous and socialistic" in character.[19]

Fermi and Segrè were much more worried about how these conflicts produced delays and uncertainties in the proceedings on the slow neutron patent. The Atomic Energy Act made their negotiations with Bush and Lavender ineffective, and forced them to wait for the AEC to be operational. Segrè was furious, especially with Lavender, who in his eyes was "extremely uncooperative."[20] Bush tried to reassure Giannini, who in turn informed Segrè that Bush was "sincerely interested in proceeding with this matter and in settling it before he leaves his office," consistently with his political stance at the time.[21] However, Bush failed to do so.

In that year, the attorney Lawrence Bernard assumed control of all legal proceedings relating to Fermi's patent, thereby becoming a chief actor in negotiations with US government representatives. A practiced lawyer in Washington, DC, he had formerly been a counsel at the Department of the Treasury and a partner in the law firm Sullivan, Bernard, and Shea (John L. Sullivan was an assistant secretary of the Navy).[22]

The Ungenerous Response

Section 11 of the Atomic Energy Act recommended the establishment of a Patent Compensation Board (PCB) to negotiate and settle awards for patents dealing with the production of fissionable materials. In January 1947 the Patent Policy Panel was established to provide recommendations on the PCB activities, and Bernard was informed that all documentation relating to the slow neutron patent had now been transferred to the newly established board.[23] The transfer allowed, nearly two years later, the filing of Giannini's application on patent number 2,206,634. As it covered "the basic process used in research and development leading up to the production of atomic energy and the production of the atomic bomb," Giannini argued that the patent was of "continuing importance in the production of fissionable materials and atomic weapons." One million dollars was now asked as just compensation for past uses, and one hundred thousand dollars a year "for the remaining life of the patent" (expiring in 1957).[24] Unofficially, however, Bernard let AEC officials know that his clients would be happy to settle the case for the lump sum promised by Bush in 1945.[25]

Pontecorvo had been involved in these proceedings. On 28 April 1948 he met with Rasetti, Segrè, Giannini, and Bernard in Washington, DC, in an attempt to solve the conflict mounting between Segrè and Giannini on the patent administration.[26] As by then Pontecorvo knew about the sum that was being requested, he was kicking himself over the 1942 agreements with Scherbatskoy and Fubini. He thus claimed that the agreement with the latter should be reviewed, as the patent was now worth much more than his earlier estimate, eight thousand dollars. Fubini tried to accommodate Pontecorvo's requests, but their relationship suffered because of these business matters.[27] Even when he moved to England, Pontecorvo continued to pursue these negotiations. Writing to Amaldi in March 1949, he stressed that he was a little pessimistic about them.[28] Yet he also promised to one of his brothers, Giovanni David, that as soon as the case was settled he could help him with his business.[29]

The AEC response to the claim presented by Giannini and Bernard shattered those expectations. In June 1949, the agency's Office of the General Counsel (OGC) analyzed the legal circumstances of the claim. Its deputy general counsel, Bennett Boskey, drafted a response that cast serious doubt on the possibility of a reward. A Harvard graduate, Bos-

FIGURE 4.2. Bruno Pontecorvo, circa 1948. Courtesy of AIP Emilio Segre Visual Archives, Physics Today Collection.

key had joined the US Army during World War II. Responsible for legal and intelligence work, he had advised the State Department on the management of German-owned patents. He was also the War Department's representative to the Joint Congressional Committee Investigation of the Attack on Pearl Harbor.[30] Boskey's understanding of the slow neutron patent case was informed by these activities, as he pointed out that three of the inventors (Amaldi, D'Agostino, and Trabacchi) were Italian citizens. Although this did not disqualify them from receiving compensation, there were "certain circumstances arising by virtue of the particular nationality." The Treaty of Peace ratified by the Italian government on 10 February 1947 stated that "the United States was not obliged to return industrial properties to Italian nationals," who "were not entitled to any patent with respect to inventions of war materials."[31]

These objections added to technical issues that also ruled out the possibility of compensation. Following a report by the PCB head, Roland Anderson, the response indicated that the slow neutron method was not essential to the production of fissionable material or usable in the development of atomic energy. The patent disclosed a process for the utilization of the slowed-down neutron principle, but Boskey denied that the AEC used it to produce radioactive isotopes, plutonium included.[32] Finally, attention was drawn to Fermi's status as GAC member: he was a government employee, so applying for compensation put him in a conflict of interest.[33] Boskey's report seemed to set the stage for refusing the claim.

Jurisprudential matters aside, Boskey's response looked ungenerous to the applicants, especially because of Fermi's activities and celebrated merits during the war. Not only was Fermi's claim supported by senior AEC officers, but even those who played a part in the legal proceedings as external advisers stressed that the compensation was primarily a matter of generosity (because of Fermi's activities). In 1952, Hector Holmes, the lawyer consulted as external adviser on the patent controversy, stressed exactly this point:

> It seems to me that the Board, or any tribunal before whom this might come on appeal, would view the Fermi invention and weigh the matter of award and compensation with generosity. Here we have one of the important discoveries in this vital field and a product of the most highly skilled research. I believe that the feeling would be wide spread, both in the world of science and industry, and with the public, that Fermi should be generously compensated [. . .].[34]

So why was Boskey now so zealous in analyzing the case of Fermi's patent? Some of the reasons are now known. In March 1948, a memorandum sent by Anderson to the OCG highlighted that of all the cases docketed before the PCB, that of Fermi's patent (technical issues aside) was by far the most commendable, as the patent had merit.[35] But another PCB report noted that settling a few meritorious cases would lead to having to deal with more unmeritorious ones as "a policy of unwise and excessively generous determinations of awards might invite too many claims, burden the budget, and possibly attract criticism that public funds are being wasted."[36]

Economic issues aside, important security matters hampered the positive conclusion of the proceedings. It is worth noting that Boskey's response was produced in the climate of security hysteria that typified the US administration at the dawn of the nuclear age. Issues concerning national security featured prominently in these dealings, defining the legal case's trajectory as well as the life path of one of the inventors, as we shall now see.

FBI Files and AEC Proceedings

In the 1940s the FBI had put together information on Pontecorvo that evidenced the possibility of connections with communist militants. The first report was filed in 1943 and was followed by another in 1949. Although this documentation did not reveal connections with espionage networks, it could be compromising in McCarthy's times.

The tightening of national security had been a preoccupation in almost every area of Truman's administration but, not surprisingly, it featured especially heavily in the management of atomic energy. Section 10 of the Atomic Energy Act allowed FBI vetting operations on employees, contractors, prospective contractors, or prospective licensees—indeed, anyone who might come into contact with secret data.[37] AEC administrators applied these security measures with eagerness, at times going beyond the legal requirements. For instance, the Brookhaven National Laboratory and other AEC facilities required FBI clearance for all scientific personnel even though they were not working on classified projects. Between 1948 and 1949, FBI checks were requested even in the context of nonrestricted AEC fellowship programs.[38]

The administrators' zeal in dealing with security was a response to accusations by anticommunist campaigners such as HUAC chairman Parnell Thomas. In 1947, following his departure from the JCAE (of which he was a member in 1946), Thomas accused the AEC director,

David Lilienthal, of being lax on atomic security. Thomas's article, published in the magazine *American*, specifically referred to protocols in the management of atomic patents, claiming that they were now available in the US Patent Office and that anyone, even Soviet spies, could obtain classified information in that way.[39] Thomas's frenzied accusation had repercussions for Lilienthal, who resigned in 1949. The administrators' zeal also informed the proceedings on AEC patent cases. Section 11 of the Atomic Energy Act made no specific reference to security matters. Yet the 1948 PCB regulation, which followed the publication of Thomas's article, indicated that the board "may issue any general and specific order, directive, or further regulation which it determines to be appropriate pursuant to Section 10 of the act to assure the common defense and security," and that "in the conduct of hearings the Board shall ensure compliance with the security regulations."[40]

These new rules did not imply that FBI security checks should be requested for all who claimed compensation. But vetting was necessary if a court case ensued. In those circumstances checks were needed, as claimants might appear as witnesses and classified subjects might therefore be discussed in public hearings.[41] The applicant's position was in all respects the same as that of an AEC employee, and just as with personnel in the AEC fellowship program, political circumstances motivated the PCB to be more scrupulous in demanding checks *before* going to court. An amicable settlement rewarding applicants who had previously been the subjects of FBI investigations could expose the PCB to political attacks if restricted information about the compensation fell into the public domain.

Most certainly the PCB and the OGC got hold of restricted security information on the applicants, including the first FBI file on Pontecorvo. In an article on the legal issues associated with atomic patents, Boskey openly asserted that one of the most controversial aspects of the PCB proceedings was the question of exactly how to deal with the claimants' security circumstances, as "indubitably some applicants not only will lack a security clearance at the outset but also would be ineligible for clearance under the Commission's personnel security clearance criteria."[42] Presumably Boskey was aware of Pontecorvo's circumstances, as the counsel was also responsible for the AEC Personnel Security Review Board. In the summer of 1949 urgent meetings were arranged to review security procedures and policy regarding applicants.[43]

These meetings took place just before Pontecorvo's second FBI file became available, and in the same period Fermi's former coworkers became aware that the AEC had been investigating their private businesses. In December 1948, Bernard received a letter from the agency in which information on the nationality and residence of the claimants was requested. The letter went on to ask for additional information on "the reported transfer by Pontecorvo of a portion of his beneficial interest in the patent to a Mr. Fubini." Segrè, as representative of the inventors, was surprised to find out that the AEC had managed to obtain some information about the deal. He then wrote Pontecorvo that he had better provide the details they were requesting, as "*hanno il coltello dalla parte del manico*" (they have the whip hand).[44]

Segrè the Whistleblower

If Segrè was sympathetic to Pontecorvo in his correspondence, he was, in fact, very anxious. He feared that because the patent proceedings made him one of Pontecorvo's associates, AEC officials might have suspicions about his loyalty just as the United States was being hit by anticommunist hysteria. The recent findings about knowledge obtained by the AEC on his former colleague's agreement with Fubini showed the agency's prowess in information gathering. Segrè wondered if the commission knew about Pontecorvo's relatives. He also worried about the financial aspects of the patent proceedings and the possibility of reaching an agreement in the presence of these hindering security circumstances. He thought the presence of communist acquaintances in Pontecorvo's family might hamper an amicable end to the controversy.

In 1949, Segrè informed AEC officer Robert Thornton about Pontecorvo's communist relationships. Segrè claimed that Pontecorvo was unreliable because "several members of his family in Italy were communists" and "had influence on him." He also alleged that Pontecorvo may have accepted the position at Harwell, rather than other job offers in the United States, "for the wrong purposes." In fact, the person referred to as "Informant B" in the 1949 FBI report was Segrè—and the government agency whose name the report did not disclose was the AEC, for which Thornton, "Informant A," worked as an adviser.[45]

Was Segrè after something specific on Pontecorvo? It is difficult to say, given the lack of archival documentation on the subject. But it is surprising that someone so close to Pontecorvo could not find evidence more incriminating than his ties to relatives during his stay in Paris ten years earlier. Of course, the information Segrè gave to Thornton (but not the information he offered to the FBI) pointed to influence exercised more recently by Pontecorvo's family members (to be discussed in more detail in chapter 5). Yet even so, the speculative tone of Segrè's accusations was clear evidence that in fact he did not know for sure why Pontecorvo had moved to Harwell, or what exactly he was doing there.

Segrè's choice of Thornton as a confidant is understandable. Ironically, Thornton was another one of Lawrence's "cyclotroneers," as was Hugh Paxton, the scientist who had acted as referee in Pontecorvo's 1943 vetting process. A British-born scientist educated in Canada, in the late 1930s Thornton had worked as Pontecorvo's assistant and, like Paxton, he had supervised the installation of cyclotrons in other academic departments before returning to Berkeley in 1945.[46] Segrè had been acquainted with him since his own arrival in California, and he considered him a "close friend." This might have led Segrè to confide in him about the vexed issue of Pontecorvo's family connections.[47] Aside from his duties as a physicist, Thornton also became a AEC manager responsible for declassification matters, especially on the electromagnetic separation process that was pioneered at Berkeley using the "calutron."[48]

The revealing information about Pontecorvo's security circumstances given by Segrè and relayed by Thornton was never attached to the PCB documentation. The lack of security documentation in the PCB files can be understood as aligning with the existing FBI disclosure policy. The PCB routinely acquired information and made it available to other governmental agencies, but recommended that it not be published so that investigations could continue and without their methods being revealed.

For instance, Boskey made reference in his paper to the infamous case of Judith Coplon, whose circumstances were not that different from Pontecorvo's as far as legal proceedings were concerned. Arrested by the FBI in March 1949 as a suspected soviet spy, Coplon was eventually released because the incriminating evidence was available only in FBI documentation that Hoover did not want to make available in proceedings. Hoover contended that the production of those documents in open court could lead to "an unprecedented outburst of criticism."[49] One of the issues at stake was the FBI attempt to protect information obtained through the decryption of a Soviet code, Venona, which had become a key tool for counterintelligence. This compliance with FBI disclosure policy explains the inconsistency between the emphasis given by Boskey in his paper to security matters, and the PCB papers' lack of documentation that would show that security played a part in the board's activities.[50]

Thus, both Boskey's response to Giannini's claim and the PCB proceedings regarding the slow neutron patent were informed by analysis of the security issues regarding the claimants in general and Pontecorvo more specifically, especially in light of Segrè's doubts. As the response directly addressed not these circumstances but other problems—such as Fermi's alleged conflict of interest, the nationality of some inventors, and the patent's validity—it created a deliberate stalemate in which no decision was taken.

Nothing, however, prevented the information from reaching Cockcroft at Harwell. We have seen that Harwell's director learned that Pontecorvo and his family were communists from an American delegate at the fourth declassification conference that took place at the AERE. The American delegation at the conference included seven AEC members, one of whom was Robert Thornton, the man responsible for declassification matters.[51] When Thornton returned from the UK, he sought to discuss these matters with FBI officers and gave assurance that the Harwell director would take action.[52] As we have seen in the previous chapter, this is exactly what Cockcroft did in the following months by arranging, together with James Chadwick and Herbert Skinner, Pontecorvo's move to the University of Liverpool.

Fermi's Conflict of Interest

In the same period when the declassification conference took place, the PCB negotiations on the slow neutron patent reached a turning point. On 19 February 1950 Anderson met with Giannini and Bernard in Washington, DC. The patent assignor and his counselor, who had carefully prepared for the occasion, believed that Boskey's negative response would not hamper the chances of settlement. Franco Rasetti, Fermi's closest collaborator in Italy, had by then replaced Segrè as the inventors' representative. He helped to prepare for the meeting by providing Bernard with copies of the pamphlet issued by the Nobel Prize Committee on the occasion of Fermi's reception of the prestigious award, and nuclear physics monographs crediting his work on neutron-induced reactions. Being in the dark regarding the role that security matters and Segrè's action had by then played in the dispute, Giannini believed that the meeting "laid the foundation for some increased activity in the processing of our claims." The settlement "was in the offing."[53] In the following months, however, protracted silence from the AEC negotiators revealed that this was not the case. Further inquiry about Fermi's alleged conflict of interest revealed that it was a complicating factor.

In 1949, Fermi wrote directly to Boskey's superior, the AEC General Counsel Joseph Volpe, to seek clarification about the accusation of conflicting interests contained in Boskey's response. In turn, Volpe sought to find out about the accusation from experts at the US Department of Justice, and in November he informed Fermi that it had merit. The accusation so exasperated Fermi that he even considered leaving the GAC.[54] The problem was exacerbated by Glenn Seaborg's predicament, which was similar to Fermi's. Seaborg was a GAC member and a claimant before the PCB for the patents on the synthesis of plutonium, which he had filed jointly with Segrè and other researchers. This ruled out ad hoc solutions. In January 1950, Volpe tried to find a way out by suggesting that the JCAE was considering an amendment to the Atomic Energy Act that would exempt GAC members from the provisions barring patent claims. The "Fermi amendment," however, could not be rapidly approved.[55]

Writing to Bernard in April 1950, Fermi argued that his only reason for not resigning was that it would damage rather than help the PCB proceedings. He thus opted for a smoother exit, letting his appointment expire the following summer.[56] By then, these private business matters had been overlapped by more serious issues, such as Fermi's political stand against the development of thermonuclear devices. In late 1949, just when the conflict of interest accusation was unfolding, Fermi and Isidor Isaac Rabi wrote the GAC report emphasizing opposition to thermonuclear weaponry: "an evil thing in any light." As the report failed to gain consensus among GAC members, Fermi decided to leave the committee.[57]

Bernard could also register the disappointment of senior AEC members, such as Brien McMahon and Sumner T. Pyke, with the whole situation; they believed that Fermi and his coworkers ought to be compensated. The lawyer even tried to gain some entrance into the decision-making process by informing Lewis Strauss, who was a former US Navy admiral and a commissioner (in 1953 he became the AEC director). Interestingly, even Strauss's intervention was of no avail. As Bernard wrote to Giannini:

> Either they [AEC administrators] are running out a bluff or Volpe and Boskey have convinced them that our claim is without merit. [. . .] As you know, John [L. Sullivan] and I met with Admiral Strauss and went into the case thoroughly with him [. . .]. Strauss seemed very sympathetic and said he would look into the entire matter at once [. . .]. A subsequent check-up indicates that no action was taken by the Commission. Since that

time the only development has been a renewed activity on the part of the Commission relative to the proposed bill to permit suits on patents owned or assigned by members of the General Advisory Committee.[58]

Giannini Strikes Back

As the summer of 1950 approached, Giannini decided to openly challenge the PCB's decisions (or indeed its lack of decision) by appealing to the US Court of Claims. In his eyes the issue of compensation was straightforward. The temperamental entrepreneur saw it solely as a matter of business, and was not aware of the proceedings' security implications. He perceived that the invention for which he held rights was being used unlawfully by the AEC and its contractors, and that it ensured growing revenues, of which nothing was being given to the legitimate owners.

The slow neutron process was fundamental in the production of plutonium, which by then had become an essential part of the US defense system. It was also thanks to this process that radioisotopes for various medical and industrial uses were produced and exported abroad. In 1949 the AEC possessed a portfolio of 109 radio-elements, of which 94 were produced in reactors using the slow neutron process. In 1949, the number of isotope shipments abroad had tripled the 1947 figure, reaching the number of 2,200 shipments.[59] On top of that, Giannini speculated that the process was being used by firms in the oil industry for well logging work, either independently or in contracts with the AEC, which from his viewpoint once again posed the problem of reestablishing fair conditions of exploitation by paying royalty fees. He asked Pontecorvo for advice on these dealings, and was told that the process was now widely used in the oil industry.[60] In light of all this, Giannini felt that he should have a fair share of present and future profits.

The lawyer Bernard presented him with the proposal of a petition. Giannini understood that unless the AEC administrators were openly challenged, they would continue to be unwilling to agree on a settlement. The plan of filing a petition with the US Court of Claims had been already suggested in 1949, but had been opposed by Fermi and Segrè.[61] Conceivably, Bernard and Giannini's decision was informed by the knowledge that from July 1950 Fermi was no longer a GAC member; they could now sue the AEC without prompting accusations about conflicts of interests or causing troubles to him. Yet they made a decision to file a suit without alerting the inventors about their intentions. As most of them, including Pontecorvo, were by then on summer vacation, they could not be easily informed. Bizarrely, they filed a petition for a sum, $10 million ($7.5 million for previous and present uses and $2.5 million for future ones), that massively exceeded anything requested of the commission so far. Giannini and Bernard obviously sought to create a sensation in the press, believing that making the American public aware of the patent controversy might force the commissioners to capitulate.

Such a decision was undoubtedly very controversial, especially in the inventors' eyes. Only Fermi could be alerted, as he was not on holiday, but was working at Los Alamos. He openly manifested his disappointment at the decision:

> I had no idea that the royalty that you propose to ask in the Court of Claims was as high as you indicate and I have the greatest misgivings that asking for such a high figure, even only as a bargaining point, will put all of us in a very unfavorable light before public opinion and ultimately decrease greatly our chances of a just settlement.[62]

Giannini replied to Fermi, stressing that it was mainly because of Bernard's experience with similar controversies that such a "high figure" was being considered:

> . . . as a layman, I have had a reaction identical to yours. On the other hand, it is my feeling that in these matters once one has placed his confidence in a law firm, one should also go along with their judgment. Mr. Bernard tells me that he has given considerable thought to the amount for which the petition was filed and has decided upon it based on his experience [. . .] . As for public opinion, I believe that the properly outlined background of us all, covering past work and accomplishments, [. . .] will do a job for us. [. . .] Mr. Bernard felt that this was not the time to do anything but to be extremely factual, answer questions as put to us by the press, and seek a minimum of publicity.[63]

In particular, Fermi, who had clashed with Giannini more than once, feared that the moment chosen by the Italian entrepreneur and the lawyer was a wrong one. In that summer, McCarthyism was rampaging and the scientists' opposition to nuclear proliferation was interpreted as lack of loyalty, if not evidence of an anti-American sentiment. Thus, Fermi feared that the press could read the petition as another attack on the US government by the atomic scientists, rather than a just claim on a due settlement.

Moreover, Giannini did not know about the personal circumstances of all the inventors involved. Thus he could not imagine that one of them, Bruno Pontecorvo, was not keen that his "properly outlined background" was being publicized by the press in this precise moment, especially as the Harwell inquiry on his security circumstances had been carefully kept concealed. A widely publicized petition might have let restricted details into the public domain with consequences that could not be easily foreseen. Giannini's attempt to strike back at the AEC ultimately seemed to realize Segrè's prophecy of two years earlier. Giannini could end up metaphorically "breaking his neck" in being so up-front and demanding with the US government.[64]

Italian Scientists Sue the US Government

On 22 August 1950, the news of Giannini's suit appeared in the *Washington Post*, although Pontecorvo's name was not mentioned. His name appeared instead in the *New York Times*. The following day the news crossed the Atlantic and reached Italy. The Italian newspapers *Corriere della Sera* and *Il Messaggero* were quick to report the news without, however, offering an explicit commentary.[65]

Different, however, was the take of *L'Unità* (the PCI-controlled newspaper), which alleged that the Americans had cheated the inventors: "The US Government condemned for defrauding Enrico Fermi. The Italian scientist has accused the American administration of building nuclear weapons without paying rights."[66] The article twisted the argument so as to lend further support to the newspaper's campaigning against nuclear proliferation. Not only had the US government decided to prioritize the military uses of the atom, but the US nuclear research program was based on a swindle of those who owned the relevant intellectual property rights. The interpretation aimed at blurring the boundaries between private affairs regarding patents and political issues regarding nuclear disarmament—exactly what Fermi had feared.

Pontecorvo read the news about the ten-million-dollar suit at the seaside resort of Circeo on the day after his thirty-seventh birthday. Already frightened by the witch hunts and aware that his employer knew about his communist acquaintances, he certainly did not welcome the publicity deriving from his involvement in a suit against the US government. Not only did the news cast serious doubt on the possibility of his receiving the financial reward he yearned for, but it suddenly brought his name into the public arena. Furthermore, it did so in connection with a serious accusation aimed at the US government—that of fraud—and a request for reparation in an extremely high amount.

Understandably, this was too much for a scientist who was already suspected in his workplace because of family connections that fueled doubts about loyalty and spying activities. By then he certainly feared that fresh news could connect the two separate issues of the filed petition and the security inquiry. A suspect scientist was now requesting an award from the US government, or accusing it of dishonesty. Most certainly Pontecorvo worried that a public scandal of major proportions was about to ensue. He presumably feared also that if the news appeared in England, it might instigate further security investigations. Conceivably, he concluded that his public image as a scientist would suffer because of this, making it impossible for him to continue his professional activities. The impending scandal was a cloud hanging over his head, and the circumstances convinced him of the urgency to make tracks. Only eight days after the claim was filed, Pontecorvo fled to the USSR.

The Clumsy Flight

The lawyer Lawrence Bernard was one of the few who spotted important connections between Pontecorvo's defection and the case of the slow neutron patent. He suggested that if the AEC administrators had awarded compensation more readily, then Pontecorvo might not have fled.[67] On the other hand, Bernard could not possibly have known that one important reason for the delay in compensation was Pontecorvo's security circumstances, which the "stool pigeon" Segrè had relayed to the commission. By contrast, most commentators who dealt with the Pontecorvo affair after the defection failed to notice these connections, presumably because the news of Giannini's petition had come out of the blue. No one, especially in Europe, was aware of the long-standing dispute over the patent that had occupied Pontecorvo, his former co-workers, and US science administrators for quite some time. The fact that the details of this dispute were by and large the subject of confidential correspondence did not help either. The commentators understood that something important had happened after 22 August 1950, but they associated Pontecorvo's erratic behavior to other circumstances, especially Gillo's visit to the Circeo resort on Bruno's birthday. Montgomery

Hyde's well known work *The Atom Bomb Spies* offers the following treatment.

> On 22 August, Gilberto and Henriette [his wife] arrived in Circeo from Rome, bringing Anna [the sister who had traveled with them to Italy] who had come on from Milan after she had seen her parents. The occasion for the family reunion at Circeo was no doubt Bruno's birthday . . . . This date seems to have been a turning point in Bruno's affairs. Hitherto the holiday had been a happy-go-lucky affair where everything went off well and everyone enjoyed themselves. Thereafter things began to go wrong, and Bruno's movements became erratic.[68]

Similarly, the *Times* journalist Alan Moorehead—the only correspondent with access to restricted MI5 information at the time—stressed that "after the visit of Gilberto on 22 *August*, everything changes."[69] Yet what made Montgomery Hyde and Moorehead believe that everything changed were in fact events taking place the day *after* 22 August. This was the day when the Italian newspapers (including Gillo's favorite, *L'Unità*) publicized the inflated news about Giannini's petition.

Everything changed on 23 August 1950. Bruno Pontecorvo suddenly decided to return to Rome. Though he was an experienced driver, he had an accident with a cyclist. The accident was not serious, but the scientist was now convinced that he ought to leave the car in Rome for repairs rather than fix it in Circeo. When he returned to Circeo without his car, he learned that his children had developed sunstroke. Two days later he returned to Rome to get the car and send a telegram to his parents claiming that due to the children's problems and the car's need for repairs, he had to change his plans and return to England rather than join them in Chamonix as planned. Pontecorvo returned to Circeo once again to retrieve his children and wife, and then headed to Ladispoli (the other seaside resort) to get his third child, Antonio, who had been staying with his other sister Giuliana. For Pontecorvo to remain on schedule, the family should have returned to England immediately. Instead, Pontecorvo returned once again to Rome. On 27 August the family stayed in the small house of Giuliana Tabet. But by then Bruno had become fidgety and decided to sleep in his car rather than in his sister's house.[70]

Dealing with unexpected problems could explain Pontecorvo's erratic behavior. But why did he want to reach Rome so urgently on 23 August

1950? And if he felt that his car could be repaired only there, why did he not immediately retrieve his family and bring them to Rome, presumably saving his children from more sunstroke? Why did he decide to return to Rome *once again* after bringing Antonio from Ladispoli, rather than head north, given his urgent need to return to England?

Herbert Skinner's book review of *The Traitors* alleged that behind Pontecorvo's erratic behavior was the fear of exposure: "Presumably, in Italy in the 10 days following Aug. 22/1950, Pontecorvo was threatened with exposure in England and therefore agreed to give his services to Russia."[71] Indeed. But nothing, aside from fresh news regarding the patent case, could really promise (or threaten) to expose him.

Interestingly, when on 23 August Pontecorvo went to Rome, he did not just have his car repaired. For instance, the physicist Giovanni Fidecaro was surprised to see him at the University of Rome. Years later Fidecaro alleged that, car problems notwithstanding, the visit was difficult to justify. In fact, he said, Pontecorvo "disembarked in Rome on the Friday following the *Ferragosto* week," and "nobody leaves a holiday place on Friday, normally waiting until the next week, unless something very urgent has to be dealt with."[72]

More eccentric decisions followed. After claiming on 25 August that he could not visit his parents in Chamonix, he stayed in Rome for another week. On 29 August 1950 he finally decided to leave the capital with his family, but not for England or Chamonix. They went to the agency of Scandinavian Airlines (SAS) and booked tickets for Stockholm, where Marianne's mother lived. Interestingly, they asked for four round-trip tickets and a single one-way ticket. At this point Pontecorvo's wife appeared "noticeably very depressed and miserable," which would seem unusual for a daughter about to visit her mother after not seeing her in a long time.[73]

Following a heated discussion with his wife, which was also noticed by the airline's personnel, Pontecorvo changed the booking to five round-trip tickets. But the purchase wasn't completed until the following day, on 30 August, when Pontecorvo returned to the agency to pay. Notably, the SAS personnel had refused a payment in Italian lire, forcing the scientist to return that day with the considerable amount of six hundred US dollars in cash.[74]

If previously Pontecorvo had been quick to communicate his amendments to existing plans, this time he made no announcement to anyone about his unexpected visit to Sweden. One might consider the misleading telegram he sent to his parents on 25 August as a case of misjudgment. But the postcard sent to AERE, Harwell, on 31 August 1950 is clearly to be regarded with suspicion, if not as openly deceitful:

> Had a lot of fun with submarine fishing but I had plenty of car trouble. I will have to postpone my arrival until first day of conference (7th). Can you tell E. Bretscher [Egon Bretscher, head of the AERE research team]? Hope everybody has prepared his talk and done good work at Chamonix. I am sorry I have missed Chamonix but I could not make it. Goodbye everybody. Bruno.[75]

The postcard did not mention the travel to Sweden, and as justification for the missed appointment with an AERE team in Chamonix it cited car troubles that had been sorted out by 25 August, a week earlier. The closing sentence, "Goodbye everybody," sounds woolly even for someone who is not that well versed in the English language.

Goodbye it was. On 1 September 1950 the Pontecorvo family flew to Stockholm on a SAS airplane that also stopped at the airports of Munich and Copenhagen. It is unclear exactly what happened next. The family definitely landed in the Swedish capital but did not visit Marianne's mother, even though she lived not far from the airport. Pontecorvo asked for accommodation in the airport hotel, but he did not turn up after making the booking. Allegedly he and his family slept in a house belonging to the Soviet embassy in Stockholm. The following day they flew to Helsinki, where a man and a woman were waiting for them. A car belonging to the Soviet embassy, customarily plated as being assigned to diplomatic corps, took them over to the Soviet military base of Porkkala, in Finnish territory. Some have claimed that from there they departed on a Russian train for Leningrad. Others have suggested that they were taken aboard the ferry *Belostrov* on 2 September, reaching Leningrad on 5 September.[76]

The ten days after 22 August quite obviously anticipated Pontecorvo's defection. Whatever plans he had made earlier were now suddenly changed in ways that are difficult to justify even in the light of his extravagant approach to holiday making or unexpected events. The changes are equally difficult to understand in light of a deliberate attempt to deceive, for the postcard sent to Harwell on 31 August is the only document showing falsehearted intentions. Ultimately, the available evidence suggests that after 22 August 1950, Pontecorvo was genuinely undecided about what to do—as well as irritable, dissatisfied, and irrational.

This account obviously does not fully explain the episode of Pontecorvo's defection. Chapter 6 will address important questions, such as who helped Pontecorvo in arranging his trip to Sweden, who provided the funding, and who helped him cross the Iron Curtain. But Pontecorvo's dissatisfaction indicates that whatever prospect was made available to him before his defection, he made his decision only after a week of extreme anxiety precipitated by Giannini's petition. Clearly, it is not so common to see patents and patent controversies featuring in matters of atomic espionage. But our narrative has shed new light on the important implications of the controversy over the slow neutron patent. It is true that the dispute involved more than one inventor, and that among all of them only Pontecorvo took the extraordinary step of leaving the Western world. Yet he was also the only one who knew about (and was the protagonist of) an ongoing security investigation.

The publication of Pontecorvo's name could have reignited the security investigation at Harwell. It easily could have created a scandal of major proportions, suggesting that a communist atomic scientist was now ready to challenge the US government in court. Carrying the weight of such an accusation was not so easy in McCarthy's times; it was much easier to lighten it by disappearing from public (and private) eyes. Whatever might happen in the United States, Britain, or any other Western nation that could threaten to expose him, in the future Pontecorvo was not about to be easily found.

5

Play It Up or Down? Confronting the Pontecorvo Affair

> I am accordingly anxious to let sleeping dogs lie and I have some hope that the matter may soon be forgotten. [. . .] My concern is to ensure that as far as possible the Pontecorvo case shall not blight the prospects of the negotiations on the Pentagon's new plan for tripartite co-operation.
>
> Oliver Franks to Roger Makins, Secret,
> 2 November 1950, in "Security: disappearance of
> Professor Bruno Pontecorvo," CAB 126/307, TNA.

Nearly two months of silence followed Pontecorvo's departure to Russia. His AERE manager, Egon Bretscher, was the first to nurture suspicions about his absence. On 20 August 1950 he sent a telegram to Potecorvo at his father's house in Milan (with a copy sent to Amaldi in Rome) urging him to meet colleagues in Chamonix. But Pontecorvo failed to turn up. Bretscher thus waited in vain for him to attend the Harwell conference on nuclear physics. Now he became more anxious, as Pontecorvo's postcard of 31 August stated that he would be back on 7 September. After the conference, Bretscher decided to alert Cockcroft, who in turn asked Harwell security officer Henry Arnold to investigate the mysterious disappearance.[1]

Arnold very discreetly began gathering information. On 17 October 1950, he met Bruno's sister Anna. She helped reconstruct the movements of Bruno and his family, having travelled with them for most of the holiday. But she was unable to offer details on what had happened after 28 August. By then she had left Italy to return to Britain. A week before Arnold's meeting with Anna, Herbert Skinner contacted Guido Pontecorvo.[2] While informing the biologist of the lack of news about his brother, Skinner wrote, "Would you kindly keep the matter as quiet as possible as we do not want the press to break out with some sensational story."[3]

This is exactly what happened three days after Arnold's meeting with Anna. On 20 October 1950 the Rome correspondent of the press agency Reuters found out that British intelligence agents had inquired about Pontecorvo with the Italian police. On the next day Salvatore Imme, chief of Rome's police department, announced the scientist's disappearance during a press conference. Imme erroneously claimed that Pontecorvo had left for Prague, but on the same day Finnish authorities confirmed that he had actually reached Helsinki a month earlier. A further inquiry by the Finnish Ministry of Interior eventually revealed that Pontecorvo had entered the country without a visa and his passport had been confiscated. From 23 October 1950 the reconstruction of his flight was available to several newspapers in Britain, Italy and the United States.[4]

This was the beginning of the Pontecorvo affair. From that day on, the mysterious disappearance became the subject of both public and secret exchanges between politicians, journalists, diplomats, and intelligence operatives. But, surprisingly enough, it was also almost the end of the affair. Though it did not take too much time to realize that Pontecorvo was in Russia, intelligence gathering was insufficient to find out how he had defected and who was responsible for the cleverly disguised travel. Three years after the beginning of inquiries, the investigators knew little more than they had a few months after the escape. Five years on, they could only speculate. Fifty years later, the case is still ambiguous in many respects. Why has Pontecorvo's defection remained a mystery for so long?

As this chapter shows, the inquiries that followed Pontecorvo's flight and typified the case were affected by two opposing tendencies that combined to greatly reduce their efficacy. On one side, British diplomats and intelligence personnel sought to play down the whole case, so the investigations on Pontecorvo's flight to Russia were never prioritized. On the other, some journalists speculated on Pontecorvo's past and present activities, suggesting that he was an atom spy. The proliferation of journalistic revelations further muddied the waters, thus hampering the investigators' work.

These tendencies to both play up and play down the case can be explained in the light of vested interests. British diplomats were engaged in negotiations about nuclear cooperation that promised to break down if nuclear security in Britain was associated with yet another failure after the Fuchs case. The leaders of US and British intelligence agencies feared that the defection could only cast a negative light on their organizations, and thus agreed to release as little information as they could. The AEC administrators had nothing to gain from challenging the British wish for protracted silence, which in fact helped them to regain control of the controversy on the slow neutron patent. By contrast, several journalists on both sides of the Atlantic sought to use the case in an attempt to inform the debate on existing and future security policies. What follows is a look at the manipulations, revelations, and omissions that typified the affair, and an explanation of why Pontecorvo's defection is still one of the most mysterious episodes in Cold War history.

Play it Down! British Diplomats Urge Continued Silence

By the time the news of Pontecorvo's defection appeared in the newspapers, a strategy was already in place to minimize its impact. The initiative of the Italian police to arrange a press conference was a major drawback for British government officials who sought to avoid sensationalism. By then, however, high-ranking diplomats had already prepared the ground for greatly reducing the case's significance.

Downplaying the Pontecorvo affair was deemed critical to an important negotiation between the United States and Britain on the exchange of nuclear information. Among those who outlined the strategy to downplay the case was the British ambassador in Washington, Oliver Franks. The offer of a less disruptive version of facts came from Roger Makins (figure 5.1), the undersecretary of state at the Foreign Office, and Michael Perrin, the director of the Department of Atomic Energy (D.At.En.) at the British Ministry of Supply. Nuclear diplomacy had been critical to these administrators' career development. In 1948 the Oxonian philosopher Franks was nominated British ambassador to the United States and left his position as permanent secretary of the Ministry of Supply. In 1946 Perrin took responsibility for the D.At.En. within the same ministry after having assisted Wallace Akers in Tube Alloys. From 1943 Makins represented Britain at the CPC, eventually becoming its chairman.[5] He served at the British embassy in Washington from 1945 before joining the Foreign Office two years later. In 1952 he

FIGURE 5.1 Roger Makins busy giving instructions on the phone, date unknown. Reproduced courtesy of the *Daily Express*; from Donald Seaman and John S. Mather, *The Great Spy Scandal* (London: Daily Express, 1955), 61.

replaced Franks as British ambassador, but continued to be alert to the management of nuclear matters. Later he became director of D.At.En.'s successor: the United Kingdom Atomic Energy Authority (UKAEA).

In the late 1940s, nuclear negotiations with the United States were vital to British interests. We have seen that the McMahon Act curtailed access to nuclear science and technology, thereby hampering cooperation on nuclear energy matters. This slowed down the advancement of the British nuclear program, which entailed the construction—in great secrecy—of an independent nuclear deterrent. Makins and Franks sought to convince their counterparts at the AEC and the US Department of State to restore cooperation, especially when it appeared that uranium shortages were affecting the US nuclear program. As Britain's supply was less depleted, the diplomats felt they had something that could make the Americans more eager to cooperate. In 1948 a bilateral agreement, the Modus Vivendi, enabled the British scientists to get access to a limited amount of information. But after that, the British negotiators were after a more comprehensive treaty—a plan the Fuchs affair disrupted.[6] The joint handling of the Fuchs affair by American and British authorities paved the way to new diplomatic exchanges in which

Franks and his assistants shrewdly used support from the US Department of State to bargain with the AEC for new conditions.[7] Moreover, following the explosion of the Soviet device Joe-1, AEC administrators realized that technical cooperation with the United Kingdom was necessary, as the British could assist in monitoring and detecting Russian nuclear explosions.[8]

The defection of Bruno Pontecorvo thus came in a period dense with diplomatic activity, and by then the British negotiators were just about to launch a new initiative. The lack of news on the Italian-born scientist stirred Franks, Perrin, and Makins into action. On 20 October 1950, the day of Imme's press conference in Rome, a secret cable was sent from the Cabinet Office in London to the British Joint Services Mission (BJSM) in Washington. The cable resumed a strategy set out by Makins to minimize the case's impact. It stressed the attempt to "keep secret any knowledge of his [Pontecorvo's] real movements" and to "do everything to play down the story."[9] It recommended that a straight "no comment" be offered as a reply to all journalists in the United States who from the next day might seek information about the defection. It also recommended that the US State Department and the AEC be alerted as a matter of urgency and asked to "reply to any press queries on same lines as we are" to avoid encouraging criticism.

Franks did not welcome the news of Pontecorvo's mysterious disappearance. Through the BJSM he asked for instructions on what exactly should be said about it. In particular, he wanted to know how to respond to queries on the management of Pontecorvo's security circumstances, and why the scientist was allowed to leave the United Kingdom if he was under suspicion.[10] Franks feared extremely negative reactions in the United States, since the FBI was so thorough on national security matters that it even confiscated the passports of scientists traveling abroad if necessary. Two days later, the Cabinet Office provided the information Franks had requested, thus emphasizing the difference between the US and British legislative systems. No legal power prevented an individual holding a British passport from leaving Britain.[11]

The exchange between London and Washington paralleled Perrin's preparation of a summary for the minister of supply, George Strauss, for the forthcoming session on the affair scheduled in the British Parliament. If Franks and Makins had been carefully monitoring the flow of information with Washington, Perrin took responsibility for arranging a portrait of Pontecorvo's career that could deflate the affair's impact. Strauss had to face a barrage of questions in the House of Commons—

especially from a Conservative MP, Frederick J. Erroll, who was deputy chairman of the Parliamentary Scientific Committee. Using the information Perrin had provided, Strauss managed to reply with acumen to Erroll's questions, suggesting that the episode was upsetting but not that disruptive after all. Of course Pontecorvo had access to secret information, but "for several years his contact with secret work has been very limited."[12] As for Pontecorvo's security status, Strauss repeated what had been already cabled to Washington: Pontecorvo's vetting had been satisfactory, and there was no means of retaining him in Britain. When the debate took Strauss away from familiar ground and into troublesome waters, the meeting was quickly adjourned. Lord John Hope, another Conservative MP, tried to keep the debate alive by claiming that the disappearance should be discussed as a matter of "urgent public importance." But the speaker ruled it out.[13]

Perrin's words, in combination with the silence of those who, like Cockcroft, knew the importance of Pontecorvo's past research activities, informed other governmental meetings. On the day before Strauss's parliamentary session, MI5 Deputy Director General Guy Liddell briefed Prime Minister Clement Attlee, repeating Perrin's version of the facts: "D.At.En. has expressed the view that for several years PONTECORVO had hardly any contact with secret work, having been mainly concerned with cosmic ray projects."[14]

Strauss and Liddell might not have known, but Perrin certainly knew that there was sufficient scientific documentation at Harwell proving that the point about limited access to scientific information was phony. It was exactly the problem of access to restricted or even top secret information that had led MI5 to recommend Pontecorvo's dismissal a few months earlier.[15] And a relevant number of classified scientific documents produced between 1943 and 1949 had been written by the Italian-born scientist. Perrin opted for ignoring this detail and making publicly available only what was already known about Pontecorvo's career. A 1948 *Nature* note on his AERE appointment turned out to be handy, as it suggested that Pontecorvo was now "conducting fundamental atomic energy research in nuclear physics."[16] Perrin used the article to create the same reassuring portrait of Pontecorvo's activities. In the meantime, Makins made sure that the same message was relayed to all British representatives abroad. On 24 October 1950 a top secret telegram was sent from the Foreign Office to forty-six British embassies and consulates reiterating that "Dr. Pontecorvo was engaged in work of a non-secret nature at Harwell, and, although he may be of use to the

Russians in the field of basic research, it is not thought that he knows anything of value regarding atomic weapons."[17]

In the days that followed Strauss's parliamentary session, the Pontecorvo affair produced some attention especially because of the news published in British tabloids. The coverage in other newspapers was thinner. The BJSM continued monitoring the American media, noticing satisfactorily that the "press reaction over here has so far been quiet."[18] Little criticism was raised by American authorities. The story was not headlined in newspapers and the AEC had only shown passing interest. During a press conference arranged on 24 October 1950, Gordon Dean (who in 1949 had replaced Lilienthal as the AEC chairman) highlighted that Pontecorvo's flight was "difficult to forecast." If US journalists wanted more information on his career, they should consult the British nuclear agency.[19] US politicians were silent as well. Many congressmen had left Washington DC because they were campaigning for the American by-election, due to take place on 7 November 1950. Thus no parliamentary sessions on the Pontecorvo affair were scheduled. Finally, the FBI informed the British authorities they had no intention to publicly comment on it, unless forced by the action of avowed anti-communist congressmen such as Joe McCarthy.[20]

Given the lack of interest in the affair in the United States, Franks recommended continued silence. On 2 November 1950 he sent a letter to Makins, informing him that only the *Chicago Tribune* had indulged in hostile comments. Franks's chief concern was that interest in the case not be revived in the future.[21] These plans could be accomplished only by showing that Pontecorvo's vetting had been accurate, that his defection could not have been prevented, and that his expertise was unlikely to be significant to the Soviet atomic program. It is exactly this portrait that Perrin and Makins skillfully manufactured in the following week.

On 6 November 1950 Strauss and Perrin met once again on the occasion of a second session on the Pontecorvo affair in the House of Commons. Roger Hollis, director of Section C (Security) at Scotland Yard, also attended the meeting. Documentation assembled by Perrin for the occasion reiterated the key points made previously, and emphasized that Pontecorvo had never had contact with atomic weapons work.[22] Erroneously (though it may well have been deliberate), Perrin claimed that Pontecorvo had not been responsible for work on the nuclear reactors ZEEP and NRX either, but had carried out only "the preliminary experiments." When Cockcroft had replaced Von Halban in 1944, Pontecorvo "withdrew from technological work and took up work in cosmic rays." Yet we have seen that exactly in that period, Pontecorvo began advising Cockcroft on crucial scientific matters such as the design and shielding of heavy water piles, also traveling frequently to Chicago to gather data on neutrons and discuss issues related to pile physics. Pontecorvo's work on the NRX did not stop in 1944 but continued until 1947, when the pile went critical. Nevertheless, Perrin stressed that Pontecorvo "was unlikely to have expert knowledge of important technological features of the pile." Finally, Perrin's report said nothing whatsoever on Pontecorvo's wide range of studies on the production of fissile materials and geophysical prospecting.[23]

Other aspects of the Pontecorvo affair that could have aroused suspicion, especially with regard to security management, were also concealed. For instance, on 21 October 1950 the FBI informed officials at the British embassy that they had reports indicating that Pontecorvo was a communist sympathizer—information that had been passed on to British security as early as 1943. Hollis felt that MI5 was unprepared to deal with this issue, and recommended that the information be hidden away. Writing to Makins, Perrin stressed that evading American questions on this point was going to be "a very tricky business," and recommended that British diplomats keep the information under wraps.[24]

However, Strauss's parliamentary session on the Pontecorvo affair focused more on security issues than on Pontecorvo's scientific profile. By then the Italian newspapers had revealed that Pontecorvo was the cousin of Emilio Sereni, a prominent member of the PCI executive committee. Strauss conceded that this element had not emerged in Pontecorvo's vetting: "One finds a great deal by screening but it so happens that the fact that he had communist relatives abroad was not in possession of security officers."[25] Not only did Strauss say nothing on the FBI information, but he was fibbing about Pontecorvo's security review, which had been instigated *exactly* by the knowledge that Pontecorvo was associated with PCI militants.

These omissions and misrepresentations undoubtedly played a part in the successful strategy designed by Makins, Perrin, and Franks. The danger of reviving interest in the case existed. For instance, on 13 November Makins was informed by one of Franks's assistants that fresh news had appeared in the American press about the loyalty of British scientists, and that "the whole question of British atomic security [. . .] will be raised when Congress re-assembles shortly."[26] It was exactly for this reason that Makins became more alert to any attempt to revive the Pontecorvo story. For example, on 24 November the British chargé d'affaires in Helsinki, A. Kellas, informed the Foreign Office that the

Finnish authorities were interested in hushing up the whole case, as Pontecorvo had entered Finland without a visa. Kellas was seeking authorization from London to "point out [. . .] more severely the error of their ways," especially as the Finnish police had been instructed by the Finnish prime minister to avoid inquiring further into the case.[27] But the Foreign Office did not share Kellas's wish, and it replied *on the very same day* that "so far from wishing to frustrate any Finnish effort to play the matter down, we have in fact an equal interest with the Finnish authorities in discouraging any further publicity. Interest in the case appears now to have died down and it would not be to our advantage to do anything that might tend to revive it."[28]

The following autumn, international events kept the press in the United States and Britain busy as the Korean War, which had broken out in June 1950, reached a stalemate once China entered the conflict. Drawing attention away from other issues, the Korean conflict strengthened the political alliance between the United States and Britain, and helped to reduce the impact of the Pontecorvo case significantly. Yet if the British attempts to downplay the affair can be easily understood in light of the existing interests on the negotiations, it is more difficult to figure out why no one tried to challenge the UK's official version of facts. Why had AEC administrators so readily accepted what the British officials had claimed?

Play it Down (Reprise): Giannini the Patriot

On the other side of the Atlantic, there was another important reason to keep the Pontecorvo affair quiet. The defection put Giannini and the Italian physicists who had filed a petition against the AEC in a very awkward position, thus greatly diminishing their bargaining power. In turn, this allowed the AEC to regain control of the proceedings on the slow neutron patent after Giannini's petition had temporarily put it out of hand. Bruno Pontecorvo's disappearance was undoubtedly a preoccupation for his former coworkers, who feared that they could be accused of complicity. Segrè worried that he could be under suspicion even though he had proven his loyalty by denouncing Pontecorvo and his acquaintances as suspicious. On 21 October 1950, when the news of the defection broke out in the USA, he expressed some grief in a letter to Amaldi: "Qui le cose non vanno affatto bene e per di più vanno male per pura stupidaggine o malvagità senza alcuna necessità estrinseca" (Here things are not well and furthermore they are going badly because of sheer stupidity or wickedness that has no obvious reason). In a handwritten postscript he added that he had just read in the newspaper about the mysterious disappearance of "π," and that he was not optimistic about where he was.[29] He then informed Fermi about the need to confer in person about "βρουνω."[30]

Segrè's connection with the missing Pontecorvo angered some of his colleagues. On 24 October the future Nobel laureate Luis Alvarez visited Segrè in his office, making "derogatory remarks" and claiming that asking for a reward because of the slow neutron patent had been improper. Alvarez concluded by asking that Segrè let him know when "Pontecorvo writes from Russia."[31] The visit made Segrè more alert to "whispering campaigns and defamatory statements [. . .] since Senator McCarthy is going on a rampage against scientists."[32] He had braced himself for the worst when Giannini also appeared apprehensive and suspicious. Giannini phoned Segrè on the day of Alvarez's visit, and on the next day he issued the following press release:

> I am today instructing my attorney to eliminate and disinvest my company of any interest in the $10,000,000 damage suit now pending in the US Court of Claims [. . .] This action is a result of the surprise and shock resulting from the amazing disclosure that Bruno Pontecorvo [. . .] has reportedly disappeared into Russia under circumstances that are highly questionable. [. . .] I am an American citizen first, and my interests and interests of my company are concerned primarily with the principles of freedom and democracy. Regardless of what sum of money is involved, we do not care to be associated with anyone whose principles or beliefs in our form of Government is in any way open to question.[33]

Giannini's patriotism was undoubtedly informed by the need to save his company's remunerative contracts with the US Army, to which he provided technical equipment. As the Pontecorvo affair threatened these deals, he returned the ownership of the slow neutron patent to the inventors to eradicate any doubt of complicity.

The inventors, especially Segrè, disagreed with Giannini on the best course of action. If Segrè was receiving no sympathy whatsoever from Alvarez or Giannini, he was being reassured by AEC administrators in San Francisco, "who took an entirely different attitude and were extremely friendly and reasonable."[34] Indeed, after Pontecorvo's defection Segrè had immediately contacted Thornton, to whom he had previously relayed information on Pontecorvo's acquaintances, presumably reiterating the point that he had been correct in denouncing these connections.[35]

This attitude was instrumental in convincing the inventors that the affair did not necessarily preclude the possibility of a settlement. On 3 November 1950 Rasetti visited Bernard, who confirmed that the best strategy to pursue was for him and Fermi to informally approach the AEC commissioners and ask for a relatively modest sum.[36] As Fermi was no longer a GAC member, the plan did not entail any conflict of interest. The strategy was formalized at a meeting on Thanksgiving Day (23 November 1950) in Chicago, when Segrè, Rasetti, and Fermi also decided to inform the inventors in Italy.

We now know what made the AEC administrators so friendly to Segrè. Not only had he offered Pontecorvo's head on a plate to demonstrate his own loyalty and that of his co-workers, but in the wake of Pontecorvo's disappearance he confirmed a willingness to negotiate rather than sue the commission. This willingness entailed an opportunity for the commission's administrators to keep controversial patent matters (which might have envisaged the analysis of classified information) behind the AEC's closed doors rather than battle them in the media or courts of justice. A few months earlier, Gordon Dean had been extremely anxious about the possibility that the case against Los Alamos experts David Greenglass and Harry Gold, who were charged with atomic espionage, would end with the disclosure of restricted information in court, especially through the cross-examination of witnesses.[37] This explains why Dean had been so happy to accept the version of facts provided by the British authorities and to invite US journalists to rely on that account of the Pontecorvo case in the pursuit of their investigations.

Towards the end of 1950, Bernard was informed about the AEC's intention of amicably settling the claim. but he was aware that such a positive conclusion could be reached only after a thorough investigation of Pontecorvo's disappearance.

Play it Up and Down: the JCAE Enquiry on Atomic Espionage

Atomic espionage was a particularly sensitive topic in the United States. In 1948, HUAC launched an investigation into these matters. Two years later the AEC's overseeing congressional body, the JCAE, took responsibility for a new inquiry into the Fuchs, Pontecorvo, Gold, and Greenglass cases. In the early months of 1951 the JCAE's chairman, Brien McMahon, ordered an investigation by the committee's staff under the supervision of his executive director, the lawyer William L. Borden. It aimed at putting together "the salient facts about the various individuals involved in atomic espionage." The final report, published in April, contradicted Perrin's summary by suggesting that Pontecorvo's flight represented the loss of a "human storehouse of knowledge about the Anglo-American-Canadian atomic projects" and "a first-rate scientific brain." It also highlighted for the first time how the defection was informed by, and had important implications for, the controversy about the slow neutron patent.[38]

William Borden (prosecutor a few years later in the judicial case against Los Alamos director J. Robert Oppenheimer) took responsibility for requesting from Fermi testimony on his pupil's mysterious disappearance. In March 1951, Fermi filed a statement before the commission. Like Perrin, he put together already known facts while omitting important details. He provided plenty of information on the physics group he had directed in Rome in the 1930s, but avoided mentioning important exchanges of information that had occurred in the context of the Manhattan Project, when Pontecorvo had met him on several occasions. Actually, Fermi declared that he "did not remember [instances] in which he took up with me any subject connected with atomic technology." He also declared that he did not know what had led Pontecorvo to cross the Iron Curtain.[39]

The JCAE's deputy director, Harold Bergman, interviewed Rasetti, who could offer no details about Pontecorvo's work in the wartime nuclear programs. Rasetti claimed that Pontecorvo had fled to Russia because of his "sincere faith in their system, [but] not for the express purpose of giving them atomic secrets." Like Fermi, he denied that Pontecorvo had ever discussed politics with him.[40]

Though Pontecorvo's former coworkers offered a reassuring picture, the final JCAE report played up Pontecorvo's role in the atomic projects and his potential for propelling the Soviet nuclear program. Alarmingly, the report concluded that every scrap of information known to Pontecorvo was now available to the Soviet Union, including details of the Canadian NRX (the reactor of most advanced design and performance) and the plutonium-producing Hanford reactor. The report also highlighted the fact that Pontecorvo's recent work had entailed research on tritium, a substance intimately related to the hydrogen bomb. The report's authors stressed that Pontecorvo was considered by some of his associates to be a more outstanding physicist than Fuchs.[41]

These conclusions may have further disrupted the proceedings on the slow neutron patent. Their potrayal of Pontecorvo's potential to boost the Soviet nuclear program promised to indefinitely halt the negotiations between the AEC and the inventors. In fact, the JCAE report set the conditions for a more open dialogue between both parties, for several reasons. First, the defection had made it apparent that Pontecorvo had links with communist militants, thereby eliminating a problem that had hindered the proceedings before his defection—namely, the possibility that Pontecorvo's communist associations might be revealed *after* a reward was authorized. Second, Pontecorvo's links had been exposed by one of the claimants, Segrè, who thereby had shown his loyalty. Third, the offer of assistance by Fermi and Rasetti to the JCAE enquiry showed, if not their loyalty, at least their wish to cooperate with US authorities in solving the mystery. These events made rewarding the Italian-born scientists less problematic, especially if a means could be found to exclude Pontecorvo from the proceedings. Not surprisingly, the PCB negotiations, which had halted in January 1951, started off again the following April, exactly when the JCAE report was published.

Purchasing the Slow Neutron Patent

On 3 April 1951 Bernard was able to approach Anderson with a new proposal. Fermi had already discussed the offer with senior AEC officers, and had made them aware of the inventors' intentions. Bernard obtained a sympathetic hearing.[42] Giannini reconsidered his decision to relinquish his rights. So Bernard was now in a position to offer a settlement for a sum in the range of $450,000. Anderson replied that the AEC was interested in a settlement and appreciated that the inventors wished the same. But he stressed that in view of undisclosed "difficulties and deficiencies," only a much lower figure could be agreed upon.[43]

In the second half of 1951 the PCB chairman, Roland Anderson, reconsidered his position on the opportunity to settle the claim. Together with the OGC lawyers, he reviewed the legal defenses, coming to the conclusion that Bernard's new offer was very advantageous to the commission. The assumptions contained in Boskey's response were hardly tenable, as Fermi was no longer in a conflict of interest. The patent's validity presented them with a legal cul-de-sac. No prior art had anticipated the invention, and Anderson's assumption that the process described in the patent differed substantially from the one used in nuclear reactors proved weak.

Anderson even sought advice from the Argonne and Oak Ridge National Laboratories on these matters. But the experts he consulted replied that it "would involve months of calculations, aside from any practical experimentation" to spot a difference between what the patent stated and the process occurring in atomic piles. Moreover, who would dare witnessing against Fermi? "Very few scientists in this country are of the stature of Dr. Fermi, and of those considered to be his peers, all would hesitate to contradict any interpretation of Dr. Fermi [. . .] in view of his position as Nobel Prize winner."[44]

Finally, Anderson understood that the patent's value might have led the AEC to pay a sum far exceeding the one originally requested by Giannini. When he considered the costs associated with the production of plutonium alone, Anderson realized that "even if *a one percent or less than one percent figure* were employed the sum would far exceed the proposed figure for settlement" [i.e., $1.9 million; my emphasis].[45]

If Anderson had opted for challenging the claimants once again, it might have led to a costly court case, whereas an amicable settlement would entail paying much less of what was due. As the case set a precedent in the reward of similar patent cases (it was the first of about fifty docketed compensation claims), it also enabled the commission to judge the merits and values of other patents accordingly, thus paying less than what it should have paid in the future. An external PCB adviser emphasized that the sum requested did not shock him; the award was "a very favorable one" for the commission.[46]

Towards the end of 1952 Bernard informed Rasetti that the AEC now appeared willing to agree to a settlement. The days turned into months and the delay exasperated Segrè. But the AEC had made a final decision to pay the inventors, and it was now making arrangements with the US Treasury. In the summer of 1953 the settlement was effected. It allowed the inventors and Giannini to pocket around three hundred thousand dollars. Fubini, who had an agreement with Pontecorvo for 50 percent of his share, was also rewarded. Not surprisingly, the only one who did not receive a penny in royalties was Pontecorvo. Anderson and Bernard agreed that his share would be returned to the commission and put in a frozen bank account, since the scientist was missing.

To summarize, the JCAE portrayal of Pontecorvo's defection emphasized its dangers, thereby casting serious doubt on the reassuring words proffered by British administrators. Yet in doing so, it actually cleared Pontecorvo's former colleagues from the accusation of complicity, which in turn helped the AEC to settle the claim on the slow neutron patent with profit and out of the public eye. Thus, as the parliamentary commission played up Pontecorvo's role in atomic espionage, the AEC could take a more watchful stand on these developments, analyzing the best way to protect its own interests in terms of information disclosure and legal and financial concerns attached to patent controversies. And even if the results of the JCAE enquiry contradicted the findings of the British officers who had analyzed the Pontecorvo affair, the security investigation that anticipated it had seen the Western intelligence services agreeing on an important pact of silence.

Play It Again, (Uncle) Sam: The Stott Papers

Since the security actions of MI5 and the FBI had been flawed in important ways, both agencies agreed to remove their mistakes from public scrutiny. Their efforts to learn the reasons for these flaws drew precious time and energy from the enquiry on Pontecorvo's defection. In fact, MI5 and FBI agents spent more time trying to verify and show that their vetting had been sound than finding out where Pontecorvo was.[47]

Since the beginning of the Pontecorvo affair, Scotland Yard had been able to count on the FBI's silence and its promise of a "no comment" to inquiries from US journalists. Yet the US agents continued to bombard their British colleagues with questions about their handling of the case. The 1943 report containing allegations about Pontecorvo's political profile had not been taken into account at Scotland Yard. The FBI wanted to know why, and a major security crisis ensued. On 21 October 1950 the FBI liaison in London, John Cimperman, met with Jim Skardon, the MI5 officer responsible for Fuchs's confession, and asked whether Pontecorvo had been transferred immediately after intelligence information had revealed his communist relations. Skardon "hedged on this," but said that the scientist had agreed to move to the University of Liverpool once his associations had been uncovered.[48]

On that day Geoffrey T. D. Patterson, Cimperman's equivalent at the British embassy in Washington, decided to write to the MI5 director-general, Percy Sillitoe, to stress his disappointment with the security management of the whole affair. By then the embassy personnel had not only worked with alacrity towards Anglo-American nuclear cooperation, but had also set the conditions for a fruitful collaboration on security matters. These had already been negatively affected by the Fuchs affair, and the British agents had to explain why the scientist was allowed to participate in secret research projects despite his previous association with the German Communist Party. For some time MI5 leaders feared that the FBI would release restricted information on the Fuchs case.

The intervention of Percy Sillitoe, who met J. Edgar Hoover and convinced him to agree on a pact of disclosure, helped to sort these matters out. After the Fuchs case, each agency would thus agree with the other on what should be made public.[49] Now Patterson feared that the Pontecorvo affair could threaten the pact. His top secret message to Sillitoe pointed out that "the Embassy were most concerned because it looked at first sight that British intelligence had *again* allowed themselves to be a target for US criticism" (my emphasis). Now that the Fuchs case was a dim memory, and "everything in the garden was rosy, [there was] another case which can, to put it bluntly, upset the applecart."[50]

In the days that followed Patterson's letter, Scotland Yard had to solve a far more important mystery than that of Pontecorvo's flight: Why had the 1943 FBI communication not been received, or—an even more worrying possibility—why had it been ignored? Liddell's briefing with Prime Minister Clement Attlee offered an opportunity to verify the whereabouts of the Stott papers (named for the BSC officer who had received them). But on 23 October, the report was yet to be found.[51] Liddell speculated that perhaps it had been destroyed at the end of the war when the organization was dismantled, or that it may have been passed on to the FBI representative in London, who may have decided to take no action. In any case, Liddell reassured the disappointed prime minister that further checks were being carried out.[52]

In the following week Scotland Yard made a major effort to find the Stott papers rather than to find Pontecorvo. Yet nothing was found at the MI6 headquarters, and nothing had been previously passed on by the FBI liaison in London to MI5 or MI6. Arnold also confirmed that Harwell's knowledge of Pontecorvo's communist sympathies was the result of the recent inquiry of March 1950. Pontecorvo's file had previously been labeled "Nothing Recorded Against." In 1948, the CIA had been asked its opinion on the occasion of Pontecorvo's naturalization, and the agency's reply had not ruled it out.[53] When MI5's Roger Hollis asked whether some of the undestroyed BSC records could be consulted, one of the BSC's former officers advised him that it would be a waste of time as he was reasonably sure that there were no adverse records.[54]

At the end of October, a report on the missing papers was drafted by Dick White (MI5 Section B, counterespionage) and made available to most governmental agencies dealing with the Pontecorvo affair, including the Foreign Office and the FBI. However, Liddell understood that the inquiry on the Stott papers could cause trouble and prevented its publication: "I am afraid but there is nothing in the latter document which would be suitable for any form of press release." He thus recommended saying nothing about the role played by the FBI in the whole story, fearing that this could draw them into making a statement.[55] When Hollis met with Strauss and Perrin, he reiterated the MI5 recommendation. The possibility that Pontecorvo's vetting might have occurred in the presence of documents showing his communist connections was thus shrouded in secrecy.

These fears appeared to be justified. On the other side of the Atlantic, Hoover made his promise of continued silence conditional upon receiving an explanation about the "slip" of his British colleagues. Hoover continued to ask for explanations up until 27 October 1950, when he suddenly gave up. The reason for his abrupt change was the realization that the FBI also had something to hide, so far as information management was concerned. The federal agency had never passed on to MI5 the report assembled in 1949 after Segrè's admissions to Thornton. Cimperman revealed the existence of this second dossier during the exchanges that followed Pontecorvo's defection.[56] And once again, the report appeared to be missing from the MI5 files. This time, however, when Scotland Yard asked for further details on *when* the second report had been forwarded to London, it appeared that the document had never been sent.

On 27 October, just as Liddell and Hollis were preparing to face criticism from their US colleagues, Patterson informed Sillitoe that Hoover was "raising hell" with the FBI espionage section. He thus concluded that "from their friendly manner and lack of direct criticism I get the impression that their position is not as strong as it well might be and that they are as anxious as we are that Pontecorvo's activities should be forgotten ASAP by the press and Congress."[57]

On 1 November 1950 Sillitoe flew once again to New York to meet with the impetuous FBI director. He now obtained from Hoover the promise not to give publicity to the 1943 reports "unless this became absolutely necessary," in which case Hoover would warn the British services beforehand.[58] Returning to London the following day, Sillitoe had to face his unhappy prime minister, who this time was "rather critical" about the whole affair. Unconvinced of Sillitoe's claim that finding out about Pontecorvo's actions was impossible because MI5 had "no magnet to find the needle in the body," Attlee was only partly satisfied when Sillitoe alleged that Pontecorvo had been blackmailed—blackmailing being the KGB's "typical technique."[59]

As for the Fuchs case, the management of security circumstances by British and US intelligence would be undisclosed for the next fifty years. Sillitoe and Hoover agreed on a nondisclosure strategy that could save the reputations of both agencies. The FBI had no interest in divulging the fact that its second report had never been passed to MI5. Scotland

Yard had no interest in publicizing the fact that the first FBI report had never been found. So no agency had an interest in making any public statement on the management of Pontecorvo's vetting. In this way, the FBI effectively agreed on the only key point that MI5 wanted to divulge: that "Pontecorvo was screened by the British security authorities in North America."[60]

Hoover Takes Revenge

The blunder, however, stirred the FBI towards investigating Pontecorvo's acquaintances in the United States. At least three American citizens were targeted. We know about the investigation not directly from FBI files, but because the AEC's Division of Security was kept updated on the inquiries.

Bruno's brother Paolo, who worked for Raytheon in Waltham, Massachusetts, was interrogated by FBI agents. From 1948 he had been the holder of a security clearance that allowed him to access restricted scientific data; it was withdrawn after his brother's defection.[61] Yet Paolo was not particularly hurt by the revision of these security measures; he actually wrote to his father that everyone was being considerate about his position.[62] By contrast, the circumstances of Neufeld and Scherbatskoy, Pontecorvo's associates in Tulsa, were more problematic: the FBI put them under surveillance immediately after the affair broke out. Neufeld, employed at the AEC Oak Ridge Laboratory, was shadowed when he went to Tulsa on vacation. He spent most of his time in Scherbatskoy's laboratory.

The FBI was heavy-handed with the Scherbatskoy. After Pontecorvo's flight he had planned to visit his father in Paris and then travel to other European countries, but this looked suspicious to the agents. Moreover, the 1949 FBI report had indicated Scherbatskoy as the subject of a pending espionage investigation because of his contacts with Amtorg Trading, the company ostensibly established in 1924 to favor trading activities between Russia and the United States, and later revealed to be a focus of industrial espionage. Scherbatskoy finally claimed that his family members were forbidden from visiting the USSR—information that did not square with FBI data showing that his father had requested a Soviet visa in 1945.[63]

These inconsistencies warranted protracted investigations, but nothing more emerged in the following weeks. Yet the agents sought to find out connections. Scherbatskoy and his secretary were investigated "rather strenuously" with methods the engineer judged improper.[64] In the meantime, the US Department of State asked the FBI to confiscate Scherbatskoy's passport, fearing that he could leave the country. He wanted to visit his ill father in Paris. His attempts to convince the Department of State to let him go were no avail—even despite the intervention of Eleanor Roosevelt, the wife of the late president and a US delegate to the United Nations.[65]

Thus Scherbatskoy could not travel abroad for two years. When his passport was finally returned to him, he was allowed to travel only to France. Until 1954 he was not allowed travel to any country outside the Western Hemisphere.[66] He alleged in various circumstances that the FBI considered him responsible for Pontecorvo's defection and for recommending the scientist for government work during the war years. As he had provided Pontecorvo with a letter of recommendation on the occasion of his employment in Tube Alloys, the FBI seemed to believe him responsible for deceiving the security authorities. He protested innocence and wrote to his lawyer that "if the FBI, Scotland Yard, General Groves, and the thousands of professional investigators with millions of dollars at their disposal were unable to find out that Pontecorvo had communist leanings, it is very unfair to assume that I knew it all the time and should have been more careful."[67]

In 1952 the US Department of State, consulted on these matters by Scherbatskoy's lawyer, indicated that an intervention by Fermi, who was in a position to recommend Pontecorvo for governmental work, might have helped in reviewing the case.[68] It is unclear whether Fermi ever produced such a document. But the secret dialogue between the British and American intelligence services suggests that Hoover wished to find someone culpable for the handling of Pontecorvo's security circumstances exactly when the FBI and MI5 blunders in this management became apparent. Actually, we have seen that Pontecorvo's recruitment by Tube Alloys was facilitated by Rapkine and Paxton's references; Scherbatskoy's recommendation played no part in it.

Play it Up: Journalists and Security Policies

If British diplomats and intelligence operators had been eager to let the Pontecorvo affair be forgotten, some journalists sought to use the flimsy information made public by governmental agencies to denounce the laxity of security procedures and forge Pontecorvo's image as an atom spy. In particular they sought to demonstrate that the scientist had been secretly smuggling information before his defection, and they alleged that the flight had been orchestrated by Soviet intelligence. Even if these allegations were never proven, they overlapped governmental inquiries about the efficacy of existing security policies. As such, they offered fresh evidence and analysis to reformers who sought to enforce more restrictive measures, especially in Britain, to counter atomic espionage.

It was especially the coverage of the Pontecorvo affair by *Daily Express* science correspondent Chapman Pincher that lent support to the image of Bruno Pontecorvo as a spy. A graduate of London King's College, Pincher had been a correspondent for the tabloid since 1946. He had been responsible for impressive scoops that had unnerved the British government, such as the revelation that Cockcroft would return to Britain before Tube Alloys was completed—a detail that marred relations with Canada.[69]

Between 21 and 27 October, reports about Pontecorvo regularly featured in the tabloid headlines, and they immediately cast doubts on the version of facts provided by government officials and more restrained newspapers such as *The Times* and the *Manchester Guardian*. Pincher's articles did not disclose more than what was already known, but they added some spin to the information available. When on 24 October it emerged that Pontecorvo was employed in Tube Alloys, Pincher stressed the existence of a connection between Pontecorvo and the atom spy Nunn May.[70] Three days later, Pincher claimed that Pontecorvo had never been screened before being employed in secret work. He went on to argue that the US and Canadian services had relied on a vetting by British security that had never occurred because Pontecorvo was not in Britain.[71]

These sensationalist accounts came right in the middle of a very tense period for British governmental authorities. The 1947 reform of the US loyalty and security system, advocated by President Truman, had led British politicians to consider a similar initiative in their own country. The government had also considered the adoption of tighter security measures. In June 1947, the Cabinet Committee on Subversive Activities (also known as GEN 183) was set up to consider *positive vetting* (or "purge procedures"). These measures aimed at extending the intelligence agencies' power to investigate individuals employed in sensitive posts.[72] The new measures enabled checks not only against filed information, but also through active investigations into an individual's private life and personal contacts. Before 1950, Attlee had felt that positive vetting compromised civil liberties, and thus postponed a final decision on its adoption. Even the Fuchs case had not convinced him to seek a security reform. On 5 April 1950 the new Cabinet Sub-Committee on Positive Vetting was established under the chairmanship of civil servant John Winnifrith, and it included among his members MI5's Roger Hollis, who was to carry out further analysis of security policies. In June 1950 the Tripartite Security Conference was organized, and US authorities were now advocating new security provisions in Britain.[73]

Even if the governmental proceedings were kept secret, the exposure of real and fictional flaws in Pontecorvo's screening put added pressure on those responsible for reforming the national security system. They suggested that the introduction of tighter security measures was indeed very urgent. Not surprisingly, the case overlapped existing debates on governmental controls of foreign citizens, which in turn called existing regulations into question.

On 26 October the *Daily Mirror* alleged that many foreign-born scientists had given up their passports to British authorities to prove their loyalty to Britain. The following day the journalist Ronald Bedford, drawing on the Pontecorvo affair, contended that the price of British citizenship was too cheap, as these foreign scientists had not even been asked about their political activities before receiving naturalization.[74] Five days later, Pincher's article "Perturbed Men" indicated that some of the foreign scientists occupying key posts in British atomic research were now frightened because the government doubted about their loyalty. The accusation even propelled the "otherwise shy and reticent" refugee scientist Rudolf Peierls to counter the claims with factual evidence, listing no less than five inaccuracies in Pincher's article. "I am not perturbed," replied the German-born scholar in a press statement that was also signed by other foreign academic colleagues.[75]

As Peierls mobilized the unperturbed naturalized researchers, the GEN 183 met once again. Attlee was still unconvinced about introducing positive vetting, but he authorized it for use in exceptional circumstances.[76] In February 1951, Pincher published a new account on Pontecorvo: A fervent communist militant, the scientist had been recruited by Soviet agents and was in contact with them until his defection, making available details on atomic weaponry that Fuchs was unable to produce. The Soviets had ordered him to leave Britain because their intelligence had concluded that in Liverpool he could no longer be used as a spy.[77] Pincher's account came out just before the publication of the JCAE report on atomic espionage which, if it did not share Pincher's allegations about spying activities, did also indicate Pontecorvo as a scientist who had played a key role in the development of applied nuclear research.

Pincher's account was clearly inconvenient for the reputation of Scotland Yard. But his allegations became far more threatening when they were coupled with Kenneth De Courcy's study of Pontecorvo's role in the Soviet atomic program. De Courcy, a British politician and diplomat, had been in and out the British corridors of power since World War II.[78] In 1951 he was responsible for an impressive number of revelations about atomic espionage. His claims caught the attention of US congressmen through the *Intelligence Digest*, of which he was editor and publisher. This publication was avidly read in Congress, especially by Republicans, and it heightened their fears about infiltration by communist sympathizers into Western government establishments. In February, De Courcy claimed that Pontecorvo had been implicated in most advanced work on atomic and thermonuclear devices and military devices that involved cosmic rays. He was now working on a cheaper atom bomb, and Stalin was waiting for Pontecorvo to complete the new bomb before launching new war plans. De Courcy's sensationalist account went so far as to suggest that Pontecorvo had a say in the timeline for these plans.[79] Unsurprisingly, De Courcy's revelations were ridiculed by British intelligence. Yet Patterson knew that De Courcy had credibility in the United States. He thus wrote to Sillitoe that a request of information from US congressmen would draw the FBI in. In turn, the FBI could ask MI5 for clarification on the current state of affairs in the Pontecorvo case.[80]

Whether or not De Courcy and Pincher were believed on both sides of the Atlantic, Attlee capitulated a few months after these reports appeared. On 27 August 1951, positive vetting was finally approved. This is not to say that the depiction of Pontecorvo alone fostered the introduction of purging practices. Other events also contributed to the reform. In May 1951 two prominent British civil servants, Guy Burgess and Donald MacLean, defected to the Soviet Union. In July 1951 a second Tripartite Conference on Security took place. In the conference, US delegates who had been influenced by De Courcy's inflated claims stressed the need for Britain to move to a new security system. If the members of the Sub-Committee on Positive Vetting struggled to endorse more restrictive security measures, the journalistic coverage of the activities of Pontecorvo and his foreign colleagues, which overlapped pressure from the political arena, helped them to take these difficult decisions. Sensationalist accounts on Pontecorvo continued to appear in newspapers for another two years in connection with his research activities in Russia. The clamor generated by these accounts counterbalanced the silence sought by British diplomats and intelligence managers, who in the meantime stealthily worked towards producing a version of the affair that could justify the conduct of Her Majesty's secret services.

Play It Out: The Inquiry on Pontecorvo's Disappearance

Owing to the British diplomats' request that the affair be quickly forgotten, the investigations into Pontecorvo's flight were carried out not with great enthusiasm, but only in an attempt to bring the inquiry to an end. The case was assigned to Ronald ("Ronnie") T. Reed, whose lack of biographical detail tells us a great deal about his ranking at Scotland Yard, as well as about the amount of attention given to the case. This is not to say that Reed was not thorough in running the investigation. In fact, he quickly sought to make sense of the scanty information available. But when the inquiry was closed, there was still very little known that could help to unravel the mystery. More significantly, Reed was assigned the task of looking into the Pontecorvo case without delving further into sources of evidence in the United States or getting in touch with the US governmental agencies' informers; particularly Segrè. The end result was a lame investigation.

On 26 October 1950 Scotland Yard agents broke into Pontecorvo's house in Abingdon, confiscating everything that could help Reed in his effort. The scientist's private correspondence was transferred to Scotland Yard and examined there.[81] But nothing was found in the house that could raise suspicion or indicate that significant changes were bound to happen in the Pontecorvo family's near future. Everything seemed to have been left as if the scientist had intended to return to Britain, and valuable documents and personal papers had not taken away.

The private correspondence found in Pontecorvo's house included letters to and from Scherbatskoy, which alerted Reed to the importance that geophysical research played in Pontecorvo's activities. Reed even sought to inform Patterson about the content of this correspondence, presumably in the hope of instigating FBI inquiries into these matters. Yet the papers did not appear to be compromising, nor did they reveal anything about the flight to Russia.[82]

The conspicuously largeness of Bruno Pontecorvo's family immediately captured Reed's attention, as he believed that one of its members might offer clues about the mysterious flight. Those who lived in Britain—the brothers Guido and Massimo, and the sisters Anna and Laura—were examined most carefully, but they appeared baffled by the disappearance. Bruno's parents said they could not believe that he had willingly decided to go to Russia, and went as far as to suggest that he had been kidnapped.[83] Reed did not believe them, and the family's correspondence continued to be tapped.

The investigators immediately focused on Bruno's sister Giuliana and her husband, Duccio Tabet, as potential accomplices. Giuliana, a PCI militant and a prominent member of the pacifist organization Partigiani della Pace (Peace Partisans, or PP), was interrogated by Italian police authorities. The couple's house in Rome was besieged by journalists looking for fresh information. In a letter to her brother Guido written on 26 October 1950, Giuliana controversially claimed that Bruno appeared "calm and happy" before leaving, and did not appear to have "tremendous plans." But her political affiliation led the intelligence officers to believe that she was lying.[84]

Giovanni David Pontecorvo was interrogated twice, but he could only confirm his parents' impression that Bruno had not gone to Russia of his own free will. On 2 November Reed interviewed Anna Pontecorvo, and her words seemed to confirm the hypothesis that the flight had been the result of a sudden decision. She said that Bruno had not brought much money with him, which did not square with his payment of a hefty airfare fee (the news "foxed her completely"). Bruno's wife, Marianne, had wanted to go to Sweden but her husband had ruled it out. Anna, who had been already questioned by Arnold just before the Pontecorvo affair went public, was taxed very carefully on the point of changes in Pontecorvo's attitudes. Guido was also questioned, especially because to the investigators he appeared to be the most balanced. He proved to be extremely cooperative, yet he could offer no suggestion as to why Bruno had defected. On the whole, Reed could only conclude that the disappearance had caught Bruno's family members by surprise.[85]

Pontecorvo's colleagues added little more that could help Reed solve the case. On 31 October Wilfrid Basil Mann, a former colleague at Chalk River and now a British delegate to the United Nations, wrote a report that MI6 forwarded to MI5. It is important to note that Mann's opinion was highly regarded at Scotland Yard because even while working in New York, he was also the MI6 atomic intelligence liaison with Britain. But Mann's report described Pontecorvo as "unfettered, uninhibited and an extrovert" who cared exclusively about cars and wealthy living.[86] Only Samuel Goudsmit, the Dutch-born American physicist who famously headed Alsos (the mission that enabled the Allies to get information about the German atomic energy project), indicated that he had the impression that Pontecorvo "had communist tendencies" when he met him in Paris.[87] The information squared with what had been secretly asserted by Segrè, but it said nothing else about the ten years that followed Pontecorvo's stay in France. On 15 November 1950 Arnold met with Henry Seligman, the head of the AERE Isotope division and one of Pontecorvo's closest acquaintances at Harwell and in Canada, to discuss his relationship with the Italian scientist. Seligman indicated that Pontecorvo was nonpolitical. Egon Bretscher, who was also at the meeting, confirmed Seligman's statement.[88]

By mid-November, Reed had to admit to having very little in his hands by which to understand what had happened to Pontecorvo. More worryingly, journalistic activities were proving very disruptive to his efforts. On 15 November, when the MI5 agent had just started to put together the evidence, the London *Evening News* claimed that Bruno's sister Laura, living in Britain, had also disappeared. A naturalized British subject, she had worked as a nurse in London and had left Britain on 7 September 1950. This news, however, was fabricated. Laura had simply decided to stay in Italy because she feared hostility in her workplace.[89]

As Reed learned from newspapers about Laura's circumstances, British journalists in Italy were busy playing the "spooks' game." *Daily Express* correspondent Roger Waters sought to find fresh evidence on the case by claiming to be an MI6 agent dispatched to Italy.[90] Waters's action was especially embarrassing because of the defective diplomatic relationship between Italy and Britain, which also hampered ongoing investigations. Italy and Britain belonged to the same defense alliance (i.e., NATO), but their governments had conflicting views on the postwar administration of territories such as the city of Trieste (which was occupied by Anglo-American forces) and the former African colonies (Libya, Somalia, Eritrea). Presumably, this is the reason why the Italian authorities had not complied with the British wish to hush the affair.

These quarrels played a part in Waters's attempt to fool the Italian police. He possessed several identities, as he was known to the Italian Carabinieri as "Agnese Ferguson" of the London Press Club–affiliated organization Circolo della Stampa, based in Milan. But he had also led them into believing that his journalistic activities were a cover-up for intelligence operations—quite the opposite of the facts. On 26 October 1950 Waters arrived at Lake Como and asked for assistance at the local police station, claiming to be working for the British secret police. Waters thus was able to visit the place were Pontecorvo had been camping and interview a few people, thereby gathering information. In particular, Waters learned that while at Lake Como, Pontecorvo had spoken with a Czech, saying, "I dare not go back. I should be sent to prison if I did."[91] Reed was unimpressed by Waters's antics.

Scanty information, deceptive journalistic activities, and the need to conclude the inquiry as quickly as possible undoubtedly all helped to prevent Reed from unraveling the mystery. On 1 January 1951, four months after the mysterious flight, Reed submitted the report that effectively put an end to the MI5 investigation into Pontecorvo's defection.

The eighteen-page-long typescript "The Case of Bruno Pontecorvo" analyzed Pontecorvo's family relations, his life and career in Canada, his research in Britain, and his mysterious departure in September 1950. Reed's balanced and accurate précis argued that it was impossible to ascertain the cause of Pontecorvo's departure, and that conceivably the decision to leave was taken out of the blue. Something had happened during the sojourn in Italy that had led him into changing his carefully assembled travel plans, but Reed could not figure out what it was. He could only conclude that Pontecorvo's sudden change of plan had received funding from a source that provided him with the sum necessary to flee the country by plane. Existing theories suggesting that Pontecorvo was a spy or had been kidnapped did not "fit all the facts." It was much more likely that Pontecorvo had been "tricked into visiting Finland," or had been promised a job of outstanding importance in Russia and went to Finland to find out about it. Giuliana, Gilberto, or some other communist family member might have convinced him of such an opportunity.[92]

The circumstances of Giannini's inflated petition, as well as those of the slow neutron patent, played no part whatsoever in Reed's report, in contrast with the JCAE enquiry on atomic espionage published a few months later. The investigative elements contained in the FBI files, including Segrè's revelations, were not even considered. Of course, Reed might have known about these elements. Yet if this was the case, he was certainly instructed to avoid looking into those matters, because they might have compromised the carefully assembled version of facts previously provided by the Ministry of Supply. Instead, Reed's investigation supported that explanation with additional evidence, thereby confirming that the British services' action had been thorough.

If Reed's report put an end to the MI5 inquiry into the affair, it did not stop the proliferation of real or fictional accounts that propelled Pontecorvo into a parallel dimension, transforming him from real person to fictional character. Soon "Dr. Pontecorvo" became the quintessential atom spy.

Pangh de Co Co and Sawbridge: A Play

In the months that followed Reed's assessment, two opposite portraits were produced. In one it was unclear why Pontecorvo had gone to the

Soviet Union, even if he was most certainly there. In the other, Pontecorvo had been an atom spy for a long time. Both these portraits offered substance to more literature that reiterated their claims or used them to fabricate fictional characters.

Reed's study was consistent with the MI5 agenda. It showed that their vetting had been efficient, and that Pontecorvo's flight could not have been prevented. This made the agency eager to find a way to publicize the report without giving away information it did not want to divulge. By then, MI5 managers also felt that the agency should offer a version of facts antithetical to that given by the likes of Pincher and De Courcy. Reed was especially annoyed by Pincher's coverage, as in his opinion it had simply taken one of the many available hypotheses and turned it into theory; he called Pincher's article of 1951 "sheer speculation."[93]

The MI5 director-general, Percy Sillitoe, instructed his staff to make available a "sanitized" version of facts to the wider public. The opportunity to do so came when one of the most prominent journalists in Britain, the Australian-born correspondent Alan Moorehead, decided to write about the atom spy cases. During the war, Moorehead had successfully covered military campaigns for the *Daily Express* before becoming one of the most popular reporters for *The Times*. His book on atomic espionage, *The Traitors*, was a best-seller. MI5 personnel provided a significant contribution to his writing by allowing him to access classified information, including Reed's report. In 1951 Reed was also asked to fill a questionnaire prepared by the Moorehead.[94] The material enabled Moorehead to publicize and embellish Reed's findings with colorful expressions such as the following: "The picture of Pontecorvo as a traitor simply does not fit the facts: it would be as rational to believe that Einstein was a secret baby-killer or that Stalin was, in reality, a fox-hunting gentleman from the shires."[95]

Yet the reassuring portrayal, designed to make MI5 look good in the public eye, was unconvincing. Fellow journalist Rebecca West labeled Moorehead as "an apologist writing under instructions from the Atomic Energy Division."[96] Her disappointment was so acute that she went on to inform the FBI, complaining that Moorehead was being granted privileged access to security information.[97] But because of the FBI/MI5 agreement on information management, she could get no sympathy from the other side of the Atlantic.

When Pontecorvo's former colleague, Herbert Skinner, wrote a book review on the monograph, he lashed out at Moorehead's claim that "nothing had gone wrong." In particular, he rejected both claims put forward in Moorehead's *The Traitors*: that the secret services had been thorough in their operations, and that Pontecorvo's knowledge was of no use to the Russians.[98] Skinner's review rang alarm bells at Scotland Yard, as it promised to reveal details the agency did not want out in the open, and to ruin its well-designed propagandistic operation. As Skinner had alerted Arnold about his intention to publish, the head of MI5's section B2a, James C. Robertson, felt compelled to convince him "that it would be wiser if he [Skinner] dropped the idea of writing on this subject at all."[99]

Skinner gave up, but between 1951 and 1953 the Pontecorvo affair continued to be in the headlines. Sensationalist accounts grew in number and in their degree of hype, offering details that in a few days were proven wrong or dismissed. Reed and other intelligence officers paid little attention to the journalistic flood of nonsense, limiting themselves to collecting the most intriguing pieces. On 12 December 1950, for instance, tabloids claimed that Pontecorvo had sent a postcard to his parents from Russia, proving that he was now there. But a few days later, the news was revealed to be fabricated. One year later, news appeared in the press that Pontecorvo had been arrested in Russia and accused of spying for Britain. By then Pontecorvo was supposedly working on a joint Soviet-Chinese nuclear program, and was called by the Chinese "Pangh de Co Co."[100] Reed found the phonetic connection amusing, but the revelation itself did not *sound* right.

These fictional reconstructions added to illustrations by tabloid artists that characterized Pontecorvo as a petty criminal. For instance, one depicted him with a pencil mustache, even though no photograph has shown him with hair on his face (figure 5.2). Since the end of World War II, the British public had associated this type of mustache with "spivs"—dealers of stolen goods resold at cheaper prices. This was partly because of a famous comedian, Arthur English, who wore one and, from 1949, had featured as a character in the popular BBC radio program *Variety Bandbox*. This deliberately misgiving imagery conveyed the notion that Pontecorvo, like thousands of other Italian migrants populating Britain at the time, was prone to illicit activities and dodgy deals—a prejudice not uncommon among the British populace.

The character fabricated in journalistic accounts was ready to become fictional once and for all. This is exactly what happened in 1952 when the writer and civil service commissioner Charles Pierce (C. P.) Snow portrayed the British atomic project and its protagonists, including the alleged traitors, in his novel *The New Men*. Snow's roman à clef was skillfully assembled to portray the scientists' effort to harness

FIGURE 5.2 Pontecorvo portrayed as a "spiv." No other photograph shows the Italian-born scientist as wearing a moustache, pencil or otherwise. Reproduced courtesy of the *Daily Express*; from Donald Seaman and John S. Mather, *The Great Spy Scandal* (London: Daily Express, 1955), 81.

atomic energy. One featured character, a scientist named Eric Sawbridge, was an atom spy. Snow would later claim that: "There was never a Sawbridge case, but several rather related cases."[101] Yet readers might have noticed that "bridge" was the English translation of the Italian *ponte*, and "Ponte" was Bruno's nickname at Harwell. Snow's literary work united three real scientists—Nunn May, Fuchs, and Pontecorvo—in a single fictional character. In doing so, it evoked the fear that the program to harness nuclear energy in Britain was experiencing problems because of the presence of these three scientists, and that their unreliability had caused severe disruption of the scientific endeavor. Snow's work, in conjunction with far-fetched journalistic inquiries, offered a character that further increased the distance between the fictional spy Dr. Pontecorvo and the real scientist of the same name. The caricature was accompanied by the fictionalization of other people close to him, such as his wife Marianne. The caricature, in fact, was made even more convincing by the presence of an equally suspicious lady at his side.

I Spy with my Little ... Wife

The conflicting portraits of Pontecorvo—the spy and the clumsy defector—included a similar view of the women surrounding him as likely suspects. A sort of morbid curiosity of the atom spy's "ladies" typified these accounts, which in turn contained embedded prejudices and preconceptions that lent support to new intelligence and journalistic inquiries. Aside from the attempts of journalists, intelligence officers, politicians, to play the case up and down, something else was at work that made it impossible to find out about Pontecorvo: the gender bias that typified the investigations.

In the eyes of investigators and journalists alike, women looked like "temptresses." It is worth noting that exactly when the atom spy cases were unfolding, the BSC agent turned novelist Ian Fleming was busy writing the first story of his James Bond series (published in 1953), which centered on a femme fatale named Vesper Lynd, who was one of Bond's lovers but also a Soviet agent.[102] The message was clear: women could be powerful tools of deception and betrayal both in public and private affairs. This stereotype informed atomic espionage enquiries as well. For instance, MI5 agents feared that Erna Skinner, the wife of nuclear physicist Herbert Skinner, played a key role in the smuggling of restricted information. This was because she was Fuchs's hostess and mistress at the same time.[103] The Pontecorvo affair elicited a similar approach in both intelligence and journalistic work. For instance, the journalists who assembled in front of the house of Bruno's sister Giuliana when the affair broke out offered her conspicuous amounts of money to pose for a photograph which would suggest that, aside from being involved in a betrayal case, she was "young and beautiful." This element, in conjunction with her communist militancy, enabled them to fashion a much better spy story.[104]

While looking for evidence corroborating the femme fatale conjecture, intelligence personnel gave extraordinary emphasis to anything that appeared avant-garde. The agents were interested in beauty, political radicalism, and dangerous eccentricity. For instance, an MI6 report of 1951 contained detailed information on an episode of two years earlier in which Gillo Pontecorvo's wife, Henriette Niepce, and Laura Pontecorvo had been arrested in Saint Tropez in the south of France. The arrest had been caused by indecent exposure, as the two were sunbathing with their tops off.[105] In the running of the Pontecorvo inquiry, details were made available by MI6 about Lydia Cassin, Lew Kowarski's secretary at the French Commissariat à l'énergie atomique. According to the agents, Cassin was "openly sympathetic to communism" and had seen Pontecorvo during her trips to England.[106] British services kept her and Kowarski under surveillance for a long time, making sure that anything unusual was promptly recorded. Yet she attracted the agents' interest because of her *sexual*, as much as her political, inclinations. In 1953 she was shadowed when she invited two English girls to her flat in Paris. As the agent duly noticed, "all three women [. . .] are said to be lesbian."[107]

This gender bias led the agents into areas of investigation that had nothing to do with atomic espionage. It was also the source of allegations about Pontecorvo that did no good to the inquiry. This is because these agents started looking at the behavior of Marianne Nordblum, Pontecorvo's wife, as suspicious. When the scientist and MI6 operative Wilfrid Mann wrote his report for Scotland Yard, he provided plenty of details on the relationship between Pontecorvo and his wife as if they were exactly what the agency had asked for. Marianne, Mann argued, was not well-liked because she was "cold, off-hand and impolite."[108] By contrast, Bruno was an exuberant man who in June 1947 had given a lift to two attractive young women of the Chalk River Chemistry Division while returning from a conference in Boston. His wife, unhappy about the episode, had left their house and withdrawn money from the bank, but was eventually convinced to return home.

Mann gave this description of a Pontecorvo family crisis because he believed that it could help in the investigation. His colleague Henry Seligman felt the need to describe the same episode to Arnold, almost suggesting that the scientist was so gullible that he could do anything, even give secrets away, if corrupted by a beautiful woman.[109] The fact that Pontecorvo's wife was considered by his colleagues to both brusque and beautiful elicited more allegations. For example, Skinner openly alleged that Pontecorvo's wife had coerced him into going to Russia. Skinner's statement defies logic, but it shows how powerful the gender bias was in atomic espionage affairs: "As confirmation of the 'guilt' theory of Pontecorvo, I think that the strange, shy and emotional behavior of Mrs. Pontecorvo [. . .] is strong supporting evidence."[110] The fact that this allegation was proffered by Skinner adds, however, a noteworthy contradiction. The secret agents involved in the Fuchs affair had suspected Skinner's wife, because she had made her husband a "cuckold." Yet Mr. Skinner suspected Mrs. Pontecorvo, exactly because she, in contrast with his own wife, had never been so "charming" with his colleagues at the laboratory.

If temptresses could be said to have played such a vital role in the unfolding of atomic espionage cases, then it is also true that women in general could be portrayed as dull housewives of no consequence in the spying game. For instance, substantial evidence of Pontecorvo's presence in Russia was offered for the first time at a Scandinavian "sewing party" organized in Moscow by a Mrs. Crapp, the Swedish wife of an archivist working at the British embassy. A Finnish guest at the party told the story that Pontecorvo had visited her country's embassy and requested the return of the passport that had been confiscated from him when he entered Finland without a visa in September 1950. His request was rebuffed and he was told "*Nitchevo*" (Nothing doing). From the diplomatic file written to report the episode to London, one gathers that Mrs. Crapp truly disappointed the secret agent who wrote it. He considered her actually a little useless because "in the hubbub of feminine chatter and her pre-occupation with her duties as hostess," she did not pursue the matter further.[111]

The proliferation of secret accounts discussing the activities of wives, sisters, and mistresses shows that while the wider public believed that an investigation of the unlawful smuggling of restricted scientific information was up and running, secret agents were dedicating precious time to shadowing activities in an effort to find out about the private lives of these women, their relations, and even their sexual habits. The activities of these agents overlapped those of journalists who were eager to link spies, scientists, and temptresses together in gossiping write-ups. This suggests another reason why the Pontecorvo case continued to be a mystery: preconceptions about women played a significant part in the running of investigations, and ultimately affected their outcome.

This bias contributed to the proliferation of fabricated accounts depicting Pontecorvo as an atom spy. As the accounts multiplied, they made it more difficult for investigators to find out why the scientist had defected and who had been responsible for his flight. Nor were these reasons and accomplices really sought by diplomats, security agents, or science administrators, who were actually more eager to withdraw information from the public domain and release only the few details that would make government officers look good in the public eye. As a result, all the official versions of fact suggested that Pontecorvo's defection was a despicable episode of marginal interest. In fact, the conflicting interpretations by journalists and government officers gave substance to the two diametrically opposite portraits of Pontecorvo—the notorious spy and the clumsy defector—that have existed side by side for the last fifty years.

The inquiry of the affair suffered substantially because of these ways of dealing with the defection. Indeed, even if Pontecorvo had conceived and improvised his flight in a hurry, he accomplished it thanks to someone who could provide logistical and financial support. Since the beginning of the investigations, that person's name had been written in newspapers and security records. But unsurprisingly, given the lack of real interest in solving the case, it took nearly three years for Reed to discover it.

6

A Political Motive

Dr. Bruno Pontecorvo, [. . .] who disappeared in 1950 while on holiday from Harwell, disclosed in an article in the Moscow newspapers "Pravda" and "Izvestia" to-day that he had been working in the Soviet Union since that time. This was the first news of Dr. Pontecorvo since he disappeared. He said in the article that he had been granted "refuge" with his family [. . .].

"Dr. Pontecorvo's Atomic Post in Moscow. 'Given Refuge' in Soviet Union: Work for Industry," *The Manchester Guardian*, **1 March 1955**

The mystery of Pontecorvo's disappearance lived on for nearly five years. On 1 March 1955 was finally exposed what a few already knew and many suspected: the scientist was in the Soviet Union. What caught almost everyone by surprise, however, was the justification given for the defection. In a revelatory article for the Russian newspaper *Pravda* that marked his comeback, Pontecorvo advocated a ban of atomic weapons and launched an appeal to world scientists, asking them to renounce the military uses of nuclear energy. He argued that his colleagues in Western countries should look at the Soviet Union as an example. There, he said, atomic knowledge was used "in the interest of peace, progress and bettering the material well-being of people." By contrast, the military applications of the atom in the Western Hemisphere had made him "ashamed of his profession" and convinced him to leave. Pontecorvo pointed out that the security hysteria mounting in Britain five years earlier had given him another important reason to go away. The situation had become intolerable to him due to "direct questioning and systematic blackmailing" by police authorities. He had thus sought asylum in Russia, where he was given the opportunity to work in the Institute of Physical Problems at the Soviet Academy of Sciences.[1]

Pontecorvo's sympathy for pacifist matters came as a surprise, mainly because in no circumstance before his flight had he openly asserted such a position. By stigmatizing the military uses of atomic energy, he thus appeared to be championing the protest against these applications, whereas no open or confidential documentation produced before 1950 confirms that he was at the forefront of, or even committed to, the battle against nuclear proliferation. In turn, his pledge in *Pravda* elicits legitimate questions that to this day have yet to be answered. How could the discrepancy between his lack of public commitment to the antinuclear cause before his defection and his later strong political stand against it be explained? Most importantly, why wait until five years after his defection to let everybody know about this commitment?

No final answer can be given to these questions unless new archival evidence, especially from the former Soviet Union, is made available. Yet Pontecorvo's untimely denunciation suggests the need to examine the circumstances of his defection in combination with an analysis of the activities of political organizations that since the late 1940s had campaigned against nuclear proliferation from within the communist camp. This chapter presents the results of this assessment.

Reed's inquiry noted that the most likely explanation for Pontecorvo's flight was that he had been "persuaded by Giuliana [Tabet], Gilberto [Pontecorvo], or some other communist [in his family] that he ought to work for Russia." What Reed did not fully explore, however, was these family members' commitment to pacifist activities. As we have seen in chapter 5, Giuliana Tabet had a prominent role in the Peace Partisans (PP), an Italian pacifist organization at the forefront of the battle against nuclear weaponry. Pontecorvo's cousin, Emilio Sereni, was an influential member in the executive committes of both the PP and the PCI. Just before Pontecorvo's defection, the PP campaign was thriving. The organization's leaders sought to target atomic scientists because they realized that these experts played an important part in the debate on nuclear proliferation. As harbingers of nuclear power, they could potentially make more of a difference in the debate than other social or political cohorts.

This brings to the fore another hypothesis about Pontecorvo's flight, notwithstanding the critical circumstances already examined in this book. The defection was conceived by some Italian pacifists in an attempt to show that unless nuclear proliferation was halted in the West, atomic scientists could resort to radical acts. Pontecorvo's eccentric move aimed to prove it, and Sereni had sufficient clout to arrange the flight using his pacifist and communist connections in Russia's satellite countries. In fact, orchestrating the defection of a prominent nuclear scientist was consistent with the political agenda of international organizations that sought to align the actions of pacifists and communists in Western Europe.

The pages that follow thus shift the focus of our inquiry from diplomatic relations between Western allies to those between the Soviet Union, its satellite countries, and the network of communist organizations supporting them in Western Europe, in order to highlight the importance that these relations played in the whole Pontecorvo affair.

Peace Partisans

Emilio Sereni had been a presence in Pontecorvo's life since his adolescence, and a very important acquaintance during his residence in Paris. But World War II abruptly halted the relationship between the two, as Sereni became a leader of the Italian resistance movement and Pontecorvo was involved in nuclear programs on the other side of the Atlantic.

In 1950 Sereni capitalized on his wartime experience by becoming a prominent politician within the PCI. Arrested during the war, he was condemned to eighteen years imprisonment, but in 1944 he managed to escape prison and was assigned a commanding role in the partisans' struggle to free Italy from Fascism. Chairman of Lombardy's National Liberation Committee (the resistance political arm), Sereni eventually contributed to the postwar coalition governments (of which the Communists were part, together with Christian Democrats and Socialists) as minister of war relief (1946) and of internal affairs (1947). When Prime Minister Alcide De Gasperi, who was also secretary of the Christian Democratic Party, put an end to this alliance, thus forcing the Communists out of government, Sereni became responsible for the PCI's cultural activities, while also continuing as a member of Parliament. By then he was recognized as a prominent intellectual who enriched the party's debate with reflections in fields as diverse as agriculture, agronomy, science, and energy policy (figure 6.1).[2]

From 1948 the controversies associated with the development of atomic energy took center stage in both cultural and political life.

FIGURE 6.1 Emilio Sereni, date unknown. Picture in the public domain.

Communist reformers such as Sereni believed that mobilizing forces against the atom bomb and for peaceful uses of atomic energy could revamp the party's activities, giving it visibility now that it was forced to operate from outside the government. In that year a new political organization, the PP, was born to counter the deterioration of international relations by promoting mass protests and pacifist activism.

Moreover, by the end of the war there was a strong component within the Italian resistance that had refused to accept the order to hand all weapons to the Allied forces, as they wanted to fight for revolution. By contrast, the PCI leaders that had instigated and supported the PP establishment, including Sereni, sought to recast the activism that had typified the resistance movement within the context of the new challenges presented by the Cold War.[3]

The struggle against nuclear proliferation propelled the PP into action. In 1948 the leader of the Italian Socialist Party, Pietro Nenni, became its president. A national PP committee was set up, which included

Sereni and Pontecorvo's sister Giuliana Tabet among its members. Several local committees were also established to coordinate action in Italy. Campaigning against nuclear weapons helped mobilize a growing number of party militants and sympathizers. The protesters were persuaded that only mass campaigning could stop a new world conflict in which atomic weaponry would be used.[4]

Sereni also played a key role in coordinating action at the international level. On 25 August 1948 he attended the first international Congress of Intellectuals for Peace that took place in Wroclaw (formerly Breslau), Poland. By then the movement of intellectuals against nuclear proliferation included renowned intellectuals (Albert Einstein, Bertrand Russell), artists (Pablo Picasso, Pablo Neruda, Renato Guttuso), and performers (Charlie Chaplin). In April 1949 Paris hosted the first World Peace Congress. Its organization was a result of the efforts of British scientist John Desmond Bernal and Pontecorvo's master, Frédéric Joliot-Curie. Sereni was present at the meeting along with delegates from seventy-two nations, thereby helping to promote the protest at an international level. The congress made possible the establishment of the World Peace Committee, whose main goal was to promote world peace and the banning of nuclear weapons.[5]

This contribution of prominent European scientists to peace campaigning was frowned upon by their own countries' governments. For instance, Joliot-Curie's activism was fatal to his career as a science manager. He advocated that peaceful applications of atomic energy be separated from military ones, but on 9 May 1950 the French prime minister, Charles de Gaulle, asked him to resign as director of the French Commissariat à l'énergie atomique. The French government, like those of the United States, Britain, and the Soviet Union, had plans to develop an independent nuclear deterrent.[6]

In the following autumn the PP organized the first World Peace Committee meeting in Rome. In the same period it also began campaigning locally in Italian cities. Its picketing against the unloading of NATO military equipment in Italian ports in an effort to halt the rearmament of West Germany was one of its most successful actions. On 15 March 1950 a second World Peace Congress opened in Stockholm and its forthcoming appeal became a key political tool in the hands of pacifists seeking to end nuclear proliferation. The appeal demanded the unconditional prohibition of atomic weaponry and the enforcement of stricter international control. It also indicated that any government using atomic weaponry in future conflicts ought to be considered guilty of a crime against humanity.[7]

American and Soviet Atoms

Pacifists based in Italy and other European countries understood that it was critical to their success that scientists, especially atomic scientists, let their views on nuclear proliferation be known. Not only could they express an opinion on the military uses of atomic energy, but some of them could actively seek to enforce their view by hampering new research on atomic weapons. Scientists were thus believed to be at the forefront of the pacifist movement. In the speech that marked the end of the World Peace Congress in Rome, Joliot-Curie argued:

> We know that science is a leading force of our times, a key factor in stabilizing or destabilizing world peace and an essential feature in the development of the quality of life across the world. Far too often scientific applications have been used for warfare activities, to serve a minority that has profited in the past and still illicitly profits from it. The men of science cannot be accomplices with those regimes that use science for illegitimate purposes.[8]

There is no doubt that Sereni and Tabet shared Joliot-Curie's viewpoint, and that Pontecorvo could not avoid contemplating his own position in this debate. He had played a key role in the design of vital processes for the production of atomic energy, and he was still contributing to it. But he had not yet openly asserted his commitment to exclusively pursue the peaceful applications of atomic energy.

Pontecorvo also had plenty of opportunities to discuss these issues with prominent scientific leaders who were campaigning against nuclear proliferation, and who could illustrate the decency of their political proposition. Just one month after the World Peace Congress in Paris, he went to Paris, where he met with Joliot-Curie. The following November he visited another celebrated nuclear physicist, Patrick Blackett, in Manchester.[9] Blackett had openly avowed the need to reduce the impact of nuclear weaponry on international relations and bring peace back to the agenda of postwar world politics. He also believed that the use of atomic weapons in Japan ought to be considered the first act of the Cold War rather than the last event of World War II. His words created enmity in the United States, and in 1951 he and his wife were even questioned by US security personnel in Tampa, Florida, while returning from a conference in Mexico City.[10]

Pontecorvo was thus surrounded by, and friendly with, outspoken critics of nuclear proliferation. Most certainly he never took part in the PP initiative in Italy. Nor did he ever speak at one of the international conferences against nuclear proliferation. There is no evidence of him considering these issues even in private correspondence. The only piece of communication in Pontecorvo's correspondence showing an interest in the pacifist cause is a copy of a support appeal sent to him in 1947 by Einstein, which he apparently did not answer.[11] Yet there is no doubt that Pontecorvo had concealed his disquiet concerning the uses of atomic energy. Conceivably, it was the pending security enquiry at Harwell that, from March 1950, had made him wary of openly supporting the peace campaign.

The existence of these political tensions can be seen as a key factor in motivating Pontecorvo to envision a future career in Russia. The protest organized by the PP in Italy, and the World Peace Congress in other countries, helped left-wing peace campaigners forge two opposing portraits of the superpowers' approaches to the controversial uses of nuclear energy. The US government was seen as warmongering, whereas Soviet Russia was portrayed as needing to catch up with US military might while intending exclusively to develop the peaceful applications of nuclear energy.

This portrayal was based on a selective interpretation of the international negotiations on atomic energy begun in 1946. The US government had instigated attempts to establish an organization devoted to international controls of the uses of atomic energy, especially through the activities of its UN representative, Bernard Baruch. Yet it was Baruch's Soviet counterpart, Andrei Gromyko, who had made the organization of such a body conditional on the prohibition of atomic weaponry and the destruction of stockpiled fissile materials.[12] Commentators were divided on the true intents of these ambassadors. Most understood that Gromyko's real concern was to slow down the pace of the US nuclear program, consistently with the Soviet Union's military agenda. Stalin believed a new world war inevitable, basing this assumption on a Marxist reading of the conflicting development of the imperialist and communist blocs. He therefore advocated a crash program to build nuclear weaponry so as, in the best case, to delay the beginning of such a war by using nuclear diplomacy or, in the worst case, to be prepared for nuclear warfare.[13] Yet leftist campaigners believed that Gromyko's proposal was more radical than Baruch's, partly because it indicated the destruction of fissile material as a nonrenounceable precondition for negotiations.

Moreover, a lack of information on the Soviet Union nuclear program, and on Stalin's mindset on the inevitability of war, combined with the Soviet Politburo's official endorsement of the peace campaign to win the hearts and minds of European militants.

Surprisingly enough, some pacifists did not doubt that Soviet Russia was exclusively interested in the peaceful atom even after the 1949 explosion of its Joe-I. After the test, Soviet authorities sought to prevent any criticism by suggesting that the detonation of this device was exclusively aimed to forcing the United States to negotiate the banning of nuclear weaponry. The test showed exclusively that the USSR would not be intimidated by Western warmongers. In the meantime, Soviet Foreign Minister Andrei Vishinsky championed a UN resolution for "the unconditional prohibition of atomic weapons." *Pravda* argued that "all supporters of peace welcome the Soviet Union's possession of atomic weapons as a victory for the cause of peace, because they firmly believe in the peaceful policy of the U.S.S.R."[14] So only Dwight Eisenhower's 1953 speech "Atoms for Peace" redressed this imbalance in the propaganda war between the US and Soviet administrations, by showing for the first time the US commitment to privileging the peaceful uses of atomic energy.

There is no doubt that the campaign presented Pontecorvo with the dilemma of whether to continue working for nuclear institutions in Western countries, thereby taking responsibility for controversial uses of the atom, or instead sponsoring the peaceful applications of nuclear energy, and even extolling the merits of the Soviet Union. This dilemma may have caused him to make extraordinary decisions—especially during the summer of 1950, when the peace campaign was gaining momentum and, conversely, Pontecorvo's career seemed close to reaching deadlock.

A Working Hypothesis

Following the Stockholm congress of March 1950, the PP's activities in Italy intensified, reaching their peak the following summer. Sereni drafted the PCI's working plan to gather support for the Stockholm appeal signature campaign. Not surprisingly, he understood that he would find a particularly responsive audience among intellectuals and scientists. He recommended supporting the actions of university deans and research center directors who had vowed to sign the petition, to try getting the petition signed by all the personnel in their organizations, and to go beyond ideological divides and reach out even to those who would normally resist engaging in political campaigning.[15] Yet the PCI plan also asserted that intellectuals and scientists known to be close to the party's position should make blunter decisions by renouncing any working commitment with military or nuclear organizations. As the protest campaign gained momentum, PP leaders such as Sereni and Tabet could register an enormous success. This made them even more eager to circulate their opinions among those whom, like Pontecorvo, they knew well.

The Stockholm peace appeal campaign began in Italy just before Pontecorvo's arrival in the country. Organized by Sereni and his collaborators, it opened on 5 May 1950 with a press release, prominent intellectuals signing the petition, and the PP president, Pietro Nenni (national secretary of the Italian Socialist Party), speaking to a big crowd in Rome. Three days later, militants of local PP committees collected signatures at public events such as festivals and carnivals. The campaign ended on 30 September 1950, in the same month as Pontecorvo's mysterious departure, when the signed petitions were sent to the national committee.[16] And if the Stockholm appeal had been signed by a conspicuous number of Italians since the very beginning of the campaign, towards the end of June unprecedented levels of support were reached. There is no doubt that the Korean War bolstered collection activities. Eighteen million people signed the petition (35 percent of the Italian population). This was an incredible result for Italy's left-wing organizations, especially because the Christian Democrats had at first followed the Vatican's official position of giving no support to the campaign. Many Catholic groups, however, signed the appeal, showing that pacifism could cross ideological divides. The petition was a resounding success at an international level too.[17]

This response made Sereni and Tabet enthusiastic about the campaign's potential, and eager to involve anyone who could be sensitive to these issues. When Pontecorvo visited Italy on vacation, he weighed the campaign's success against the preoccupations deriving from his precarious professional and security positions. We know that Giuliana Tabet solicited her brother to decide against continuing to work in Britain. Pontecorvo had not disclosed to her the content of the security investigation, but he had confided that the planned move to Liverpool had not made Marianne particularly happy.[18]

These concerns informed a very unconventional plan: seeking asylum in a country championing the peaceful applications of the atom. A plan that was quickly conceived and rapidly executed in ten days, from 22 August to 1 September, enabled Pontecorvo to obtain refuge in Russia. The swiftness of the move indicates that suggestions about it had been made earlier, either in conversations with Sereni or with another member of Pontecorvo's family. Yet Pontecorvo considered it as a possibility only when he saw his professional career sinking to a low. News from the United States regarding Giannini's petition added further anxiety. Yet the news brought grist to Sereni and Tabet's mill, as they confirmed that the US government had first used the atomic scientists to obtain nuclear weapons and then cheated on them by failing to reward them financially for their efforts. The present monopoly on nuclear weaponry was the result of a trick.[19]

If the plan to migrate to the Soviet Union was an answer to Pontecorvo's professional and personal preoccupations, it was also consistent with the political goals of its promoters. It fit well into the PP's political campaigning, as the militants could use Pontecorvo's flight to demonstrate that prominent members of the Western scientific community had become unwilling to continue working for their governments given the security purges, the lack of recognition for past research activities, and the impossibility of influencing policy making on nuclear matters.

Such a political motive only could become reality if an organizational arm existed that allowed clandestine migrations across the Iron Curtain. We shall now see that not only was Sereni an organizer of political campaigns, but also a key player in the organization of covert actions. He could muster significant political clout to turn the working hypothesis of Pontecorvo's flight to Russia into a well-designed operation.

From Motive to Execution: Cominform and Sereni

Sereni is one of the most controversial figures in the Italian postwar political scene, and opinions are still divided regarding his activities. Some historians have emphasized his ability to rejuvenate the party's work through the pacifist initiative. Others believe that he was an intellectual who would hardheartedly conform to the party line—or even a Stalinist, who would not object to violence if perpetrated for the benefit of his political organization.[20] Such difference of opinion is understandable not just in light of the ongoing historical debate, but also because of the lack of detailed knowledge about the role he played in a number of clandestine party operations. In particular, Sereni's only partly understood contribution to the Communist Information Bureau (Cominform) reveals his role as eminence grise behind the PCI's covert activities, including Pontecorvo's flight.[21]

Established in 1947 at Slazrska Poreba, Poland, the Cominform aimed at coordinating political activities between the communist parties of Russia's satellite nations, as well as those of Italy and France. These operations coincided with Stalin's effort to promote a political strategy in allied and non-allied countries that would be consistent with Russia's geopolitical interests. In Italy this strategy materialized in the effort to provide financial support to the party organization and arrange communication and propaganda activities that were consistent with Moscow's agenda.[22]

The Cominform had supported the PP action in many ways. The first World Peace Congress of April 1949 was organized thanks to financial support from Moscow amounting to one hundred thousand dollars. In the meantime, peace campaigning became a prioritized item of discussion at the Cominform executive committee meetings. In June 1949 Stalin drafted a key document arguing that broadening the peace partisans' action represented the main priority for communist militants in Western and Eastern Europe.[23]

After the Stockholm peace congress, the Cominform took responsibility for coordinating the campaign among communist parties. Several leading members of the Soviet Politburo, such as Mikhail Suslov—one of Stalin's closest collaborators, *Pravda*'s editor in chief, and a member of the party's Central Committee—took responsibility for convincing the PCI's leaders of the importance of peace campaigning. For instance, in November 1949 he ignored their objections about resistance from the militants and pressed them to consider pacifism as a main political objective. After the Cominform meeting of April 1950, the PCI executive committee decided that supporting and invigorating the peace campaign was going to be the party's priority for the foreseeable future.[24] Between 1949 and 1950, Sereni continued to demand more commitment. He resented the national trade union's approach to peace campaigning, as it was not attracting enough workers. He also criticized the level of support the party had offered.[25] These critiques made him unpopular with other party members but, not surprisingly, they were consistent with Suslov's attempt to align the party's actions to Moscow's wishes.

Sereni's activities within the Cominform led him to travel frequently to meet "comrades" in Russia and its satellite countries. These trips also coincided with the movement of documents, funding, and personnel that the PCI clandestinely transferred to and from Moscow.[26] Prague was one of the most visited places, as the Czech Communist party functioned as a connecting unit in clandestine operations between the PCI and the Communist Party of the Soviet Union. In 1948 restricted documents produced by the PCI executive committee in previous years were transferred to Prague, and from there to Moscow. A number of Italian communists found refuge in Czechoslovakia, especially those who were charged for crimes connected with their political activities. Some of them helped set up clandestine radio stations transmitting from Prague to Italy. Requests for financial assistance from Moscow and transfers of money were also managed through the Czech capital.[27] These connections turned out to be very profitable when Pontecorvo considered the possibility of moving to Russia.

A Question of Money (Dollars, Actually)

Reed's investigation had laid out an important line of inquiry that was never followed after December 1950, when his report was archived. The MI5 officer did not manage to find anything about those who had instigated, planned, and executed Pontecorvo's move to Russia. But he alleged that such support had come from an unknown source and was necessary because of the Italian physicist's sudden decision to flee.

Reed understood that the payment for the flight to Scandinavia was the key investigative element in his inquiry of the Pontecorvo affair. Six hundred dollars in cash was the only available circumstantial evidence that the flight had been arranged with someone's financial assistance. We have seen in the previous chapter that Bruno's sister Anna was surprised that he could pay the airfare because by the end of the vacation he had no money left. In her journalistic investigation, Rebecca West also indicated that Pontecorvo and his family could not have left Britain with more than £205 (approximately six hundred US dollars).[28] Because of the circumstances of this payment, Reed decided to follow the money trail: "The dollars were in 100-dollar bills, which are said to be scarce in Italy except perhaps amongst Americans. They are also no doubt available at Embassies and Legations, or even perhaps at Communist Party Headquarters."[29]

Could Pontecorvo have gone to the Soviet embassy himself and received the money there? In light of the evidence available, this seemed unlikely to Reed. Could Pontecorvo have collected the money from some prominent Communist Party official in the aforementioned headquarters? The scientist's family relations made this more conceivable, as "through the good offices of Communist Party headquarters or the Soviet embassy, they provided the money to go to Stockholm."[30]

The payment in dollar bills was indeed an important lead into key aspects of the case that were unknown at the time of Reed's inquiry. Historians have now ascertained that most financial transactions between the Soviet Communist Party and its satellite organizations, including those in Western Europe, took place by using one-hundred-dollar bills—often using the contacts that the Cominform had established. This was deemed necessary in order to ensure that the currency could be widely used internationally, and that any attempt to track down the source of funding would fail. In fact Reed could only speculate about the source of funding for Pontecorvo's flight, but he couldn't go further to identify the origin of the money handed over to the SAS offices in Rome.

One-hundred-dollar notes were routinely used when the PCI needed funding. Since 1947, six hundred thousand dollars in notes had been given by Stalin to Pietro Secchia, the PCI deputy chairman, and safely stored by Giulio Seniga, the party official responsible for treasury activities. This money represented what party officials dubbed the "Moscow Gold" (*Oro di Mosca*). When in August 1949 PCI officials put forward a request for financial assistance to Moscow, the point about dollar bills was reiterated. Correspondence between Czech colleagues who forwarded the request reveals that the contribution "should be exclusively in *bank notes* because any other method would not be safe enough. The best way would be *Dollars*, even if British Sterling or Swiss francs might do as well" (my emphasis).[31]

Following Seniga and the dollar trail, we can infer something important about Sereni and the Pontecorvo affair. A former partisan, the party treasurer Seniga stayed with the political organization up until 1954. In that year, after Stalin's death and the shakeup within the Italian party, he left grumbling, promising to create a sensation. He threatened to take a "suitcase full of *dollars*" to the Italian Parliament to reveal the foreign origins of PCI funds. He then fled Italy for Moscow, along with the dollar bills he had amassed in previous years.[32]

In the 1990s Seniga revealed the existence of a secret committee responsible for the party's clandestine operations. He also revealed how the covert funding bestowed by the Soviet leaders was used in a number of those operations. Seniga stated that one of the most successful operations of this committee was Pontecorvo's flight to Moscow—a flight, he added, that had been engineered by his cousin Emilio Sereni. Seniga went on to reveal that three party leaders were responsible for discussing confidential details with the Soviet embassy in Rome: Palmiro Togliatti, the party chairman, and Pietro Secchia with his brother Matteo. Following the indications of these three men, the secret committee used its logistical arm to allow the transfer of men and materials across the Italian border. The organization also had houses in the Dolomites, the mountainous region of the eastern Alps bordering Austria, at its disposal. They functioned as outposts for clandestine migrations, and were used to supply counterfeited passports.[33]

According to Seniga, Pontecorvo left Italy with the support of the party's secret committee, which provided the needed funding and arranged the entire operation. The fact that Seniga has been discredited by other former party members might induce one to doubt his revelations. Yet they are strikingly similar to those that a yet-to-be-identified source offered to Reed in 1953.

The Revelation That Reed Believed True

Sereni's name was not new to Reed, as it was mentioned from the very beginning of the Pontecorvo affair and throughout the run of the inquiry. Imme's press conference of 21 October 1950 first reported that Sereni and Pontecorvo were cousins.[34] Reed also highlighted Sereni's details in the description of Pontecorvo's family prepared for his final report. Yet he never went on to investigate Sereni's circumstances after the report was filed.

By 1953, Reed had abandoned hope of finding out more about Pontecorvo's flight. He continued to collect amusing newspaper clippings about "Pangh dc Co Co," but he never seriously considered reopening the case, constantly dismissing the reliability or significance of fresh news and claiming that the new information had no bearing on the known facts. Yet in April 1953 he reviewed the case and suddenly changed his position. He thus annotated Pontecorvo's security file: "I now think that it may have been Emilio SERENI who persuaded PONTECORVO to defect."[35]

This change of mind depended on revelations from an untested source whose identity is still secret. James C. Robertson jotted a few lines in a secret note to Reed: "This information rings true."[36] In turn, Reed analyzed the content. Its source was acquainted with Sereni's daughter, as she was his wife's English teacher. According to the source, Sereni had left for the USSR a few days before Pontecorvo, and stayed there for a month to introduce his cousin to Soviet officials and be his companion there. The source was also close to leading party members and diplomats. He argued that in private talks held with Matteo Secchia after the defection, he had been informed that Pontecorvo had reached Moscow by traveling through Austria and Czechoslovakia. The Czech embassy official Milan Matusel had also indicated to him that Sereni was the main sponsor of Pontecorvo's flight. Sergei Mikhailov, of the

Russian embassy in Rome, asserted that Western intelligence services could not find Pontecorvo.[37]

It is impossible to say whether these revelations prompted new investigations at Scotland Yard. The lack of further detail suggests that this was not the case. But the presence of significant omissions and cuts in the archival papers leads us to believe that more information may have been put together at the time, even if it is not yet available to historians. Nevertheless, these revelations fit well with those offered by Seniga. They indicate Sereni as the instigator of Pontecorvo's defection, and they also highlight the possibility that other prominent party officials (such as Pietro Secchia) may have known about the operation or even played a part in it. More generally, both sets of revelations highlighted the existence of a clandestine network responsible for migrations to Russia and its satellite countries that was at work in Pontecorvo's attempt to defect. However, some of the points made by the untested source were untrue—such as the claim that Sereni introduced Pontecorvo to Soviet authorities, or that his daughter taught English; although she was a language teacher and gave private lessons, she did not know English well enough to teach it.[38] More significantly—a detail that caught Robertson and Reed's attention—there was no doubt that Pontecorvo had gone to Russia through Scandinavia, and not through Austria and Czechoslovakia.

Who was this source? Only a few Soviet officials are known to have worked and defected so early in the history of the Cold War, and only one of them began his activity as CIA informer: Piotr (or Peter) Popov, a GRU official. Popov stayed in Vienna for some years before returning to Moscow, where he was exposed and given the death penalty.[39]

The mole certainly knew about an episode in Sereni's family life that occurred during this period, in which Sereni's daughter, Lea, and her partner, Antonio Natoli, were protagonists. Natoli fled Italy at the beginning of 1950 after being convicted for crimes related to his political activities. He was followed by Lea, who was in financial troubles before reaching him in Prague. Lea was also a language teacher, which squares with the leaked information. Yet Popov never left Vienna—and although Lea might have sojourned in Vienna before going to Prague, there is no evidence of this. Alternatively, the source might have stayed in Rome—in which case nothing can be said about his identity. In any case, there is no reason to believe that the untested source was deliberately deceitful as he revealed the identity of key players in the PCI's covert actions, such as Matteo Secchia, whom he indicated as being the main courier between Rome and Prague. In light of the revelations by Seniga and the unknown source, together with analysis of the movements of the protagonists in the Pontecorvo affair, it is now possible to lay down a realistic description of what happened in the summer of 1950.

Conjectures on the Pontecorvo Affair

Sereni and Pontecorvo ostensibly met on only a few occasions before the flight to Russia. In April 1949 Sereni was in Paris, where he attended the World Peace Congress; Pontecorvo was in Paris the following month. On 30 May 1950 Sereni was in London, where he met with the scientist and peace campaigner John D. Bernal; Pontecorvo lived less than a hundred miles away from the capital. These meetings allowed the cousins to discuss their successes and shortcomings. If in April 1949 Pontecorvo was an interested observer of the peace campaign, by May 1950 he was aware of the dilemma that the peace campaigners posed to nuclear scientists. In the same period he also had private quandaries, such as the outcome of the pending security investigation, and the proposal to move to Liverpool, which did not please Marianne at all. If, especially during the second meeting, Sereni put forward an eccentric proposal such as that of going to Soviet Russia; Pontecorvo was likely to promise his cousin that he would think about it.

It is during Pontecorvo's vacation, however, that the proposal to defect materialized while he visited places near the Dolomites that were a focus of clandestine PCI activities. He also met Sereni again. After staying at a campsite in Menaggio from 31 July to 6 August, Pontecorvo and his family crossed the border to reach Landbeck, Austria, and returned to Italy two days later. On 12 August they went to Milan to visit Pontecorvo's parents.[40] So Pontecorvo and his family were in the Dolomites for four more days, between 8 and 12 August, and Sereni was also there during the same period. On 10 August 1950 Sereni left Rome, arrived in Bolzano the following day, and then went to the mountain resort of Pera di Fassa.[41]

During their rendezvous in the mountains, Sereni and Pontecorvo might have continued the dialogue they had begun a few months earlier about the unconventional proposition of moving to Russia. Now Sereni the politician could have also showed, away from indiscreet eyes, that he possessed the means necessary to make the whole operation possible. He could have let Pontecorvo know about the outposts, the routes for clandestine Italian migrations, and even the places where counterfeit passports were produced. The two men might have conceived a plan together. In 1992 Seniga recalled that some of the outposts the PCI secret committee used were located in the area were Sereni stayed. More precisely, they were in a place called Costalunga, on the road between Bolzano and Vigo di Fassa, near Pera.[42]

On 14 August Sereni went to Prague, where a PP meeting—extended to include other European pacifists and, among others, Joliot-Curie—had been arranged between 17 and 18 August. Sereni returned to Pera on 23 August, where he stayed for another week. He was in Rome from 4 September to 2 October 1950.[43] In the meantime Pontecorvo went to Rome, where he met with his sister Giuliana, who presumably reiterated Sereni's offer and informed him about the successful outcome of the peace campaign. After 22 August, as we have seen, Pontecorvo finally decided to "disappear" in an attempt to avoid repercussions to his career deriving from Giannini's petition in combination with the pending security investigation at Harwell. He thus returned to Rome in a hurry, seeking to alert Sereni, who was now in Pera. Sereni was thus informed about Pontecorvo's decision, and he in turn informed the party officials who were responsible for clandestine operations.

The operations took place during a period that was very eventful for party managers. On 22 August the party's secretary, Togliatti, was involved in a car accident on his way back to Rome. Three days later, when the executive committee convened in the capital, only a few members were able to attend. Among them was Pietro Secchia, who acted on Sereni's behalf and took responsibility for alerting the personnel at the Soviet embassy in Rome about Pontecorvo's and Sereni's wishes. He called for the Soviets to provide logistical support to the operation. The secret PCI committee, or perhaps even Pietro Secchia himself, arranged the flight to Russia. The unknown informer of Scotland Yard argued, on the basis of what Matteo Secchia had told him, that the best and habitually most used way to reach Russia was through Austria and Czechoslovakia. This indicates that this was the route initially considered by the secret committee. However, Pontecorvo and his family had crossed the Austrian border only a few weeks earlier, and it would have been risky to cross the same border again with false documents. No doubt they could leave the country with their real passports, but that could attract unwanted attention. A decision was thus made to avoid the conventional clandestine route and instead try the Scandinavian corridor through Sweden and Finland.

This pathway had never been used for clandestine activities by party operatives, but several elements made it practicable in the Pontecorvo operation. First, the travel to Sweden was not suspicious because of Marianne, whose mother lived in Stockholm. When Pontecorvo booked the air tickets at the SAS office, he registered his three sons with the double surname Nordblom-Pontecorvo.[44] Furthermore, Finland's precarious position in Cold War geopolitics made it a free port, ideal for the operation to be carried out. In 1950 Finland was officially a neutral country, but it continued to host Soviet military bases, such as the one at Porkkala (allegedly used in the Pontecorvo operation), in compliance with the Finno-Soviet Treaty of Friendship, Cooperation, and Mutual Assistance.

Once the stealthy travel was arranged, the only problem was finding the money necessary to complete it. Enter Seniga. As he was responsible for party finances and controlling the "Moscow Gold," Secchia or Sereni solicited Seniga to give Pontecorvo the amount of dollar bills needed for paying the airfare. It is exactly because of this transaction that Seniga was eventually able to reveal that the PCI secret committee had been involved in the financial aspects of the Pontecorvo operation.[45]

Meanwhile, Sereni monitored the operation from a very safe position. He returned to Rome only after the flight was successfully accomplished, thereby concealing the role he had played in it. Sereni's personal diary shows that he was not in Moscow welcoming Pontecorvo after the flight, as was claimed by Robertson's still unidentified mole. Yet there is no doubt that Sereni was often in Moscow for diplomatic activities. He was in Prague between 2 and 6 October, and during the following week he visited the Russian capital. He returned to Rome only on 4 November 1950.[46] Thus it is conceivable that although Sereni was not waiting for Pontecorvo in Moscow, he waited for the whole operation to be successfully accomplished before meeting him in Russia, spending some time with him and also introducing him to his acquaintances there.

These conjectures, based on sound evidence assembled from different archival materials and revelatory statements offered by a number of individuals, undoubtedly approximate a solution to the mystery of Pontecorvo's defection. Yet one element still defies reasonable explanation. Why did it take so long to let everybody know that Pontecorvo was in Russia? Once again, it is an analysis of the open and covert goals of the peace campaign that offers a credible answer to this question.

Five Years of Unexplainable Silence

If Pontecorvo's move to the Soviet Union was orchestrated by Sereni to denounce the military uses of atomic energy as deviant and shameful, it is surprising that the endorsement of such a political act occurred five years after it was committed. Peace campaigners and nonproliferation protesters said nothing about Pontecorvo's flight before 1 March 1955. In fact, if Pontecorvo's example could have been used to make the case against nuclear weapons, then Sereni and his comrades missed an important opportunity.

By contrast, after the publication of Pontecorvo's 1955 article, there was a flourishing of pacifist propaganda counting the scientist as a backer of antinuclear campaigns. Pontecorvo became the scientist who had chosen the peaceful atom, in contrast with his colleagues who were complicit in the despicable crimes of the Western military and nuclear establishment. His words were thus duly reported with those of other movement leaders as if he could now represent a model of conduct in the public arena. For instance, the Italian magazine *La Nave* reported his words:

> The NATO powers are preparing for atomic warfare. They believe nuclear weaponry to be a legitimate means of warfare. Yet in the last four years that I've spent in the USSR, I have become convinced that the Soviet people, all the Soviet people, want peace, and that the USSR government is taking all the measures needed to stop war. I am begging all men who are honest, especially the scientists and the physicists who I know and whom I have worked with, [. . .] to take a position. In these days it is no longer possible to stand aside.[47]

Indeed it wasn't. However, if Sereni sought to use the Pontecorvo affair to denounce the militaristic uses of the atom in the West, then it was ultimately up to the Russian authorities to consider whether and when to use Pontecorvo's defection as a political tool of nuclear diplomacy. Given the political factors that followed the flight, these authorities kept it secret, believing that no benefit could derive from immediately disclosing that Pontecorvo was now in Russia. This is because after Pontecorvo's defection, the peace campaign lost its appeal to the wider public, and although it continued to be an important part of the struggle against nuclear proliferation, it could no longer count on the levels of support that had typified its early days. The PP meeting that Sereni attended in Prague deliberated the organization of a fourth peace congress in Sheffield, England. Yet Attlee's government was successful in imposing restrictions, through the Home Office, on the attendance of foreign delegates. Another congress was hurriedly organized Warsaw, but the late change of venue had a negative impact on participation.[48]

The Italian delegates arrived at the meeting divided on a number of political and organizational issues. In December 1950 Sereni reported on the campaign at the party executive committee in Rome, arguing that the PCI had done too little too late to make it a success. Once again Sereni's criticism attracted the party leaders' attention, but this time Secchia did not support his claims, indicating instead that the party had other matters—pacifist protest aside—that it ought to look into and prioritize.[49]

The Warsaw congress led into the organization of the World Peace Council, of which Joliot-Curie was elected president by acclamation. Yet nuclear disarmament no longer featured as the key issue of the "pacifist international." The Korean War had an impact on these decisions, as the conflict was not fought, as expected, with nuclear weaponry—even if their use was considered by the US military. It was mostly biological warfare, a new feature of the conflict in Asia, that now attracted the attention of Joliot-Curie and others and became the new priority for the internationalist movement. Therefore, the Warsaw appeal, which in many ways was similar to the Stockholm appeal, referred to the renunciation of nuclear weapons only as one of several items, such as the elimination of chemical and biological weapons of mass destruction.[50]

Some militants did not accept these swift changes in emphasis, and abandoned the protest. Others begun to suspect that the campaign was not entirely based upon a democratic debate, and that its priorities really were responsive to the Soviet Union's political urgencies. Among others, Albert Einstein expressed some criticism of these developments, claiming that Soviet and East European delegates did not seem to have an opinion—they just reiterated the official party line.[51] In this political climate, a disclosure that Pontecorvo had gone to Russia to express disapproval of nuclear proliferation could have greatly reduced the impact of his denunciation. The movement was struggling to keep satisfactory levels of participation, and was attempting to direct the campaign towards other themes, thereby keeping the issue of nuclear proliferation in the background.

Another reason why a public condemnation by Pontecorvo did not occur immediately after his defection was Moscow's changing political agenda. The interest of Soviet leaders in the peace campaign was now less pronounced. On 22 November 1950 the Cominform delegates were to have met in Sofia, Bulgaria, to discuss the development of the peace and nuclear disarmament campaigns. But after a last-minute decision by Stalin, the points of discussion were changed to consider a review of the bureau structure. The Cominform, like the PP, was in decline.

A few months after the Sofia meeting, Stalin asked Togliatti to become its chairman in an attempt at facelifting its activities, but the PCI leader refused the offer.[52]

Furthermore, in the second half of 1950 Stalin, Suslov, and their party associates felt safer about the possibility of soon restoring nuclear parity, because the Russian nuclear program was now gaining momentum. On 8 August 1953 the Soviet leader Georgy Malenkov confirmed that Russia was in possession of thermonuclear weapons during a speech at the Supreme Soviet of the Soviet Union. Four days later a "boosted" fission bomb was exploded at the Semipalatinsk polygon to prove it. Even if the test was a mockery, Russia was close to completing its nuclear program. On 22 November 1955 a proper thermonuclear device was tested.[53]

Malenkov's speech confirmed that Russia's support of the PP was instrumental in its attempt to restore nuclear parity. Inevitably, this revelation further exacerbated the tension between the pacifist movement and its communist component. It also set the conditions for ending the Cominform's support of pacifist organizations. In 1956, three years after Stalin's death, the organization was dissolved because it had lost its main function as an engine of propaganda activities.[54] In the meantime, Eisenhower's Atoms for Peace redressed the imbalance in the propaganda war. Even if the US campaign retained a propagandistic agenda, it was effective enough to create divisions within the pacifist camp. The image of a warmongering US government that opposed a Soviet one that was eager only to boost the peaceful atom was now nonexistent.

Pontecorvo's defection occurred in the middle of this substantial restructuring of political strategies that weakened the position of antinuclear campaigners and blended their criticism of nuclear proliferation with other issues then arising in the pacifist camp. Furthermore, the Soviet leaders had no interest in immediately reinvigorating the peace campaign, because they had always perceived it as subordinate to their geopolitical considerations. The lack of a strong political motive thus induced them into waiting for better times to transform Pontecorvo into the champion of nuclear disarmament that he eventually became. As we shall see, Sereni was by then out of action because of health problems. This meant that he could hardly influence what decision was taken in Moscow.

So if Sereni wanted to reveal the reason for Pontecorvo's flight as soon as it was accomplished so as to dispel the impression that the act was a betrayal, he had to wait for approval of the revelation to be agreed upon by Moscow's top brass. If he wanted to highlight the defection as a political act that indicated the need for scientists around the world to renounce their complicity in nuclear warmongering, he had to wait for Stalin and his affiliates to consider the act beneficial to the Russian political agenda. Presumably, party officials also weighed these benefits against the negative impact the disclosure would have on diplomatic relations with Britain. Regardless of what Sereni wished or aimed to do, he had to wait until conditions were ripe for such revelations.

Awakenings in Dubna

By 1955, when Pontecorvo disclosed that he was in Russia, much had changed in the international and Soviet political landscape. Stalin's death in 1953 paved the way to a period of transition that led into Nikita Khrushchev's new regime. Khrushchev criticized Stalin's administration and sought to establish new relations with the West. In contrast with his predecessor at the Kremlin, he did not see war as inevitable, and he advocated peaceful relations between the two blocs.

The thaw in international relations that accompanied Khrushchev's action deeply affected the scientific community in general, and the nuclear physics community more specifically. Pontecorvo's surprising reappearance in Moscow anticipated a number of initiatives, including the first serious disarmament proposal coming from the Soviet Union. Some of these initiatives were aimed at establishing collaborations between scientists in the East and the West on issues of atomic proliferation. In June 1955, India's Prime Minister Jawaharlal Nehru visited Moscow State University, presenting his proposal for setting up a committee of scientists to discuss arms control. The following month, Bertrand Russell and Albert Einstein presented a similar scheme, also urging a joint scientific appraisal of new developments in nuclear warfare. These proposals led in 1957 to the establishment of the Pugwash Conferences on Science and World Affairs; which included American and Soviet scientists.[55]

Other initiatives aimed at letting scientists from the Western and Eastern blocs exchange scientific information and ideas. Between 1 and 5 July 1955 an international conference took place in Moscow on the peaceful applications of atomic energy. It was followed in August by the Geneva conference which was also devoted to these applications. Clearly, the détente in scientific international relations was interwoven with other interests, such as each nation's desire to show its scientific and technological prowess to the enemy, as well as using international meetings as a chance to gather intelligence.[56] Yet there is no doubt that

Pontecorvo could discuss his political act because of these more relaxed political relations between superpowers.

In March 1955, following the publication of Pontecorvo's articles in *Pravda* and *Izvestia*, a conference took place at Dubna, the site of the institute for nuclear research that he had indicated as his new workplace. During the conference he was asked whether he had renounced British citizenship. The scientist took an ID card from his pocket, showing that he had been a Soviet citizen since 1952. He spoke in Italian, but his words were simultaneously translated into Russian and English. He claimed to have agreed to a press conference because "a free exchange of opinion would be beneficial. This is why I am happy to be among you for a friendly conversation."[57]

Pontecorvo restated the key points he had made in his two articles for the Soviet newspapers, and when asked whether he was working on the military applications of nuclear energy, he replied, "The correspondent is mistaken if he thinks that I am working in the field of application of atomic energy for military purposes." He went on to claim that he was working in high-energy physics analyzing meson production and diffusion. Soviet physicists held "the first place in the world" in this type of investigation.[58]

At no time during the conference did Pontecorvo explain why he had decided to defect in the summer of 1950. Nor did he clarify how he had done so. Nothing was proffered about his disappearance, the instigators of his flight, or his many changes of travel plans during the week after 22 August 1950.

The conference did not remove all doubts about Pontecorvo's conduct of five years earlier; if anything, it aggravated some of the accusations that had been proffered immediately after the defection. In Britain, Pontecorvo was still considered a traitor.[59] The final sanction was the withdrawal of his British citizenship. On 9 March 1955 the Home Office secretary, the Conservative Gwilym Lloyd George, informed the British Parliament that he had produced an order depriving Pontecorvo of his British citizenship on the ground that "he has shown himself by act and speech to be disloyal or disaffected towards Her Majesty."[60] Information about the order was forwarded to the Russian Ministry of Foreign Affairs, which in turn passed it on to Pontecorvo. The scientist replied the following April to a ministry official, claiming that

> I am a Soviet citizen and that therefore I am not interested in the question of what decision will be taken by the British Home

Secretary concerning British citizenship. At the same time I consider it necessary to emphasise the fact that I have not shown any elements of disloyalty or unfriendliness towards the British people and that I still entertain the highest feeling for them.[61]

The order was effected the following May on the basis of Section 20 of the British Nationality Act. The Foreign Office asked for Pontecorvo to return his passport and naturalization documents. But the following year the scientist claimed that he could not return them because they were no longer in his possession.

Pontecorvo's pacifist-inspired choice did not receive the sympathy of his former colleagues. In 1956 a high-energy physics conference at the Soviet Academy of Sciences in Moscow presented an opportunity for Pontecorvo to meet with his former colleague Emilio Segrè, who had been busy completing the research on antiprotons that eventually gained him the 1959 Nobel Prize with Owen Chamberlain. The meeting was, unsurprisingly, ice-cold. Segrè later vilified Pontecorvo in his autobiography, claiming that during the meeting he was "so little Russified" that he couldn't speak Russian properly and his colleagues asked him to continue his talk in English because they did not understand him. In addition, they treated him as a "party zealot."[62] These words evidenced Segrè's grief over Pontecorvo's decision to flee the Western world. On the other hand, it was Segrè who had been responsible for instigating the security operation that eventually led Pontecorvo to leave Britain. Segrè paid the price for it by having to substantially review the financial details of the compensation claim on the slow neutron patent.

To summarize, the press conference did little to heal the wounds the defection had opened. Pontecorvo's decision met with general disapproval that was only mildly mitigated by the support received by pacifist campaigners. If Pontecorvo held pacifist beliefs, he did not fully succeed in getting them across. Nor could he count on the support of his cousin Sereni, who in that same time period had to face accusations that seriously undermined his own reputation.

Sereni's Downfall

As the peace campaign lost its luster, Sereni could not contribute to the political protest either. From February 1951 he was in bed with serious health problems. The following April, the party agreed to let him rest for two more months, as his condition had not improved. He attended the fourth meeting of the World Peace Council in Vienna, but by November he was ill again. He thus decided to go to the Soviet Union for treatment that put him out of action for the next two years.[63] By the end of 1953 he could work again, but by 1955 the PP was no longer a major political force and Sereni resigned. Nenni left the organization too, as by then the Italian Socialists had distanced themselves from the Communists and accepted Italy's participation in NATO. The PCI continued to support the pacifist campaign, but its contribution to PP activities was now negligible.[64]

Sereni could not enjoy Pontecorvo's March 1955 pledge to abandon nuclear weapons, either. On the day of his cousin's press conference, his reputation was seriously undermined by a recent investigation into his past, whose details were disclosed in the Italian Parliament. On 5 March, a retired marshal of the Italian Army, Senator Giovanni Messe, accused Sereni of partisan activities during World War II. Messe's exposé was based on a leaked 1943 document of the French services discussing Sereni's attempt to sabotage Italian troops at the front. Sereni sought to reply to the accusation, but the stormy parliamentary session was abruptly brought to a halt due to the bickering between Sereni's associates and the right-wing parliamentarians supporting Messe.[65]

Sereni's partisan actions were thus portrayed by Italian newspapers in conjunction with Pontecorvo's recent admission that he was in Russia. Some journalists, such as *Il Tempo* editorialist Vittorio Zincone, assimilated their conduct. Both cousins believed that to favor their own political organizations, it was legitimate to help their country's enemies.[66] *Il Tempo* continued to publish accusations of Pontecorvo and Sereni, recalling the time of their stay in Paris: "There [Pontecorvo] worked in Joliot-Curie's research institution . . . [also meeting with] that communist Senator who has recently been revealed to the whole nation as a deserter and as the architect of acts of violence behind the back of our troops at the front, Emilio Sereni, cousin and Parisian comrade of the traitor Pontecorvo!"[67]

Sereni eventually abandoned active political work, even if he continued to have a prominent role in cultural activities. Despite accusations and troubles, the cousins continued to see each other, facing together the few highs and many lows of state socialism. Pontecorvo's name does not appear in the diaries before 1955, but that is mainly because Sereni did not want to give away any information about his missing cousin. After 1956 Pontecorvo was a constant presence in all the diaries, which contained his postal address and phone number. He and Sereni also met on several occasions. Sereni's private diaries document those meetings, especially the ones that took place when he visited Russia. For instance, on 5 August 1957: "With Giuliana (Tabet), Duccio and Bruno." And on 30 October 1973: "Dinner with Bruno."[68]

Sereni died on 20 March 1977. When in the 1980s Bruno Pontecorvo agreed to discuss his life and career with the Italian journalist Miriam Mafai in preparation for the biography *Il lungo freddo*, he stated that only on one point was he not prepared to reveal anything: the identity of the person who had helped him cross the Iron Curtain. Pontecorvo's statement caught the journalist by surprise, as she could not figure out why he was so eager to keep that person's name secret after so many years. This chapter offers an answer. Pontecorvo wanted to avoid more attacks on his cousin after his death. He sought to protect a friendship that had begun in the Parisian boulevards, was interrupted by the conflict, and was renewed by political activism in the name of pacifism—a friendship that led Sereni to act with generosity when his cousin needed help. Surely these activities exhibited some naïveté, especially about how an act that united private and political motives could be capitalized upon to strengthen the pacifist movement. In fact Pontecorvo's defection didn't fulfill that purpose, because by the time the Italian scientist openly endorsed the pacifist cause, the political project for which Sereni had worked had fallen into a state of sheer decadence.

7

Bruno Maximovich and Professor Pontecorvo

For nights there have been various bangs which broke the sodden gloom—
Backfires, perhaps, from motor-cars but rather I assume
Our ardent youth rehearsing, with approving squeaks and squawks,
The damp November festival of caitiff Guido Fawkes.
A little out of date, I think, in this atomic age;
Were it not better done to turn a newer, apter page?
If dogs and cats should now be doped (see warning in the press)
To shield their shattered nerves and save the darlings from distress,
Could we not try a newer guy to justify their terror,
A fresh design and more in line with modern moods and error?
For I should like to see to-night across the skies as courser
Professor Pontecorvo on a crimson flying saucer.

Lucio, "Miscellany," *Manchester Guardian,* **4 November 1950.**

"Who is he *really*?" In the forty-three years that followed the defection, the question that had ignited early inquiries into Pontecorvo's flight across the Iron Curtain continued to puzzle reporters across the world. It never received a final answer. Chronicles from within the Soviet empire portrayed the achievements of the Russian citizen Bruno *Maximovich* Pontecorvo, a resourceful scientist who had pioneered the development of high-energy physics in the USSR. Opposed to the military uses of atomic energy, the physicist worked at the Dubna Institute for Nuclear Research (from 1956, the Joint Institute for Nuclear Research, or JINR) until the end of his career.

Yet the cascade of revelations that typified articles on "Professor Pontecorvo" in the West suggested something different: The professor, especially in the five years that followed his defection, was busy with defense studies, such as the development of atomic weaponry and baffling "death rays." He had also been sighted in remote places beyond the Arctic Circle, busy with mysterious research tasks, such as detecting UFOs. His previous activities as an atom spy had been now more extensively scrutinized, promising to provide conclusive evidence that he had given away scientific information while working in the West. Whether or not Bruno Maximovich was really involved in restricted research, his bogeyman alter ego, Professor Pontecorvo, continued to appear in the news. In this way he became an emblematic figure who could be used to spice up any information about nuclear science and technology that Western journalists managed to obtain from the otherwise secretive Soviet Union.

As far as Bruno Pontecorvo was concerned, Reed concluded his 1950 report with words that in retrospect sound prophetic. He argued that Pontecorvo was unlikely to find living in the Soviet Union as pleasant as he had wished: "He will almost certainly not find life in Russia as he wanted it to be, especially as he and his wife enjoyed foreign travel so much and life in European capitals."[1]

After his defection, Pontecorvo sought to become part of the local scientific community. He even adopted the patronymic Maximovich, so as to please Russian colleagues who found it too embarrassing to address him only by his first name. Whether or not he was sufficiently "Russified" (to use Segrè's slanderous term), he undoubtedly contributed to the expansion of Soviet science with experimental and theoretical contributions, especially on neutrinos. This research gained him prestigious awards such as the 1953 Stalin Prize, the Order of Lenin (twice), and, from 1958, honorary membership in the Soviet Academy of Sciences.[2]

Yet Pontecorvo did not find it so easy to accept the rigid party rules and, more generally, the political directions taken by the Soviet regime. In the fall of 1956 the Soviet invasion of Hungary shook his belief in the merits of state socialism, but ultimately he defended the choice of Russian administrators to crush the revolt. But the launch of the first Soviet satellite, *Sputnik*, a year later troubled him. He could now see how paradoxical it was to invest so much money in the country's space program while Soviet citizens struggled to get hold of basic commodities such as toilet paper and toothpaste. When, in the spring of 1968, the Prague uprising was silenced by Soviet tanks, Pontecorvo began to seriously doubt the meanings and implications of the socialist experiment taking place in Russia. After the tragic events in the Czech capital, his socialist dream was dead and buried.[3]

Disillusioned by the activities of the Soviet Communist Party, in the early 1970s Pontecorvo supported the initiative of his colleague, the physicist Andrei Sakharov, to ban antiballistic missile defense, recognizing that it increased rather than reduced the probability of escalating the Cold War into open atomic warfare. It is worth noticing that the target of Pontecorvo's criticism this time was not exclusively the Western governments, but also the Soviet politicians, who were eager to show the USSR's military prowess in the confrontation between superpowers. Not only did Pontecorvo refuse to sign a petition of the Soviet Academy of Sciences calling for a condemnation of Sakharov's campaign, but he sought to let Western scientists know about it, thereby further exacerbating his political problems.[4]

If these political frictions disenchanted Pontecorvo, then the travel limitations imposed by the Soviet regime made him very depressed. Only in 1978 was he authorized to visit Italy, the country he had left twenty-eight years earlier. The occasion of this visit was Edoardo Amaldi's seventieth birthday (figure 7.1).[5] By the time Mikhail Gorbachev's perestroika blossomed and the Soviet Union collapsed, Pontecorvo was struggling against Parkinson's disease. In 1993, he died at the age of eighty.

Pontecorvo's distancing from the Communist regime took place only after years of intense research activities that in the past have been understood as being exclusively connected with pure research on nuclear particles and interactions. Yet what is revealed here is Pontecorvo's contribution to the Soviet nuclear program in two specific areas—nuclear geophysics and nuclear reactor technologies—that, as we have seen, also featured prominently in his research interests before the defection. If we are to believe current literature, these interests literally disappeared from his studies during his days in the Soviet Union. This chapter highlights the fact that such a reading of Pontecorvo's career after his flight is unsatisfactory.

The chapter also documents the proliferation of accounts of the antics of Professor Pontecorvo in espionage literature. Even when the leakage of restricted security information allowed revelations about significant aspects of Pontecorvo's past, most of these works sketched improbable spying scenarios. What matters to us is highlighting how these projections continued to play a part in the cultural and political debate

FIGURE 7.1 Friends after all these years? Bruno Pontecorvo (left) and Emilio Segrè (center) meet again in 1978, on the occasion of the seventieth birthday of Edoardo Amaldi (right). Copyright E. Recami, courtesy AIP Emilio Segré Visual Archives and E. Recami.

in the Western world—especially during the 1980s, when the Cold War reached new heights and the spying game returned to a prominent space in political debates. The new accounts of the atom spy Pontecorvo thus helped to alert public opinion about the paucity of security regulations and the need to tighten them.

Bruno and Borodino

The Soviet atomic program greatly benefited from the existence of a security shield that made it virtually impossible for many within and outside the USSR to know anything about it. For some time (as shown in the previous chapter), this lack of information helped to promote the image of the Soviet Union as a country interested exclusively in the peaceful uses of the atom. In great secrecy, however, the technological gap dividing the country from the United States was overcome, and the Russians invested heavily in both peaceful and military uses of atomic energy. Such research was given top priority by Stalin and unlimited state funding that drew heavily on the gross national income. Fourteen billion rubles were set aside in the two years preceding Pontecorvo's defection.[6] This program developed along lines that favored large-scale installations and disrespect for environmental and safety issues. The disaster of Chernobyl's nuclear reactor demonstrated, thirty years later, its limitations and flaws.[7] More significantly, technological advancement in the nuclear field was not administered through democratic mechanisms of decision-making, thus leaving scientists and citizens unable to have any say on the directions taken. The image of the Soviet atomic program that Western peace campaigners helped to propagandize was thus inconsistent with its real features.

Stalin gave Lavrentiy Beria, director of the Narodnyy Kommissariat Vnutrennikh Del (People's Commissariat for Internal Affairs, or NKVD) the power to make decisions about the direction of research, assigning him the task of completing the atom bomb program as soon as possible. Planning was discussed exclusively between party leaders and no one was allowed to influence it. The Soviet populace suffered massively for this totalitarian approach to the exploitation of nuclear energy, as significant human and material resources were drawn upon at all stages of development, from the extraction of strategic minerals to the design and building of nuclear reactors and atomic weapons.

The key role atomic scientists played in the nuclear enterprise made Stalin and the party leaders wary of imposing constraints on their work. Yet the Soviet leader was prepared to seize any opportunity to undermine or sabotage the program from within. In the 1930s he had considered the emergence of nuclear physics, relativity, and quantum mechanics as dangerous ideological departures from the principles of dialectical materialism. He was eager to fund, encourage, and support the atomic scientists only as long as they gave priority to applied studies in the nuclear field. Their steadfastness in reaching the goal of making nuclear reactors and weapons available to the Soviet Union was all that could save them from falling into disgrace. Notably, in 1949 Stalin replied to a disappointed letter from Beria, who had criticized the attempt of some Soviet physicists to arrange a conference on nuclear physics, by saying reassuringly, "Leave them in peace, we can always shoot them later."[8]

If Operation Borodino, the codenamed nuclear program set up by the Soviets, could hardly be considered a model of democratic participation, it is worth wondering how Bruno Pontecorvo could have any influence on its development. In the previous chapter we saw that even his wish to let the Western world know about the motives of his defection was subordinated to Moscow's political agenda. Similarly, if he ever attempted to contribute *exclusively* to the peaceful aspects of atomic energy, his actions would clash with the reality of a scientific program in which a clear divide between peaceful and military applications did not exist. Even in the 1980s, Pontecorvo continued to contend that by going to Russia he had decided to use his expertise for peaceful purposes only.[9] But newly released archival materials and recollections cast serious doubts on this allegation. Surely Pontecorvo could not contribute to studies on nuclear weapons—and in fact he had never done so. Yet his expertise in all activities that anticipated nuclear weapons production, from prospecting to nuclear reactor design, was unique. He may have been excluded from classified work because of security concerns. But it is very difficult to believe that when asked to make information available about one of these activities, he could actually say no and demand evidence that the information would be used only in peaceful applications.

This is not to say his contribution was indispensable. In fact, by the time Pontecorvo landed in Russia, Soviet physicists had already come a long way in understanding the key aspects of harnessing atomic energy. On the other hand, this understanding had combined with serious disadvantages that, unsurprisingly, scientists could obviate only by developing some expertise on technical aspects that Pontecorvo had already studied. It is conceivable that an analysis of what the Soviet nuclear program could gain from Pontecorvo's presence was made before he defected. Yet there are no declassified documents proving this.

In the ten years before Pontecorvo's flight, Russian atomic scientists had already obtained outstanding results following a research pattern that did not differ much from those of other nuclear projects. In 1940, the Soviet Academy of Sciences established a committee to investigate uranium fission. In the same period, the physicists Yakov Zel'dovich and Yulii Khariton, of the Leningrad Institute of Physics, revealed how heavy water could be used as a moderator in nuclear process. Yet Germany's invasion diverted their attention towards more urgent matters, and their research on atomic energy was temporarily halted. It occupied them once again towards the end of World War II, when Stalin declared the completion of Russia's nuclear project a matter of national importance.[10]

In 1946 a team led by the physicist Igor Vasilyevich Kurchatov developed the experimental nuclear reactor F-1, which was similar in design to the one built by Fermi's team at Chicago. Information leaked by Fuchs and others did not disclose anything new to Kurchatov. Yet it helped the Soviet physicist avoid the mistakes of Fermi's team, thereby enabling him to complete the pile more quickly. In December of that year the installation at Laboratory Number 2 of the Soviet Academy of Sciences went critical. Two years later, the Russians managed to produce the key fissile material for atomic weapons, plutonium, and to build a production reactor at the atomic facility Cheliabinsk-40 in the Ural Mountains. Much as in the Manhattan Project, another research facility, Arzamas-16 (nicknamed by historians "Los Arzamas"), was entirely devoted to assembling a nuclear device under Khariton's direction.[11] The explosion of Joe-1 in 1949 confirmed that Soviet scientists had acquired sufficient scientific knowledge on atomic energy and hardly needed more when Pontecorvo landed in Russia. Yet his arrival anticipated important developments in areas of research where the Russians lagged behind, such as the development of heavy water reactors.

Tritium and Heavy Water Reactors

Truman's 1950 announcement that the United States intended to proceed towards the production of a thermonuclear device impelled the Russians to proceed along similar lines to catch up in the escalating Cold War conflict. Hydrogen bombs possess an explosive power much higher than that of atom bombs, because they exploit the far more energetic process of nuclear fusion. But they also need enormous quantities of deuterium and tritium, the two isotopes of hydrogen that are used as ingredients for these weapons of mass destruction.[12]

By 1950, Russian atomic scientists such as Sakharov, Zel'dovich, Igor Tamm, and Vitaly Ginzburg had already begun tackling theoretical and design problems associated with thermonuclear devices.[13] However, one important problem they had to face was the absence in Russia of heavy water reactors that could be used as tritium production facilities. Neutrons emitted from the pile during the reaction transformed heavy water into "tritiated" (tritium-containing) water, from which the hydrogen isotope was extracted.

The lack of heavy water reactors in Russia resulted from the absence of power plants producing its main constituent. In 1940 Kurchatov had considered the option of building an experimental reactor with heavy water, but he had abandoned the plan and reverted to using the more conservative uranium-graphite assembly due to a shortage of the precious hydrogen isotope. Only from 1944 was Russia able to produce considerable amounts of heavy water, thanks to construction of a new plant in Chirchik, near Moscow.[14] And it wasn't until 1949 that a zero-energy heavy water prototype was available to Russian experts. Designed by the Russian physicist Abraham Alikhanov, the model reactor was built at Laboratory Number 3 of the Academy of Sciences. Its successful design would lead Soviet science administrators to build a much larger heavy water reactor for the production of tritium, but that project would be completed only in 1952, at Cheliabink-40. When Pontecorvo landed in Russia in 1950, the completion of a tritium-producing heavy water reactor was still deemed urgent and necessary. Recent revelations suggest that he may have offered advice before its completion.

When Pontecorvo was interviewed by Miriam Mafai in preparation for his biography, he disclosed that he had discussed details concerning shielding problems with Russian experts.[15] As we have seen, his study of shielding problems had occurred exclusively in connection with the heavy water reactor NRX. His findings could obviously be applied to any other type of reactor, but heavy water reactors need a specific shielding design because of their considerable radiation and heat output. Indeed, the timing of Pontecorvo's arrival in the Soviet Union, along with his revelation to Mafai, suggests that the building of a heavy water reactor motivated those responsible for its design to seek advice from the only scientist in Russia who had been designing shielding facilities for heavy water reactors, as well as monitoring their instrumentation and investigating lattice problems. By then, Soviet intelligence had already gathered information about Pontecorvo's presence in Canada in connection with the construction of a nuclear reactor there. In particular, a note of 29 July 1943 indicated that "Halban with most of his team moved [from Cambridge] to Montreal, Canada, where he had been joined by Prof. Auger, Placzek, and Pontecorv [sic]."[16]

The historian David Holloway has also argued that some developments in heavy water reactors technology could take place in the Soviet Union during this time because of leaked information made available by Alan Nunn May.[17] Yet, Nunn May could only have given the Soviet experts information produced by others, including Pontecorvo. Nunn May was not a pile physics expert; his research dealt mostly with the chemical analysis of fission products. Thus, the "intelligence trail" shows that by the time Pontecorvo defected, the Soviet scientists knew that a major expert who had contributed to the design of the first heavy water reactor was now in the Soviet Union.

There is no doubt that Pontecorvo's presence in Moscow made atomic scientists eager to obtain restricted information on anything he knew about atomic energy. A memoir written by the Russian physicist Boris Ioffe has contended that immediately after Pontecorvo's arrival in 1950, a meeting of Soviet physicists with their Italian colleague was arranged, somewhat unusually, at the Kremlin. Among the participants was Ioffe's colleague A. V. Galanin, another atomic scientist responsible for reactor development. Some years after the meeting, Galanin confessed to Ioffe that the meeting (or, as Ioffe puts it, the "interrogation") was aimed at obtaining for Russian experts all the information Pontecorvo had on the development of nuclear science and technology in Western countries. Ioffe asserted that Pontecorvo was able to offer only an analysis of the basic physical principles, and was not that useful in the description of technical details.[18]

These revelations shed new light on something that had already emerged much earlier in connection with the JCAE atomic espionage investigation. The investigator's final report indicated Pontecorvo's expertise in the study of tritium as the most worrying aspect of his defection. Not only did the 1951 investigation conclude that "every scrap of information known to Pontecorvo is today known to the Soviet Union," but it stressed that his most recent studies had included work on tritium, a key ingredient in hydrogen bombs.[19] There is no doubt that Pontecorvo's unclassified study of tritium's beta spectrum examined through the proportional counting technique in collaboration with Hanna and Kirkwood, was aimed primarily at inferring data on the mechanisms of neutrino production.[20] Thus it was part of Pontecorvo's research on neutrino physics and had no immediate application. However, the JCAE conclusions provided evidence for the worry that Pontecorvo's novel experimental technique could help in developing tritium quality assessment methods. Indeed, during the 1950s the proportional counting technique was widely used to assess the purity of tritium compounds.[21] This demonstrates once again that Pontecorvo's research on heavy water reactors and their products, especially tritium, was bound to have major repercussions for the development of the Soviet nuclear program in both peaceful and military applications.

The only aspect of the history of the Soviet nuclear program that minimizes the significance of tritium production is the decision, taken in the early 1950s, to use lithium deuteride as an alternative to tritium in thermonuclear devices. This followed Ginzburg's reasoning that given the shortage of tritium and the need to produce such a device as quickly as possible, lithium deuteride offered a better prospect.[22]

Even so an analysis of the timing of Pontecorvo's flight and of the scientific, technological, and industrial shortcomings of the Russian atomic program in the same period confirms the significance of Pontecorvo's presence in Russia. Yet none of these drawbacks can compare with the one that historians have considered the "bottleneck" of the Soviet nuclear program: the shortage of raw materials. Was Pontecorvo's expertise in geophysical prospecting going to be of any help?

Prospecting for Uranium

Previous investigations into the Soviet nuclear program have highlighted the fact that the real deficiency of Russia's nuclear program was not a lack of knowledge about how to produce atomic energy, but the lack of access to uranium.[23] This is not surprising, as the establishment of the Combined Development Trust (CDT) by the United States, Britain, and Canada was aimed at setting up a worldwide monopoly on available deposits of radioactive minerals. By the time the Soviet Union decided to enter into the race to harness atomic energy, most of the uranium resources outside the country were already controlled by the CDT. In 1950 the Soviets were still desperately looking for fresh uranium deposits.[24]

By the end of World War II, the scarcity of uranium hindered Stalin and Beria's plans to quickly transform Russia into a nuclear state. When the nuclear program was originally given the go-ahead, the Soviet Academy of Sciences had put forward a request for ten tons of uranium per year. But only one ton per year was supplied, and extraction and purification operations did not begin until 1942. The slow pace made the atomic scientists pessimistic about the outcome of the entire project. In 1945 Kurchatov considered the lack of uranium the main obstacle, and he complained about it to his superiors. The only reason why reactor F-1 could be assembled and go critical was the lucky discovery by the Soviet Army of one hundred tons of uranium in a German repository during World War II. The shortage was also partly overcome by the use of uranium from Eastern Europe, which accounted for 80 percent of the program's supplies. Yet the Czech deposits—the largest available—were already depleted and could satisfy only 15 percent of the Soviet demand for uranium until 1950.[25]

In the second half of the 1940s new deposits were found in the Soviet Union, especially in Siberia, on the border with China, and in the Fergana Valley of Uzbekistan. Yet these territories were inaccessible, and finding rich veins of uranium continued to be problematic. Dimitri Scherbakov, who worked for the government commission responsible for uranium prospecting, has documented the backwardness of their methods: they used the "tactic of the broad front," trying to prospect large areas that presented different geological profiles.[26] The Russians did not have up-to-date technology like that used by Pontecorvo in the Northwest Territories of Canada.

In 1946 and 1948, British intelligence reports stressed that the "limiting factor" of the Soviet program was indeed the scarcity of uranium. It was exactly in light of this deficit that until 1952 they recommended keeping the technical details of uranium prospecting equipment classified. The Americans estimated that before 1953, the Soviets had enough uranium to build only twenty atom bombs.[27] They were so desperate that they tried to buy it on the international market. In 1948 they sought to acquire some of the Congolese uranium held by a Belgian consortium, but CDT administrators secretly arranged a deal with Belgian diplomats to thwart the Soviet attempt. The Soviets denounced the Western nations' control over uranium deposits; in 1955 a Russian journalist wrote a book on the subject, and his allegations were promptly reported in *Pravda*.[28]

The Soviets' output of uranium grew enormously in the period leading up to Pontecorvo's flight. In 1946 they had amassed one hundred tons of it. By 1950 they possessed nearly two thousand tons. Yet this accumulation was achieved by extensive mining operations for which gulag prisoners had been recruited. Scientific and technological prowess in the prospecting field played no part in the process.[29] As Kojevnikov has recently noted: "Economically, the development of the mining industry and the supply of uranium were the most challenging part of the entire Soviet atomic project, consuming the lion's share of total expenditure and manpower and determining the ultimate time schedule for acquiring the bomb."[30] Because of the Soviet urgency to progress in the field of prospecting, we have another element suggesting that Pontecorvo's presence in Russia was associated with the Soviets' attempts to learn more about how to find uranium.

Sightings

The flow of journalistic speculations that followed Pontecorvo's flight failed to impress investigators, mainly because most of the information came from untested sources. Yet between 1951 and 1953, these speculations together with intelligence reports highlighted Pontecorvo's involvement in uranium prospecting. This on its own is remarkable, because at no time after the defection had Pontecorvo's expertise in uranium prospecting been exposed. In fact, this study is the first to document the extent of Pontecorvo's involvement in dealing with prospecting problems. Thus, the speculations on Pontecorvo's prospecting can now be viewed in a different light.

On 16 March 1951, the *Daily Telegraph* published fresh news on Pontecorvo's work in Czechoslovakia. The unsigned article indicated that Pontecorvo, together with Czech and Russian scientists, was now responsible for supervising the development of atomic research in Cominform countries. Five days later, the same newspaper claimed that Pontecorvo was in Kamenice, near Prague, where fresh uranium deposits had recently been found.[31] The news overlapped secret information that British military intelligence personnel based in Trieste forwarded to Scotland Yard the following April. An "American source" had made information available "through a fairly reliable sub-source" indicating that Pontecorvo was in an area one hundred kilometers southeast of Prague.[32]

If it is hard to believe that Pontecorvo was given a supervisory role so soon after his flight, the news about his participation in uranium prospecting cannot be so easily dismissed. Pontecorvo was familiar with uranium-bearing geological structures and knew how to explore them. He was an ideal candidate for suggesting ways to improve geophysical exploration techniques and instruments. By the 1950s, the largest share of the world's uranium supply came in the form of uraninite and pitchblende from three primary veins located in Shinkolobwe, Belgian Congo (now Democratic Republic of Congo); Port Radium, Canada; and Jáchymov, on the Saxon border between Germany and Czechoslovakia. By then geologists had ascertained that these areas presented similar structures.[33]

The Russian thirst for Czech uranium made them eager to prospect new areas. From 23 November 1945, a bilateral interstate agreement allowed the Soviet Union to use uranium through a national enterprise, Jáchymovské doly, which was part of the Czech Ministry of Industry. The company continued to operate under several names until the dissolution of the Soviet Union.[34] By the 1950s, Kamenice was the location of the fastest growing uranium prospecting in Czechoslovakia. As the Jáchymov deposits had been exploited for quite some time and their depletion preoccupied the Russian atomic program administrators, new prospecting began in areas surrounding Prague—some of which, like Horní Slavkov, presented the same geological setting and mineralogy as Jáchymov.[35]

Intelligence and journalistic reports relayed information about Pontecorvo's presence in the largest mining district in the country: Příbram, seventy kilometers southeast of Prague. From 1950, this center and Horní Slavkov provided 39 percent of the total Soviet production of uranium, which from 1958 was processed at the nearby processing plant of Bytíz.[36] The information leaked in 1951 provided details about the plans to build this plant, thereby further confirming some of the earlier reports.

In the following three years, fresh news about Pontecorvo's participation in prospecting continued to appear in the press. His presence was reported in the main uranium extraction area in the Soviet Union before the Czech deposits were made available. The area he visited, Tyuya Muyun in the Fergana Valley of Uzbekistan, was a secondary deposit presenting characteristics very similar to those found in the Colorado Plateau. Thus, prospecting in that area with up-to-date instrumentation, such as that used by Scherbatskoy and Russell, was bound to have a major impact on the Russians' frantic search for uranium.[37]

Reed was never impressed by the news of Pontecorvo's presence in uranium prospecting areas under Soviet control, especially when he was sighted at the Chinese border with Russia. As the leaked information originated from the Chinese Nationalists' intelligence organization, Reed thought that Pangh de Co Co was being used for purposes of anticommunist propaganda. The signaling of his presence highlighted the collaboration between Chinese and Russians in atomic research, and revealed China's attempt to harness nuclear energy. Reed commented: "These unsubstantiated rumors are without doubt K.M.T. inspired, and are probably part and parcel of recent Chinese nationalist propaganda designed to emphasise the over-riding importance, in their view of the Communist menace in the Far East."[38]

Yet the news appeared in connection with intelligence communications and referred to an aspect of Pontecorvo's research, uranium prospecting, that was unknown at the time. The rumors appeared in the press from June 1951 and indicated Pontecorvo's presence in the Takla Markham (now known as Taklamakan) desert. This was an area that the Russians now occupied since Chinese nationalists had fled Sinkiang. The former secretary-general of the Sinkiang government, Issa Yusuf Bey Aliptekin, confirmed the information the following November, claiming that Soviet military authorities had seized the provincial capital, Urumchi, and the city of Kulja to extract uranium and build a production plant.[39] The news was further confirmed the following December, when the MI5 overseas representative in Hong Kong (also known as the security officer, or SO) forwarded confidential information to the British Admiralty's Directorate of Naval Intelligence. The secret dispatch indicated that the Chinese geologist Weng Wenhao had recently conducted a survey trip in the remote province of Kwelin (Guanxi, southwest China). Upon his arrival in Lanchow (Kansu province, central China), he was joined by Pontecorvo, with whom he flew to Tihwa in Sinkiang. The Chinese geologist, also former president of the Executive Yuan of China and founder of that country's National Resources Commission, was one of the most prominent Chinese academics working on geophysical prospecting.[40]

The details on Weng's travels were given by the intelligence operator with no estimate of accuracy. In fact, recent historical work has shown that Weng's travels to Sinkiang were "part and parcel" of Chinese Nationalist propaganda. The Chinese historian Li Xuetong has recently noted that the geologist did not leave Beijing in 1951.[41] Yet, similar information was obtained the following year when the MI5 Security Liaison Officer in Pakistan informed Percy Sillitoe that Pontecorvo had carried out prospecting in Sinkiang province.[42] And in 1953 the chairman of the Canadian Peace Congress, James G. Endicott, claimed that Pontecorvo was employed by the Chinese government in the search for large uranium deposits. In his memoir, Ioffe provides evidence of another trip to China in 1956.[43]

So was Pontecorvo really contributing to prospecting in Czechoslovakia, Russia, and China? Was he really engaged in research that could help Russia overcome its scarcity of raw strategic minerals? Recently declassified documents allow these questions to be only partially answered. Nevertheless, the analysis of developments in neutron well logging provides further confirmation of the significance of Pontecorvo's research on prospecting problems.

The Legacy of Neutron Well Logging

Pontecorvo developed oil prospecting techniques based on nuclear methods that found wide application in the United States and other countries. After World War II, nuclear well logging was further refined, becoming one of the most fertile techniques used by oil prospectors. These advancements did not occur only in the West. During the second half of the 1950s, Soviet experts managed to close the technological gap in the field of oil prospecting mainly thanks to techniques that stemmed from Pontecorvo's original ideas and experimentation. The available archival documentation does not show a direct transfer of relevant geophysical knowledge in connection with Pontecorvo's flight. But the recollections of a prominent Russian geophysicist confirm the existence of an important legacy, suggesting that the interest in Pontecorvo's newly designed technique may even have clinched the Soviets' decision to give "refuge" to the Italian-born scientist.

In the second half of the twentieth century, neutron well logging led towards a diversification of techniques based on similar principles, and greatly improved them. Two of these techniques in particular proved very successful: the Epithermal Neutron Log (EPL) and the Compensated Neutron Log (CNL). The EPL exploited the properties of new detectors that were designed to be sensitive only to epithermal neutrons—that is, neutrons that were slowed down in the presence of hydrogen-containing substances but had yet to reach the thermal state. It allowed the gathering of information on lithological structures and formations, especially in terms of porosity, by measuring the hydrogen index (concentration) of underground substances. Its key merit was its ability to indicate the presence of chlorine, the largest neutron absorber of all chemical elements, thereby enabling geophysicists to detect the presence of salt (sodium chloride) in wells. Salt can be found in water, but not in petroleum. Thus the EPL was used mainly for "remote sensing" oil/water contact points. The CNL was a refinement of earlier techniques of simultaneous well logging in which information from more than one detector was used to produce a combined log. In particular, it exploited the detection either of gamma rays and neutrons or of neutrons alone by two detectors located some distance apart in two different directions from the source of radiation, This technique was also known as dual neutron porosity log.[44]

As the history of these groundbreaking techniques is yet to be fully explored, there are several interpretations of their origins, and Pontecorvo's contribution is hardly ever fully acknowledged. For instance,

Allaud and Martin have only credited Scherbatskoy for the introduction of the EPL in the United States.[45] Part of the problem in tracking down Pontecorvo's contribution to oil prospecting on both sides of the Iron Curtain has been that previous studies of his career have indicated neutron well logging as a brief departure from his main research interests and associated it with his employment at Well Surveys between 1941 and 1943. But we now know that he continued working for Scherbatskoy after that time, actually becoming a key player in the advancement of the engineers' neutron well logging program.[46]

From 1949 Pontecorvo considered this job as one of the most important aspects of his career—one that he felt would propeled it forward, especially in light of the fast-growing demand for new prospecting methods. In two letters written in 1950, he indicated this as his main activity: "act[ing] as a consultant on a geophysical company in the U.S.A." British security agents such as Reed and Arnold understood the importance of this work, and sought to inform their colleagues in America about the status of Pontecorvo's prospecting studies.[47]

From 1948, the manufacture of new detectors was discussed in Pontecorvo's correspondence with Scherbatskoy and Neufeld. By then he understood the potential of scintillation counters as EPL detectors and was investigating several possibilities, such as using anthracene or sodium iodide crystals as scintillation elements.[48] In 1949 the description of one of these detectors was included in one of Scherbatskoy's patents; it established his firm's control over future developments in neutron well logging, and also challenged more prominent players in the prospecting field, such as Schlumberger.[49] Similarly, Pontecorvo offered advice to Scherbatskoy on important consultancy work for large US firms such as the Perforating Guns Atlas Corporation (PGAC, later Pan Geo), the first company to successfully deploy EPL apparatus in geophysical prospecting.[50] By advising on these matters, Pontecorvo showed his willingness to set out agreements for future collaboration. Yet his decision to waive rights on his inventions and sell them to Scherbatskoy might explain why his contributions to nuclear geophysics have never surfaced in historical accounts.[51] If Pontecorvo thus played a pivotal role in the advancement, and not just the *origins*, of neutron well logging, this fact has never come to light before.

Pontecorvo's activities as innovator in the oil prospecting field in the years before his defection elicit legitimate questions about his contribution to the Soviet nuclear well logging program after his flight. Those who have known his work well, such as Herbert Skinner, have noted that his knowledge on oil prospecting techniques was likely to rouse some interest in Russia.⁵² This has been further confirmed by a pioneer of nuclear geophysics in Russia, the scientist Boris Yerozolimsky, who worked in the field between 1950 and 1970.

Before the 1950s, nuclear geophysics was so underdeveloped in the Soviet Union that little had been published by experts in specialized journals. The physicist Georgy Nicolaevitch Flerov set the conditions for expanding the field. One of Kurchatov's closest collaborators in the development of fission experiments at the Soviet Academy of Sciences, Flerov was also one of the key contributors to his country's race for nuclear energy.

According to Yerozolimski, it was information about Pontecorvo's work on neutron well logging that in 1949 stirred Flerov into action. The physicist sought to invest material and human resources in developing nuclear techniques in Russia, thinking that it would lead to better oil prospecting methods, therefore increasing the country's oil capacity. We have seen that in the same period, Pontecorvo's method was attracting interest in other countries, and it is worth speculating on whether it was such interest that made Flerov eager to welcome Pontecorvo to Russia when he decided to flee the West.

In 1951 Flerov was working towards the establishment of the Petroleum Institute in Moscow. The laboratory was set up to further develop Pontecorvo's technique under the guidance of Bernard Lapuk. Among others, it included Yerozolimski, who had previously been working in Kurchatov's laboratory in Moscow.⁵³ The laboratory team was responsible for developing theoretical and experimental research. Several new detectors like those designed earlier by Pontecorvo were tested, including BF_3 counters and innovative scintillation counters that used photomultipliers. Yerozolimski was assigned the task of verifying the effectiveness of these new detectors. In the same period, an operating model well was built in the oil fields of the Volga-Ural oil-bearing region (Bashkiria-Tatarstan).⁵⁴ Like the Western geophysicists, the Soviet researchers mainly tried to tackle the problem of how to use neutron well logging to understand chlorine concentration in wells.

Following Pontecorvo's innovative method, the Soviet geophysicists eventually took the lead in its development.⁵⁵ In the 1960s a major reorganization took place that led Flerov to direct a new laboratory for nuclear techniques at the Soviet Academy of Sciences after the Petroleum Institute was shut down. That laboratory was eventually expanded to become the All-Union Research Institute of Nuclear Geophysics and Geochemistry (abbreviated as VNIIYaGG in Russian). A new technique was designed there which used pulsed neutron sources to detect of salt-bearing substances. The plan to use "pulses," derived from Yerozolimski and Flerov's intuition that neutrons were emitted like electromagnetic waves in radar, was used to obtain more accurate measurements and avoid "poisoning" effects (i.e., faulty measurements due to continuous irradiation of element targets).

A pulsed neutron generator was realized at the laboratory of the future Nobel Prize winner Il'ya Frank, another one of Pontecorvo's collaborators at Dubna.[56] Its design, a sealed vacuum tube filled with deuterium and tritium in which deuterium ions were accelerated towards the target by an electric field, was the result of a collaboration that included researchers from the Kurchatov Institute and the VNIIYaGG. It eventually opened a path towards the use of small "particle accelerators" in oil wells.[57] Yerozolimski's recollections show very clearly that Soviet researchers working on nuclear geophysics inherited the technique Pontecorvo had pioneered, including in the design of nuclear detectors and neutron sources (both pulsed and non-pulsed), in the use of isotopes such as deuterium and tritium, and in the development of particle accelerating devices. Did the Soviets also receive direct advice from Pontecorvo? Yerozolimski's words are puzzling and telling at the same time. He has argued that because Pontecorvo was a "secret person" between 1950 and 1955, nobody involved in the progress of neutron well logging, aside from Flerov, had any contact with him. After 1955, he visited the VNIIYaGG but never offered "any considerations or new ideas" on its program.[58] So despite Flerov's interest in developing the prospecting technique further, and the presence in Russia of the scientist who had effectively engineered it for the first time, we are to believe that Pontecorvo never offered, or was asked to offer, any input—which in fact seems, to use Yerozolimski's words, "really strange."[59]

Presumably before 1955, Flerov sought Pontecorvo's advice at critical moments during the advancement of the neutron well logging program—without, however, letting the Soviet researchers have direct contact with him. Thus, his expertise may have helped researchers at the Petroleum Institute first, and at the VNIIYaGG afterwards, to avoid mistakes in experimenting with the technique he was already familiar with. In the meantime, tight security kept Pontecorvo—especially in the period in which he was "missing"—away from researchers to avoid leaks of information on his or their activities. Ioffe clearly states that Pontecorvo was not allowed to leave Dubna until 1955, or to publish in academic journals. Nobody, there or elsewhere, was allowed to mention his name. Nothing appeared in the Soviet press that could confirm his presence in Russia.[60]

In any case, from the 1960s Soviet geophysicists, also thanks to Pontecorvo's technique, had effectively closed a technological gap with American experts with the invention of the pulsed neutron method. A new institution controlled by the VNIIYaGG was set up in Novosibirsk, Siberia, and prominent geophysicists such as Eugenii M. Filippov expanded the field further by experimenting with several alternative techniques based on pulsating neutron sources. Interestingly, Filippov's work was informed by the reading of innovative research carried out by Pontecorvo's former WSI coworker, Robert Fearon, who in 1963 had managed to obtain more energetic neutron sources in the development of neutron well logging apparatus. Once again, this suggests contiguity between Pontecorvo's achievements and the advancement of nuclear geophysics in Russia after his flight.[61]

During the 1960s the replication of experimental trials in the United States with detectors based on pulsed neutrons further confirmed that in the Soviet Union there had been important achievements in the field—"spectacular successes," actually, according to one British geophysicist.[62] These successes cannot be quantified, but in the ten years that followed Pontecorvo's defection, the number of research teams adopting nuclear techniques in oil detection grew from 2 in 1949 to 185 in 1960. The Institute of Economics of the Soviet Academy of Sciences calculated that the adoption of these techniques allowed savings of sixty million rubles, and that oil production markedly increased in Russia and its satellite countries. Not unlike what had happened in the US Southern states in the early 1940s, nuclear techniques allowed the reuse of wells that more traditional prospecting techniques had deemed depleted.[63] Following this progress, petroleum became a key tool of Soviet diplomacy. In the early 1960s, the USSR began flooding the European market with oil obtained from newly discovered deposits in order to weaken the NATO defense alliance. This attempt to dominate the European oil market created anxiety in Washington, for both commercial and foreign policy reasons.[64]

There is no doubt that from the 1960s, Pontecorvo's contribution to Soviet nuclear physics studies took center stage in his career, thereby limiting any other "consultancy work." In the 1970s, however, he still played a key role in establishing relations between geophysicists in Western and Soviet-bloc countries. In 1976 Scherbatskoy went to the Soviet Union to meet with Pontecorvo. We know that the two were in contact because shortly afterwards, Scherbatskoy settled the still-unfinished question of the reward for the slow neutron patent. Following the exchange, Pontecorvo sent the US Treasury a letter in which he asserted that Scherbatskoy was the owner of 5 percent of the patent royalties. The US Treasury thus paid $750 back to Scherbatskoy. It is unclear whether Pontecorvo ever claimed his own part of the royalties.[65]

To summarize, there are important connections between Pontecorvo's mysterious disappearance in 1950 and the startling changes in trajectory and interest that typified nuclear geophysics in the Soviet Union. If Pontecorvo's expertise informed its development, then its sudden rise could be more easily explained. In the light of these changes, it would be more difficult to believe that Pontecorvo's flight marked a discontinuity with previous research activities, and that by the time he was employed at Dubna he was dedicating all of his time to particle physics. It is thus likely that future historical studies, drawing especially on Russian archive material or recollections, will reveal involvement by Pontecorvo in a far wider range of interests.

The Informer's Plot

After his defection to Russia, Pontecorvo was the subject of more journalistic investigations seeking to reveal his past as atom spy. These inquiries and the resulting espionage literature paralleled the historical transitions of the Cold War and proliferated especially in the moments when tensions between the blocs were more visible. In turn, this literary production projected anxieties about the existence of security risks, suggesting that the Soviet bloc was unable to match the West's technological advancement, and could compete with it only by "filching" scientific information. The espionage "clairvoyants" indirectly, but in most cases deliberately, helped to spread the image of Western technological superiority.

This literature was by and large based on a recurrent plot. Pontecorvo had left Britain because security agents had found out about his past espionage activities, and thus he could no longer help Soviet intelligence gather restricted scientific information. In most cases the reporters sought to reveal that he was one of the many codenamed spies mentioned in encrypted intelligence material. However, it was mainly thanks to the revelations of former intelligence officers that allegations about Pontecorvo's past could be put forward. Thus, the fast growth of this journalistic literature was informed by the revelations of these agents. The informants were also playing a game of their own by deciding to reveal details that, given the agreements between security agencies, should never have been divulged. In some circumstances their agenda was to reveal corruption and inefficiency in the Western security system, and thereby inform its reformation. In other cases it was the end of the Cold War that convinced those in the know to add fresh evidence. For instance, former KGB agents sought to propagandize their ability to infiltrate Western research establishments.

The first account that promised to reveal more than what was already available on Pontecorvo was written just two years after the publication of Moorehead's *The Traitors*. Its author, the US historian David Dallin, belonged to the exiled Menshevik community in the United States and regularly contributed to the anticommunist magazine *The New Leader*. His monograph *Soviet Espionage* alleged that "in 1949, a Communist friend of Pontecorvo who had broken with the party reported Pontecorvo to the US authorities, giving a complete picture of his activities and connections. No action was taken. The information, however, was turned over to British authorities, but again nothing happened."[66] Dallin thus concluded that "inexcusable blunders" typified the action of the security services. His account shows that someone had told him about the secret exchange between the FBI and MI5 that was based on information Segrè (who, however, was not a communist) had originally offered to Thornton in 1949.

Atomic espionage literature was a popular genre in the mid-1950s. It rapidly declined as the superpowers sought to establish a new regime of coexistence, but it developed once again during the 1980s when new espionage scandals and tensions between East and West brought it back to life. With the rise of Ronald Reagan and Margaret Thatcher, the trans-Atlantic cooperation on military and nuclear matters was reinvigorated. Deterrence strategies replaced détente policies, generating preoccupations which nourished a literary production that mixed facts and fiction.

In 1980 the former intelligence agent and University of Oxford lecturer Harford Montgomery Hyde published the widely acclaimed *The Atom Bomb Spies*, which sought to reveal more about the inner ring of spies that undermined the Manhattan Project. Montgomery Hyde claimed that Pontecorvo had indeed been linked with the Gouzenko affair. When the Soviet cipher clerk Igor Gouzenko defected in 1945, he made some documents available to Western intelligence, including a notebook with the names of some "activists" involved in the spy network.[67] An auxiliary group existed in Montreal which included a "Gini" and a "Golia." Montogomery Hyde claimed that French sources of uncertain reliability identified "Golia" as Fuchs. He went on to allege: "Gini more than likely was Bruno Pontecorvo, which may appear fantastic, but then Bruno Pontecorvo was in many ways a fantastic character."[68] At no time did Montgomery Hyde attempt to clarify the reasoning behind this extraordinary deduction. Fuchs had briefly joined Tube Alloys, but he was not part of the Montreal mission. By 1 August 1946, he was already at Harwell as part of the new British atomic establishment. Golia *worked* with Gini, but Pontecorvo had never really worked with Fuchs. Fuchs's connection with Gouzenko appears to have been as bogus as Pontecorvo's. If it ever existed, it was related to the relationship between Fuchs and the Canadian mathematician Israel Halperin (codenamed "Bacon") that was discovered after Fuchs's arrest.[69] The only elements connecting Gini and Pontecorvo were thus their Italian-sounding surnames.

It was especially the study of the so-called Cambridge spy ring—uniting the two 1951 defectors Guy Burgess and Donald MacLean with their colleague, the diplomat Kim Philby, who defected in 1963—that made it possible to formulate a new investigative hypothesis.[70] The Philby case inspired many authors, partly because of the key role Philby played as first secretary at the British embassy in Washington between 1949 and 1951. In the 1980s, journalist Chapman Pincher alleged that Philby had been responsible for deceiving Scotland Yard by removing critical evidence showing that Pontecorvo was indeed a spy. Pincher had by then moved on from contributing to the *Daily Express* to publishing bulky monographs on espionage. His 1981 *Their Trade Is Treachery* revealed for the first time the existence of a 1943 FBI report that detailed the search of Pontecorvo's house in Tulsa. He presented the revelation again in the 1984 book *Too Secret Too Long*. Pincher's studies also informed American scholars working on the subject.[71] In both volumes Pincher claimed that the Stott papers had not reached British authorities because Philby had "sat on the report." According to Pincher, Philby had alerted Soviet agents that Pontecorvo had been discovered, and the agents in turn advised him to defect.[72]

Even before verifying this allegation, it is important to stress that the information being relayed was the subject of an agreement between Western intelligence agencies about its confidentiality. It is thus worth wondering *who* made the decision to inform Pincher about the FBI report going against the pact of nondisclosure between the FBI and MI5 (as discussed in chapter 5). US espionage scholar John Costello has alleged that Pincher received the information in 1980 while interviewing former MI5 principal scientific officer Peter Wright in his house in Tasmania.[73] Wright went on to write about Soviet infiltration in Scotland Yard in his 1986 memoir *Spycatcher*. The monograph famously led to suspicions against one of its key men, Roger Hollis, and only a parliamentary enquiry set up during the Thatcher government demonstrated that Wright's allegation was unfounded. The former MI5 officer also alleged that Philby's deceiving activities had affected the inquiries on Klaus Fuchs, which obviously makes it even more likely that Wright had been responsible for disclosing information regarding the management of the Stott papers.[74]

In any case, the recently released MI5 documents show that Philby could not have dealt with the handling of the FBI reports on Pontecorvo. If the communication between Western security agencies was seriously flawed, this most certainly was not because of Philby. If Philby had come to know about the early FBI notes of 1943, he could hardly have been in the position to remove them, as at the time he was employed within the Special Operations Executive (SOE) in Baulieu, Hampshire. As the case was handled by the BSC from its headquarters in New York, Philby was not in a position to either get hold of or withdraw the FBI report. If Pincher's claim is instead that Philby "sat on" the 1949 FBI document, we have seen that the FBI indirectly admitted its own responsibility for the mishandling of restricted information; this was exactly the reason why Hoover had been unhappy with his agents. Philby could not have deceived the British services even if he wanted to.

Gorbachev's glasnost anticipated the end of the Soviet Union. Former KGB officials who had played prominent roles in the communist regime fled to Western countries, and some of these people sought to enrich the existing atomic espionage accounts with their own new disclosures. It is not difficult to spot an agenda for their flow of revelations since, while the Soviet empire was crumbling, they helped to project a positive image of its intelligence services. Soviet politicians might have been incompetent, but KGB agents were shrewd and proficient. In 1990 the double agent Oleg Gordievsky, secretly working for MI6, claimed that Pontecorvo had been a spy as important as Fuchs, and had voluntarily provided classified documents and calculations to the Soviet embassy in Ottawa since 1943.[75] Yet in substantiating this claim, Gordievsky made reference to Montgomery Hyde's earlier claim that Pontecorvo was the codenamed Gini—a claim which was, by Montgomery Hyde's admission, speculative.

Equally controversial was the 1994 memoir of "Soviet spymaster" Pavel Sudoplatov, entitled *Special Tasks*. A former NKVD officer, Sudoplatov was arrested in Russia in 1953 and only released fifteen years later. He was rehabilitated following the collapse of the Soviet Union, which gave him an opportunity to discuss past NKVD activities. According to Sudoplatov, in the 1930s the organization targeted

Pontecorvo and Fermi as "dedicated antifascists." Then, when the two moved to America, the NKVD contacted them again. In January 1943, NKVD agent Semyon Semyonov received a "telephone message" from Pontecorvo containing a full report on Fermi's recent experiment, and claiming that "the Italian sailor [had] reached the new world." A full report on Fermi's experiment was also provided a short time later. Pontecorvo, according to Sudoplatov, was known to the NKVD as "Mlad" (meaning "youngster" in Russian) and Fermi as "Star."[76]

The reliability of Sudoplatov's claims has been taken for granted for some time—in turn also influencing important research on the history of the Cold War, such as Vladislav Zubok's study of Stalin's Russia—even if the nuclear physicist Edward Teller and the journalist Richard Rhodes both relayed doubts about it.[77] Sudoplatov's allegations are hardly tenable. Understanding Fermi as a "dedicated antifascist" is wide off the mark, as there is no historical literature suggesting it.[78] These allegations contain laughable mistakes as well. Pontecorvo's message to Semyonov is far too similar to Compton's secret cable to Conant in which he announced that the CP-1 had reached criticality. In his telegram, Compton had alluded to Fermi's conquest of atomic energy by comparing it to Christopher Columbus reaching America. But Pontecorvo (or Fermi or Semyonov) could not have known those details at the time, as they were disclosed only at the end of the war. Furthermore, as it stands the message no longer retains its core meaning. Columbus was a "navigator," as Compton indicated, and not merely a "sailor," as we might have to deduce from Semyonov's call, if it was ever made. Sudoplatov's allegations about the code names have also been proven false by the study of the cipher system Venona by US historians John Haynes and Harvey Khler. Haynes and Khler identified "Mlad" as Harvard graduate and Los Alamos scientist Theodore Hall, and "Star" as his friend Saville Sax.[79]

The most recent theory about Pontecorvo being an atom spy was put forward in 1998 by the historian Arnold Kramish, who claimed that he may have been codenamed "Huron" or "Guron." The Venona cable 259, of 21 March 1945, shows that KGB New York decided to send secret agent Guron to Chicago "to reestablish contact with Veksel" and Goldsmith (later identified as Hyman Goldsmith, an Austrian-born physicist at the City College of New York). Kramish has claimed that Pontecorvo may have been named Guron because it is the Russian word for Huron, the lake that lies "only" about one hundred miles from the facility at Chalk River. Timing is suggestive, because the first Soviet cable reached Moscow just four weeks after Pontecorvo visited the Met

Lab in May 1944. Kramish has, however, been clear about the speculative nature of his claim.[80]

The hypothesis has several counterarguments. Pontecorvo had already visited the University of Chicago during the time when he was working in Tulsa. So May 1944 was certainly not the first time he went there. In 1944 he visited the university four times in the context of Tube Alloys work.[81] These visits were part of an official exchange of information fully authorized and acknowledged by security and Manhattan Project managers. On 21 March 1945 Pontecorvo, in contrast with Guron, did not travel to Chicago.

It now seems clear that in the last fifty years, atomic espionage literature has promised to reveal more than it actually could deliver, filling new studies with code names that could or should have corresponded to several scientists, including Pontecorvo. What makes this literature less convincing is its attempt to use specific espionage cases to make more general claims about other issues that are presumably part of the authors' agendas. Dallin and Pincher sought to highlight flaws in the management of national security. Gordievsky and Sudoplatov wanted to portray the Soviet intelligence system as being very efficient. The disclosure of archival material has, by and large, revealed their accounts as implausible.

We are thus left with a few code names that could match Pontecorvo's profile but don't, thus renewing the mystery surrounding the scientist. Clearly, there is always a possibility that new information will finally reveal that Pontecorvo was indeed a spy. On the other hand, the fact that such information has not emerged in the last fifty years is telling. Reed concluded in 1951 that the hypothesis that Pontecorvo was a spy could be made, but that there was not sufficient evidence to support it. Fifty years later, the situation has changed little.

8

Conclusions: The Noisy Echo of Secrecy

We need to learn more about the Pontecorvo affair if we want to fully understand its significance in the history of the Cold War. This monograph reveals that important manipulations and omissions of information occurred both during and after its unfolding. Our knowledge of what actually happened in the summer of 1950 has been deeply affected by the use of secrecy. The gaps that still exist in the historical reconstruction of key events associated with the case depend by and large on this management of restricted information. Yet we now know that at no time was Pontecorvo's flight the subject of adequate investigation, that the search for the motives and implications of his defection was never prioritized, and that no effort was really made to consider its impact for nuclear security.

As Pontecorvo's flight happened at a crucial moment during diplomatic and intelligence consultations in Britain and America, the inquiry into it reflected the hidden agendas of the parties involved. Thus, scanty evidence was used to demonstrate that it was a despicable episode indeed, but of marginal consequences for nuclear security. Or, conversely, it was used to prove that Pontecorvo was an atom spy. These depictions appeased diplomats, pleased government officers, saved the reputation of security officials, and helped some journalists build their careers. Later, the Pontecorvo affair became a minor problem, a microscopic issue of no relevance to anyone. Has this volume restored its original historic proportions?

It has certainly helped to make sense of the motivations behind these distortions, thereby shedding new light on Pontecorvo's activities as a scientist, alleged spy, and defector. At the same time, the new picture of the affair has been conducive of a novel understanding of the history of nuclear physics, security, and diplomacy during the early days of the Cold War. In particular, we have seen how the withdrawal of information from the public domain featured as a key tool of nuclear diplomacy —and how, in this way, the use of secrecy has affected our current understanding of important transitions in nuclear science and security.

Secrecy as a Tool of Nuclear Diplomacy

The study of the Pontecorvo affair has offered important indications of the reasons why some information produced by government agencies and scientific organizations was made secret. The legislation concerning the protection of information through secrecy ("official secrets acts" in Britain and other Commonwealth countries; "espionage acts" in the United States) appropriately emphasizes the advantage that could come to enemies or potential enemies from its disclosure. Yet our treatment reveals that in the Pontecorvo affair, secrecy was used as a valuable tool of policy making that made it possible to shape international alliances in the nuclear field and address specific national interests, regardless of whether or not the information in question would have been useful to foes.

British government officials had much to gain from publicizing Pontecorvo's research interests in cosmic rays while omitting compromising details about his expertise in pile physics and geophysical prospecting. At stake was the future of nuclear pacts between their country and the United States. To protect these interests, British diplomats and policy makers mobilized to carefully select the information to be divulged and decide what to keep under wraps. Some of these officials—Makins, Perrin, and Franks—agreed on a strategy of information disclosure that provided a selective reading of Pontecorvo's career and the circumstances of his defection. In the meantime, their colleagues in security departments—Hollis, Sillitoe, and White—quarantined spicy details on the handling of his vetting to comply with the diplomats' wish to avoid any release of information into the public domain that could jeopardize this strategy. Fifty top secret telegrams were dispatched to British embassies to convey the same version of facts, asserting that Pontecorvo's research was going to be of little use to the Russians and, in so doing, informing the conduct of diplomats abroad. Any future attempt to revive interest in the case, or promote further enquiring, was promptly halted by the Foreign Office.

Ironically, the attempt to protect UK interests in negotiations with the United States failed exactly because of innovative prospecting technologies, such as those designed by the defector Pontecorvo. In 1951 the discovery of new high-grade uranium ores in the Athabaska region of Canada, not very far from Great Bear Lake, helped the AEC to overcome its temporary shortage of uranium. Thus US diplomats decided to further delay the negotiations with the UK, and the explosion of the first British atomic bomb in October 1952 left them in their original "state of non-existence."[1]

US government officials had an equally ambivalent approach to disclosing information and raising concerns about Pontecorvo's flight. The beginning of the Korean War led them to prioritize other issues. The JCAE enquiry sought to reveal that the impact of the defection on the Russian nuclear program may have been very disruptive indeed. Yet that inquiry was the only attempt by US congressional bodies and governmental agencies to investigate the episode. In the meantime, AEC managers were happy to let their colleagues in Britain handle the problem. Moreover, the prospect of an amicable agreement on the thorny proceedings on the slow neutron patent further convinced them to support the British quest for continued silence. Hoover and his assistants at the FBI might have had an interest in divulging information on Pontecorvo's security management that their colleagues in the UK wanted to keep concealed. Releasing the details could have revealed how strained, dysfunctional, and corrupt the British security system was, thereby informing its reformation. But then the exposed deficiencies in the FBI's own communication management convinced Hoover of the importance of avoiding a public scandal, and he agreed with Sillitoe on a pact of silence.

Finally, we have seen how the Russians matched the British officers' quest for protracted silence, and the Americans' wish not to challenge those intentions, with a similar wish to keep the whole affair secret. Actually, if we now know what strategies were put in place by British and American government officers, we still have to learn a great deal about what made the Russians eager to welcome Pontecorvo to the Soviet Union, exactly because the whole affair was enshrined in secrecy.

Pontecorvo's move to Russia was engineered by Italian communists as a political act aimed at convincing nuclear physicists to join in the struggle against nuclear proliferation and follow the example of the Soviet Union, where the atom was being harnessed for peaceful purposes only. Yet this political stand was made conditional upon decisions made in Moscow by the Soviet leadership, which was becoming less eager to support the pacifist cause, and more eager to restore nuclear parity. Thus, the only official act confirming that Pontecorvo was in the Soviet Union was the publication of his 1955 article in *Pravda*. In the five years preceding that publication, everything about him was kept secret by Soviet officials, and he was accommodated in a house in Moscow under strict police supervision.

On the whole, as the governments of three leading Cold War nations decided to shroud key aspects of the Pontecorvo affair in secrecy, their actions set the conditions for the case to be removed from the public eye and quickly forgotten. When Britain's official version of facts was challenged, especially in journalistic accounts, some intelligence officers tried to establish some control over media releases by feeding selected information to some journalists. Thus the use of secrecy meant removing information from the public domain as much as stealthily establishing some control over details disclosed in the public arena. The secret services' urgency to inform the wider public came from their understanding that simply withdrawing information could leave journalists free to speculate wildly on the case. Some, like Chapman Pincher, challenged the official version of facts produced by MI5. Others, like Roger Waters, pretended to be spooks to gather more information. Still others, like Rebecca West, exposed the flaws of security actions. Journalistic reports were detrimental to the quest of the FBI and MI5 to hush the case. Their officers feared that these speculations could inform the political debate, thereby forcing them to allow disclosure of details they wanted to keep hidden.

Thus MI5 secretly worked towards producing its own account, so as to convince the public of the efficacy of its investigations. To have the account ready as soon as possible, Reed's inquiry was hastened so as to reach a conclusion two months after the defection was announced. The agent could look into Pontecorvo's relatives and acquaintances without examining the important ones—such as, for instance, Emilio Segrè. Therefore, Reed's report highlighted some implications of Pontecorvo's flight but, not surprisingly, his conclusions were skewed towards minimizing the responsibility of MI5, which in fact was consistent with the wishes of the British Foreign Office, the UK Department of Atomic Energy, and Scotland Yard. Then, when Reed's account was completed, MI5 sought to promote a public release of his findings through Moorehead's book *The Traitors*. In this way the original source of the information in journalistic accounts was kept hidden; and the secret dialogue between the intelligence services and the reporter was concealed, but the public would still understand the Pontecorvo affair consistently with the MI5 position.

To summarize, secrecy was used in the Pontecorvo case exclusively to address issues that had major implications in the handling of national and international affairs. It was used extensively to curtail and manipulate the release of information to the public. The consequences of this management of information were far-reaching, as they affected our current understanding of Pontecorvo's role in the advancement of nuclear physics and the implications of his move to Soviet Russia.

The History of Nuclear Physics and the Effects of Secrecy

Past examinations of Pontecorvo's scientific career have mainly centered on his achievements in particle physics. These accomplishments, especially in terms of theory and experimental detection of neutrinos, have lent support to the portrayal of a scientist who was mostly interested in understanding the nature of the "subatomic" and considered applied science as a departure from his main research interests. This picture fits well with the manufactured claims put forward immediately after Pontecorvo's defection, such as the claim that his work would be of little use to Russian experts experimenting with atomic energy because he was mainly interested in cosmic rays.

Yet that picture does not match the portrait given here, which confirms the prominence of Pontecorvo's research interests in applied fields. His scientific education at the research laboratories of Rome and Paris entailed dealing with practical problems such as envisioning new and more effective methods of producing radioisotopes; investigating how these methods could be used in industry and medicine; and protecting the design of new methods and instruments with patents.

In the ten years before his defection, Pontecorvo traveled to many places in North America. He visited the skyscrapers of New York, the oil fields of Oklahoma and Texas, and the Canadian forests and mines located above the Arctic Circle. These travels made him even more aware of the potential of nuclear physics for practical applications. The neutrons continued to be a "handy companion" in these journeys, enabling him to consider their uses in prospecting as well as in the harnessing of atomic energy. From 1940, examining the interaction of neutrons with chemical substances in the strata of wells led him to understand the potential of neutron-based detectors in oil exploration, thereby pioneering the nascent field of nuclear geophysics. From 1943, following Fermi's study of nuclear piles, Pontecorvo was actively involved in the design of the first heavy water reactor ever built, also being responsible for shielding and monitoring systems and the design of nuclear detectors. These practical applications of nuclear science made him one of the few scientists responsible for highly secretive aspects of wartime and postwar atomic research. He contributed also by producing instrumentation for uranium prospecting, participating in fieldwork at locations were uranium could be found, and discussing these details at restricted meetings on prospecting problems set up in the context of the wartime collaboration between the United States, Britain, and Canada.

Nobody can deny that after 1947, when the Chalk River pile went critical, Pontecorvo's interests in basic research in particle physics became more significant. This is exactly the reason why a strategy aimed at downplaying the relevance of his defection could be put in place in 1950. But these studies combined with applied interests again, especially in terms of manufacturing more sophisticated instrumentation for radioactivity detection.

The skillfully drafted career profile prepared at the British Ministry of Supply shortly after the defection exclusively emphasized Pontecorvo's interest in cosmic rays, while hiding his almost unique knowledge and experience of all the processes that anticipate the production of atomic weaponry. Perrin emphasized, correctly, that Pontecorvo knew little about the physics and engineering associated with building a nuclear device. But he deliberately avoided mentioning that Pontecorvo had a vast knowledge of everything that anticipated it, such as searching for uranium deposits and producing fissile materials in nuclear reactors.

Was this knowledge and experience influential in the development of the Soviet nuclear program? Of course it is important to stress that *no* isolated contribution to that program could ever be decisive. Most historians of nuclear research programs, on the basis of extensive knowledge and comparative analysis of many different cases, agree that there never was a specific "secret" or piece of knowledge that had the power, on its own, to propel the nuclear research of the Soviet Union or any other nation.[2]

Yet our treatment of Soviet advances in nuclear science and technology reveals deficiencies in some key areas Pontecorvo was familiar with, such as radioactivity detection, radioisotope production, and geophysical prospecting. It was especially Pontecorvo's knowledge of prospecting methods that did not go unnoticed in Russia. Boris Yerozolimski's recollections suggest that Pontecorvo's nuclear well logging techniques were actually foundational in the establishment of nuclear geophysics in the Soviet Union. Journalistic speculations on Pontecorvo's presence in uranium prospecting areas in Czechoslovakia and China reinforce the hypothesis that his contribution to prospecting problems was critical in fostering his acceptance in the restricted circle of Russian experts dealing with nuclear science. Certainly his defection made science administrators eager to extract critical information without jeopardizing security, as confirmed by Ioffe's revelations about the scientist's trial (and Yerozolimski's recollections). Even so, the outstanding progress in underdeveloped areas of nuclear research in Russia can no longer be understood without considering Pontecorvo's presence there.

If Pontecorvo's research interests were much broader than has previously been assumed, then the impact of his defection must have been grossly and deliberately underestimated in the West. The emphasis on the "atomic secrets" that typified government releases, journalistic accounts and, later, historical reconstructions has withdrawn attention away from a threat that was much more salient. Pontecorvo's flight did not just bring the Soviets snippets of information that could help but not radically transform their atomic energy research. Instead, it promised to provide them with know-how on many different aspects of nuclear science and technology, including defective or not yet fully understood aspects of their program. Even more worrying to Western experts should have been the understanding that in the ten years before his defection, Pontecorvo had been crossing the boundaries between secret and non-secret research by readapting the knowledge of existing practices to new problems. This flexibility had the potential to propel the Soviet nuclear program by promoting a new understanding of the research tasks ahead rather than just illuminating what had already been tried in the West and was now being attempted in the Soviet Union.

The Effects of Secrecy on the History of Nuclear Security

The manipulation of secret details that occurred during the Pontecorvo affair have also led to a misleading understanding of the ways in which nuclear security was handled in the early days of the Cold War.

In the 1940s the FBI and MI5 had to deal for the first time with the problem of defending their countries from the threat of nuclear espionage. Their effort to "defend the realm" in the context of national atomic energy programs was broad, as in the ten years after the Manhattan Project they were involved with vetting, gathering information, analyzing data, reconstructing spy networks, deciphering coded messages, seizing informants, tapping correspondence and telephone communication, and interrogating an incredibly vast number of atomic scientists.

However, the Pontecorvo case shows that the investigative criteria adopted by intelligence personnel may have been unsuitable. Historians have explained these flaws as the unwanted consequence of the agents' ineptitude,[3] and they have revealed that these shortcomings depended on the enemy's ability to corrupt the agents. Our study confirms that these historical inquiries have been accurate to the extent that they have highlighted the mismanagement of individual cases. The lack of communication between British and American officers demonstrates it, as in 1943 one FBI file was presumably never received by the agent in charge of Pontecorvo's vetting, and in 1949 another one was never sent.

Nevertheless, our treatment of the Pontecorvo affair reveals that something more compelling than incompetence or bribery might have hindered the agents' work: namely, their lack of understanding of what was distinctive about nuclear security. Therefore, their strategies may have been inadequate to counter the threat posed by the spread of atomic information. The emphasis on protecting "atomic secrets" meant that enormous weight was given to circumstantial evidence on potential suspects who could relay restricted information to potential enemies, rather than to unassailable evidence showing that the crime of spying had indeed been committed. In fact, what the security services made available was evidence that exclusively corroborated "reasonable doubt" about the conduct of scientists. But making decisions exclusively on the basis of doubts, whether reasonable or not, was extremely problematic. It introduced into the inquiries considerations of elements, such as the scientist's nationality or relatives, that might have led to patent threats but were not sufficient to actualize them. There is a vast literature showing how these criteria were persecutory in character, often chastising scientists for aspects of their private lives and careers.[4]

Bias aside, however, there is something distinctive about the Pontecorvo affair. The security services' efforts to protect nuclear security almost exclusively on the grounds of reasonable doubt may have actually been counterproductive. First, the evidence was too ambiguous to clearly recommend action. MI5 had considered Pontecorvo's employment questionable as early as 1940 since he was a refugee from an enemy country. But it was difficult for wartime research organizations like the DTA, which had the compelling need to employ experts, to rule Pontecorvo out because of secondary details that might or might not have represented a security threat. The 1943 FBI report highlighting Pontecorvo's associations with communists was insufficient to rule out his employment, and even the FBI was noncommittal about recommending action on the grounds of those findings, which were relayed to the BSC. In 1950 Roger Hollis made it clear that even if the report had reached the security officer in charge of vetting, it likely would not have had major repercussions.

Second, giving too much weight to hazy information meant that security action was deeply affected by major social and political events such as the postwar witch hunts. The 1949 FBI report and the information passed on by Thornton to Cockcroft in 1950 were both inconclusive because they pertained only to Pontecorvo's associations of ten years earlier. Though Segrè alleged that these relations still had an "influence" on Pontecorvo—who might therefore have made certain career decisions for "the wrong reasons"—he offered not a single piece of documentary evidence to corroborate his claim. As this information disclosed a doubt rather than representing any factual evidence, it only showed how McCarthyism had instilled a sense of anxiety and insecurity in prominent members of American society, making them willing to relay fears rather than sound information in an effort to provide reassurances about their own loyalty and avoid persecution.

Third, foggy evidence could lend support to alternative (or even contradicting) security actions that could hardly be justified with scientific organizations, or with those who were victimized by these decisions. In 1950, evidence of Pontecorvo's relations with communist militants was sufficient to rule out his employment, despite the fact that in 1943 similar associations had been considered acceptable. Of course one might sympathize with the security services' conduct, suggesting that their privileging of circumstantial evidence had the purpose of removing potential threats before they became real. But it is equally important to sympathize with those who, like Pontecorvo, saw their employment made conditional upon circumstances beyond their control, such as the political affiliations of their relatives. Pontecorvo felt that the 1950 security enquiry was unjust because he had no influence on his relatives and, at the same time, his actions had never shown that he deserved to be removed from the British atomic program. Working under the threat of a "reasonable doubt" dismissal seriously undermined his dedication to the job, since his removal was forthcoming regardless of his conduct or performance. Pontecorvo's decision to flee to Russia, although controversial in many ways, can be interpreted as signaling these deficiencies in nuclear security.

Even the adoption of positive vetting to "purge" British scientific personnel with suspect relatives was presumably a legislative decision that could hardly address these shortcomings. Afterwards, scientific personnel felt even more ostracized, whereas the security services were burdened with having to gather more information on them. This documentation suffered from the same flaws that typified their documentation on Pontecorvo.

What measures could have been more effective in tackling atomic espionage? Recent historical research has revealed that the threat was much less salient than previously assumed, and involved fewer people than those who were hit by investigations. Some historians have even suggested that Western security action was affected by a hysteria that exaggerated the real dangers.

Could Pontecorvo's removal have been made conditional on the finding of unassailable evidence? It is worth considering that a search for indisputable proof could have been far more difficult and, at the same time, less profitable politically. Committing hundreds of agents to look for a few real spies could have been hard for security administrators to justify. But highlighting the threat posed by thousands of "potential suspects" was more rewarding politically, as it made policy makers eager to fund the expansion of the security system.

In the end, Western intelligence agencies prioritized the gathering of low-key evidence over in-depth investigations that could have returned more serious allegations. This strategy, based on privileging quantity over quality in evidence gathering, suggests that intelligence services amassed information mainly because, regardless of whether it was truly useful, it could extend their own powers and render them watertight against accusations of not doing enough to protect nuclear secrets.

Counterespionage operations conceived in this way made MI5 a key player in the wartime debate in Britain on the question of introducing foreign nationals into secret research programs. During the McCarthy era, similarly conceived operations were decisive in the FBI's effort to hamper the activities of prominent political organizations such as the FAS. On the other hand, they were far less successful in providing conclusive evidence on atomic espionage. When the search for atomic spies demanded selective information on a few individuals, the services' response was to broaden their investigations, thus obtaining too much data to be of use.

Furthermore, unassailable evidence was never really assembled on the "spy" Pontecorvo. All the services' documentation in the years preceding his defection never exposed him as having conveyed secret information to the Soviet Union. After his defection, the secret services no longer placed priority on proving that he had been a spy in the recent past. In 1950 their urgency to make sense of the evidence amassed in the previous ten years instigated new enquiries that aimed exclusively at demonstrating that they had been thorough in their treatment of the case. Astonishingly, they did not even interview the most obvious source of information, Emilio Segrè, to find out whether he knew of specific episodes of unlawful transfer of scientific information.

It may be that this understanding of the services' actions is derived from a noncomprehensive reading of archival documents and the fact that we know about the restricted FBI information only from the accounts of MI5 personnel. In light of the evidence available, however, it would be misleading to view the security agencies' treatment of Pontecorvo's circumstances as either a failure or a success. The episode seems only to show how flimsy evidence that did not result directly from atomic espionage investigations could be used to affect the lives and careers of atomic scientists. In this respect, equating Pontecorvo with Nunn May and Fuchs would be misleading. Pontecorvo should be compared instead to other scientists, such as Oppenheimer, Rabi, Condon, or Joliot-Curie, who suffered for the intensification of anticommunism in the United States and Europe.

A Tale of Two Emilios: The Affair and its Protagonists

Because of its secrecy, the Pontecorvo affair has been one of the most mysterious episodes in the history of the Cold War. Why did Bruno Pontecorvo go to the Soviet Union, thereby creating a unique international case? This volume has explained how his decision can be understood as having derived from a combination of factors. Constraints and opportunities convinced Pontecorvo to make a move that was sudden, radical, irrational, and ill-fated.

The security inquiry of March 1950 created impediments to Pontecorvo's continuing research at Harwell, and made him unhappy about his professional circumstances. The offer of a move to Liverpool, which was "warmly recommended" by British intelligence and nuclear science administrators alike, was also not ideal. Pontecorvo was tempted by the prospect of using the University of Liverpool's new particle accelerator, which was the largest in Europe at the time. But he had to face his wife's opposition.

Meanwhile, the campaign against nuclear proliferation was gaining momentum with the Stockholm appeal and the mobilization of prominent scientists, including Pontecorvo's master Joliot-Curie. Pontecorvo had not been known to be politically active. Even if he had become a Communist party member in 1939 (and whether he actually did is still unclear), in the ten years that followed, he always appeared uninterested in politics.

However, the campaign against nuclear proliferation was new in ways that could not easily be dismissed. It urged atomic scientists in Western countries to take a stand or be considered complicit in the use of nuclear weapons. The portrayal, albeit realistic, of Soviet Russia as "a promised land" where the atom was used only for peaceful purposes induced Pontecorvo to think that he could cultivate the same research interests in that country without carrying the blame for their military implications. The fact that some of Pontecorvo's family members, such as Emilio Sereni were playing key roles in the campaign made him even more alert to this prospect.

Yet Pontecorvo decided to go to Russia because of a visible threat, a cloud hanging over his head, that made him responsive to Sereni's suggestions. Giannini's much-publicized petition against the US government made Pontecorvo shaky and fearful. Especially in Italian newspapers, it was presented in a way that suggested contiguity between the pacifists' denunciation of the military use of atomic energy, and the unlawful exploitation of that same atomic knowledge. This in turn caused Pontecorvo to worry that Giannini's petition might ignite a major scandal. Details of Pontecorvo's security inquiry could be divulged, thereby ruining his already precarious career. If we have no final proof that Giannini's petition instigated Pontecorvo's flight, we do have an important confirmation. Only in the days that followed the announcement of the petition did Pontecorvo's movements in Italy become erratic, leading eventually to his mysterious departure.

One element that this book offers as crucial to the case, but cannot fully unravel, is the problematic relationship between Pontecorvo and the "two Emilios," Segrè and Sereni. The history of these three men's friendship contains elements of loyalty, betrayal, and grievance worthy of epic films such as *Once Upon a Time in America*. During their adolescence, the two Emilios engaged in heated discussions not only about the merits of Poincarè's philosophy, but also over much less theoretical issues, such as the merits of opposite political and economic systems. Over the next twenty years they continued to exert an influence on Bruno Pontecorvo.

Sereni contributed to Pontecorvo's political schooling while in Paris; Segrè played a key role in his training as a nuclear physicist in Rome and arranged his employment in the United States. Yet Segrè also undermined Pontecorvo's position in the Western nuclear establishment by relaying his own doubts about him to their peers. The proceedings on the slow neutron patent guided his action. This made Segrè wary of problems that could result from the release of information on Pontecorvo's relatives, and possibly eager to remove any suspicion about his own loyalty while being associated with Pontecorvo in the compensation case before the AEC.

While Segrè betrayed Pontecorvo, Sereni presented him with the unconventional proposal of employment in the Soviet Union. Pontecorvo's defection left Segrè extremely disappointed, as evidenced by the harsh comments about his zealotry that followed their meeting in Russia. At the same time, the defection strengthened Pontecorvo's friendship with Sereni, who became one of his chief companions in the following years. Some of those who have looked at Pontecorvo's decision to flee the West have highlighted the strength of his conviction in the merits of communism.[5] But his controversial relationship with Segrè and Sereni suggests that his political passion was not as strong as his trust (or gullibility) in friendship. Every move abroad that he made in his life was somehow influenced by suggestions and advice from these two very prominent intellectuals.

Presumably Pontecorvo saw in them the incarnation of two opposite but equally attractive models. Segrè was the scientist who had succeeded in establishing himself first in Rome and then in the United States, exploiting his experimental skills and profiting from them, even financially. Sereni was the intellectual whose radical ideas had informed the political debate in Italy. He had pioneered the analysis of the social role and responsibility of intellectuals in general, and of scientists more specifically.

The history of these relations demonstrates that the Pontecorvo affair exemplifies important tensions that typified the Cold War; especially in the entanglement of political and private matters. Thus it shows how the former tarnished the latter. None of the nuclear physicists who were involved in major nuclear programs followed Pontecorvo into the Soviet Union. Many, however, shared his wish to have a say in directing atomic energy towards peaceful rather than military uses. Some saw the administration of intellectual property rights under the new monopolistic regime established by the AEC as controversial. Many more shared Pontecorvo's disquiet over the ways in which scientists' participation in atomic projects gave way to the secret services' surveillance and overt persecution, and to project administrators' questionable decisions about their careers. What makes the Pontecorvo affair unique is how these elements combine like pieces of a jigsaw puzzle in the scientist's decision to leave the West.

The Pontecorvo Affair in Perspective

Is there a lesson that can be learned from the Pontecorvo affair? This is definitely a difficult question to answer, mainly because the events described in this book occurred in a time when there was no preparation for confronting the spread of nuclear knowledge across countries. Obviously the problem has become more acute since then, especially as the number of states that want to become part of the "nuclear club" has grown considerably. Soviet Russia was only the first to catch up with the United States; it was followed by Britain, France, China, South Africa, Pakistan, and India. It might also be followed soon by other states.

This expansion resulted from episodes that in some ways resemble Bruno Pontecorvo's flight to Russia. For instance, in October 1986 the nuclear engineer Mordechai Vanunu was hihjacked by Mossad agents after having provided the British press with restricted information on the advancement of a military nuclear program in Israel. Whether or not Israel has become part of the nuclear club remains a mystery to this day. In 2004 the Pakistani nuclear engineer Abdul Quadeer Khan confessed to have provided a number of countries including Libya, North Korea, and Iran with restricted information regarding the production of centrifuges to enrich uranium.

In recent years we have also seen how the secret management of issues such as the spread of atomic knowledge has informed international politics. For instance, when in 2001 the British government decided that an attack on Iraq would be a legitimate way to prevent that country from using weapons of mass destruction, including nuclear weapons, the decision could be evaluated on the basis of intelligence information revealing the existence or nonexistence of such a threat. British intelligence went against the final conclusions of the International Atomic Energy Agency (IAEA), which showed that the threat was negligible. The recent Chilcot inquiry on the Iraqi war has now shed new light on this episode, also showing that in fact the IAEA was correct in denouncing the lack of an immediate danger to Britain. The inquiry also highlights the existence of political motives that engendered decision-making in those critical years, and therefore suggests that the invasion of Iraq was illegal.[6]

The Pontecorvo affair tells us that the mysteries associated with the spread of nuclear and other military technologies and the unlawful transfer of scientific information cannot be immediately understood, except for some marginal aspects. Secrecy regulations, omissions of information, and far-fetched official statements impinge too heavily on our understanding of critical episodes in the history of the nuclear age and make it impossible to discern their real implications even many years after they have occurred. Only a few people have access to the restricted information on these episodes, and those few have an incentive to manipulate the information flow so as to grant more malleable explanations that will agree with a given agenda.

Our reading of the Pontecorvo affair can be used in interpreting other cases of unwanted spread of nuclear information. First, it invites us to look at those cases in terms of more than just secrets smuggled by foreign intelligence agencies. In particular, it suggests rejecting simplistic readings that are based exclusively on an analysis of the "spying game." In fact, such a reading on its own is not conducive of a comprehensive understanding of what really was at stake in the Pontecorvo case. Thus, the study of the affair presented in this volume suggests that while it is undeniable that security agencies play a key role in controlling the dissemination of restricted scientific information, other important factors also have an impact on the spread of this knowledge. Second, this study draws our attention to the ways in which foreign researchers abroad are administered and how their expertise is valued and used. Third, it recommends that we rethink the quest for control of material resources and the use of political and diplomatic relations toward that end.[7]

In any case, the Pontecorvo affair invites us to be wary of the investigative methods used in intelligence operations. If they have been praised by governments in the past because information-gathering is an essential part of modern governance, the more recent history of intelligence activities, especially in connection with atomic espionage, presents a less reassuring picture. Much remains to be learned about the real and imaginary aspects of atomic espionage and the ways in which real scientists were used to manufacture accounts that had far-reaching implications for them and for the emergence of political and social tensions in the early days of the Cold War. The circulation of restricted knowledge was decisive in distorting the public perception of facts. Crimes, guilty parties, and critical evidence were emphasized, whereas political circumstances, investigative methods, and flaws were either downplayed or never publicized. Revealingly, the new documents on the Pontecorvo affair suggest that an intelligence program genuinely aimed at tackling atomic espionage, in the way we know it and understand it, may have never really existed.

Abbreviations

Archival Sources

ACS	Archivio Centrale dello Stato (Italian National Archive), Rome
AIP	Niels Bohr Library & Archives, American Institute of Physics, College Park, Maryland
AMF	Archivio Museo di Fisica, Istituto di Fisica, Università "La Sapienza" (Institute of Physics Archive, University of Rome)
CAC	Churchill Archive Centre, Churchill College, Cambridge, UK
FIG	Fondazione Istituto Gramsci (Gramsci Foundation), Rome
LUL	University of Liverpool Library and Archives, University of Liverpool
MAE	Archivio del Ministero degli Affari Esteri (Ministry of Foreign Affairs Archive), Rome
NARA	National Archives and Record Administration, College Park, Maryland
SI	Archives Center, National Museum of American History, Smithsonian Institution, Washington
TNA	National Archives, Kew Gardens, London
UCA	University of Chicago Archives

Organizations

AEC	Atomic Energy Commission (US)
AERE	Atomic Energy Research Establishment (UK)
BCSO	British Central Scientific Office, Washington

ABBREVIATIONS

BJSM	British Joint Services Mission, Washington
BSC	British Security Coordination
CDT	Combined Development Trust (US/UK/Canada)
CIA	Central Intelligence Agency (US)
CNR	Consiglio Nazionale delle Ricerche (National Research Council, Italy)
CNRS	Centre National de la Recherche Scientifique (National Research Council, France)
Cominform	Communist Information Bureau
CPC	Combined Policy Committee (US/UK/Canada)
D.At.En.	Department of Atomic Energy, Ministry of Supply (UK)
DCOS	Deputy Chiefs of Staff (UK)
DTA	Directorate of Tube Alloys (UK)
FBI	Federal Bureau of Investigation (US)
GAC	AEC General Advisory Committee (US)
GEN 183	Cabinet Committee on Subversive Activities (UK)
GMCO	Geophysical Measurements Corporation (US)
GRU	Glavnoye Razvedyvatel'noye Upravleniye (foreign intelligence service, USSR)
HUAC	House of Representatives Committee on Un-American Activities (US)
JCAE	Joint Committee on Atomic Energy (US)
JINR	Joint Institute of Nuclear Research (USSR)
KGB	Komitet Gosudarstvennoy Bezopasnosti (national security agency, USSR)
NKVD	Narodnyy Kommissariat Vnutrennikh Del (Commissariat for Internal Affairs, USSR)
NRDC	National Research Defense Committee (US)
OGC	AEC Office of General Counsel (US)
OSRD	Office of Scientific Research and Development (US)
OSS	Office of Strategic Services (US)
PCB	AEC Patent Compensation Board (US)
PCd'I	Partito Comunista d'Italia (Italian Communist Party, 1921–43)
PCI	Partito Comunista Italiano (Italian Communist Party, 1943–91)
PNF	Partito Nazionale Fascista (Fascist National Party, Italy)
PP	Partigiani della Pace (Peace Partisans, Italy)
PSC	AERE Power Steering Committee (UK)
RCMP	Royal Canadian Mounted Police
SSC	Seismic Services Corporation (US)
TRE	Telecommunications Research Establishment (UK)
WSI	Well Surveys Incorporated (US)

Notes

INTRODUCTION

1. UK Foreign Office (FO) files "Defection to USSR of Dr. Pontecorvo," (FO 371/84837) and "Disappearance of Dr. Bruno Pontecorvo in Finland" (FO 371/86437) were originally retained under Section 3(4) of the Public Records Act (1958). In March 2002 I asked the Foreign and Commonwealth Office records manager to review the files to establish whether the secrecy conditions still applied. In May 2002 the papers were released.

2. On the Manhattan Project, see Jeff Hughes, *The Manhattan Project: Big Science and the Atom Bomb* (London: Icon, 2002). See also Richard Rhodes, *The Making of the Atomic Bomb* (New York: Penguin, 1986); Stephane Groueff, *Manhattan Project: The Untold Story of the Making of the Atomic Bomb* (Boston: Little, Brown, 1967); Lawrence Badash, J. O. Hirshfielder and H. P. Broida, eds., *Reminiscences of Los Alamos, 1943–1945* (Dordrecht: Reidel, 1980); and Lillian Hoddeson et al., *Critical Assembly: A Technical History of Los Alamos during the Oppenheimer Years, 1943–1945* (Cambridge: Cambridge University Press, 1993).

3. See for instance Richard Rhodes, *Dark Sun: The Making of the Hydrogen Bomb* (New York: Touchstone, 1995), and, more recently, Jim Baggott, *Atomic: The First War of Physics and the Secret History of the Atom Bomb: 1939–1949* (London: Icon, 2009). On the Soviet nuclear program, see David Holloway, *Stalin and the Bomb: The Soviet Union and Atomic Energy, 1939–1956* (New Haven: Yale University Press, 1994), and Alexei B. Kojevnikov, *Stalin's Great Science:*

The Times and Adventures of Soviet Physicists (London: Imperial College Press, 2004).

4. Spencer Weart, "Secrecy, Simultaneous Discovery and the Theory of Nuclear Reactors," *American Journal of Physics* 45:11 (1977): 1049–60.

5. Donald MacKenzie and Graham Spinardi, "Tacit Knowledge, Weapons Design and the Uninvention of Nuclear Weapons," *American Journal of Sociology* 101 (1995): 44–99. On replication of experiments, see Harry M. Collins, *Changing Order: Replication and Induction in Scientific Practice* (Beverley Hills and London: Sage, 1985); H. Collins and Trevor Pinch, *The Golem: What Everyone Should Know about Science* (Cambridge and New York: Cambridge University Press, 1998).

6. David Kaiser, "The Atomic Secret in Red Hands? American Suspicions of Theoretical Physicists during the Early Cold War," *Representations* 90:1 (2005): 28–60. On confusion in the United States over the term "atomic secret," see also Charles A. Ziegler and David Jacobson, *Spying Without Spies: Origins of America's Secret Nuclear Surveillance System* (Westport, Connecticut: Praeger, 1995), 25.

7. See, for instance, Brian Balmer, "A Secret Formula, A Rogue Patent and Public Knowledge about Nerve Gas: Secrecy as Spatial-Epistemic Tool," *Social Studies of Science* 36:5 (2006): 691–722; Peter Galison, "Removing Knowledge," *Critical Enquiry* 31 (2004): 229–43; Anne Fitzpatrick, "From Behind the Fence: Threading the Labyrinth of Classified Historical Research," in R. Doel and T. Söderqvist, eds., *The Historiography of Contemporary Science, Technology and Medicine: Writing Recent Science* (London and New York: Routledge, 2006), 67–80; Michael Aaron Dennis, "Secrecy and Science Revisited: From Politics to Historical Practice and Back," ibid., 172–184.

8. See Christopher Andrew, *The Defence of the Realm: The Authorized History of MI5* (London: Allen Lane, 2009). See also Nigel West, *MI5* (London: Bodley Head, 1981); N. West, *A Matter of Trust: MI5, 1945–1972* (London: Weidenfeld and Nicolson, 1982).

9. C. Andrew and Oleg Gordievsky, *KGB: The Inside Story of Its Foreign Operations from Lenin to Gorbachev* (London: Hodder and Stoughton, 1990); C. Andrew and Vasili Mitrokhin, *The Mitrokhin Archive: The KGB in Europe and the West* (London: Penguin, 1999); Pavel and Anatoli Sudoplatov, *Special Tasks. The Memoirs of an Unwanted Spymaster* (Boston, Little Brown & Co., 1994).

10. Alan Moorhead, *The Traitors: The Double Life of Fuchs, Pontecorvo, and Nunn May* (London: Hamish Hamilton, 1952); Harford Montgomery Hyde, *The Atom Bomb Spies* (London: Sphere, 1980).

11. Chapman Pincher, *Their Trade is Treachery* (London: Sidgwick and Jackson, 1982); *Too Secret Too Long* (London: Sidgwick and Jackson, 1984); John Costello, *Masks of Treachery* (New York: William Morrow and Co., 1988).

12. See Michael S. Goodman, "British Intelligence and the Soviet Atomic Bomb, 1945–1950," *Journal of Strategic Studies* 26:2 (2003): 120–51.

13. Mark Hollingsworth and Nick Fielding, *Defending the Realm: MI5 and the Shayler Affair* (London: André Deutsch, 1999) and Jessica Wang,

American Science in the Age of Anxiety (Chapel Hill: University of North Carolina Press, 1999).

14. Richard Aldrich, *The Hidden Hand: Britain, America and Cold War Secret Intelligence* (London: John Murray, 2001).

15. Septimus H. Paul, *Nuclear Rivals, Anglo-American Atomic Relations, 1941–1952* (Columbus: Ohio State University Press, 2000); Bertrand Goldschmidt, *Atomic Rivals*, trans. Georges M. Temmer (New Brunswick: Rutgers University Press, 1990); Donald Avery, "Allied Scientific Co-operation and Soviet Espionage in Canada, 1941–1945," *Intelligence and National Security* 8:3 (1993): 100–129. See also D. H. Avery, *The Science of War: Canadian Scientists and Allied Military Technology during the Second World War* (Toronto: University Press, 1998).

16. Brian Cathcart, *The Fly in the Cathedral: How a Small Group of Cambridge Scientists Won the Race to Split the Atom* (London: Penguin, 2004).

17. See for instance John L. Heilbron and Robert W. Seidel, *Lawrence and His Laboratory: A History of the Lawrence Berkeley Laboratory, Vol. 1* (Berkeley: University of California Press, 1989); Thomas Lassman, "Industrial Research Transformed: Edward Condon and the Westinghouse Electric and Manufacturing Research Company, 1935–1942," *Technology and Culture* 44:2 (2003): 309–36; Jeff Hughes, "Plasticine and Valves: Industry, Instrumentation and the Emergence of Nuclear Physics," in J. P. Gaudillere and I. Lowy, eds., *The Invisible Industrialist: Manufacturers and the Construction of Scientific Knowledge* (London, Palgrave Macmillian, 1998), 58–101.

18. See, for instance, Gerald Holton, "Enrico Fermi and the Miracle of Two Tables," in *Victory and Vexation in Science: Einstein, Bohr, Heisenberg and Others* (Cambridge, MA: Harvard University Press, 2005), 48–64; G. Holton, "Fermi's Group and the Recapture of Italy's Place in Physics," in *The Scientific Imagination: Case Studies* (Cambridge, MA: Cambridge University Press, 1978), 155–98. Details on the group's patenting activities can be found in: Emilio Segrè, *Enrico Fermi, Physicist* (Chicago: University of Chicago Press, 1970), 83–85; Edoardo Amaldi, "From the Discovery of Neutrons to the Discovery of Nuclear Fission," *Physics Reports* 111 (1984): 154–60; Laura Fermi, *Atoms in the Family: My Life with Enrico Fermi* (Chicago: University of Chicago Press, 1954).

19. See for instance Lawrence Wittner, *One World or None: A History of World Nuclear Disarmament Through 1953, Vol. 1* (Stanford, CA: Stanford University Press, 1995).

20. Miriam Mafai, *Il lungo freddo. Storia di Bruno Pontecorvo, lo scienziato che scelse l'URSS* (Milan: Mondadori, 1983). S. M. Bilenky et al., eds., *B. Pontecorvo Selected Scientific Works: Recollections on B. Pontecorvo* (Bologna: Società Italiana di Fisica, 1992).

21. See for instance Sylvan S. Schweber, *In the Shadow of the Bomb: Oppenheimer, Bethe, and the Moral Responsibility of the Scientist* (Princeton, NJ: Princeton University Press, 2000). For an analysis of the use of biographies in the history of science, see also Joan L. Richards, "Introduction: Fragmented Lives," *ISIS* 97:2 (2006): 302–6 (and all the other articles, part of a special focus session on biographies in the history of science).

22. As indicated by Thorpe in Charles Thorpe, *Oppenheimer: The Tragic Intellect* (Chicago: University of Chicago Press, 2006), chapter 1.

23. Maurice A. Finocchiaro, *The Galileo Affair: A Documentary History* (Berkeley: University of California Press, 1989). Note 1 at p. 325 presents more works discussing the appropriateness of using the word "affair" in connection with Galileo's trials. See also G. V. Coyne et al., eds., *The Galileo Affair: A Meeting of Faith and Science. Proceedings of the Cracow Conference, 24–27 May 1984* (Vatican City: Specola Vaticana, 1985).

24. David Joravsky, *The Lysenko Affair* (Chicago: University of Chicago Press, 1970). Incidentally, the affair overlapped that of his colleague Nikolai Vavilov, sentenced to death in 1940. See Mark Popovski and Mark Aleksandrovich, *The Vavilov Affair* (Hamden, CT: Archon Books, 1984).

25. Alan Sokal and Jean Bricmont, *Intellectual Impostures* (London: Profile Books, 1998).

26. S. Turchetti, *Il caso Pontecorvo: Fisica nucleare, politica e servizi di sicurezza nella Guerra Fredda* (Milan: Sironi, 2007).

CHAPTER 1

1. On secrecy and the invisibility of nuclear researchers in the post–World War II years, see Sean F. Johnston, "Implanting a Discipline: The Academic Trajectory of Nuclear Engineering in the USA and UK," *Minerva* 47 (2009): 51–73.

2. Irene Bignardi, *Memorie estorte ad uno smemorato: Vita di Gillo Pontecorvo* (Milan: Feltrinelli, 1999), 11.

3. Clara Sereni, *Il gioco dei regni* (Florence: Giunti, 1993), 114.

4. Director of the 1966 acclaimed *Battle of Algiers*, which portrayed the anticolonialist struggle against French troops in the North African city. Bignardi, *Memorie estorte*, 17.

5. Miriam Mafai, *Il lungo freddo*, 44. Lucia Orlando, "Physics in the 1930s: Jewish Physicists' Contribution to the Realization of the 'New Tasks' of Physics in Italy," *Historical Studies in the Physical Sciences* 29 (1998): 141–81, on 143.

6. The Knighthood is the national Order of Merit for Labor.

7. Bignardi, *Memorie estorte*, 12.

8. Mafai, *Il lungo freddo*, 46.

9. Ibid., 46. Bignardi, *Memorie estorte*, 13.

10. At the beginning of the twentieth century, when only 40 percent of the population met standards of basic literacy, the rate was down to 2 percent within the Jewish community. In 1938 only 12 percent of Italians aged between eleven and eighteen had access to higher education, in contrast with 85 percent of their Jewish fellows. Orlando, "Physics in the 1930s," 144.

11. Bernard L. Cohen, "Guido Pontecorvo 'Ponte,' 1907–1999," *Genetics* 154 (2000): 497–501.

12. Alberto Asor Rosa, "Il fascismo alla conquista del potere (1919–1926)," in A. Asor Rosa, ed., *Storia d'Italia. dall'unità ad oggi*, vol. 4 (Turin: Einaudi, 1975), 1358–1470, on 1465.

13. Ruth Bondy, *The Emissary: A Life of Enzo Sereni* (London: Robson, 1977), 34.

14. Bignardi, *Memorie estorte*, 13.

15. Luigi Tomassini, "Le origini," and Raffaella Simili, "La presidenza Marconi," in R. Simili and G. Paoloni, eds., *Per una storia del Consiglio Nazionale delle Ricerche* (Bari: Laterza, 2001), 5–127.

16. William G. Welk, *Fascist Economic Policy* (Cambridge, MA: Harvard University Press, 1938), 159. See also Gianni Toniolo, *L'economia dell'Italia fascista* (Bari, Laterza: 1980).

17. Anna Guagnini, "Patent Agents, Legal Advisers and Guglielmo Marconi's Breakthrough in Wireless Telegraphy," *History of Technology* 24 (2002), 171–201. Segrè argued: "Marconi was interested in Patent Applications, not scientific papers, and for this reason he was rather secretive about his procedures, a habit that has presented additional difficulties for anyone trying to reconstruct his way of methodology." E. Segrè, "Preface," in Giancarlo Masini, *Marconi* (New York: Marsilio, 1995 [1975]), 10.

18. Renato Giannetti, *Tecnologia e sviluppo economico italiano, 1870–1990* (Bologna: Il Mulino, 1998), 47–51.

19. *Il popolo d'Italia*, November 28, 1932 (cited in Masini, *Marconi*, 327).

20. Renato Giannetti, "Il CNR e le politiche per la ricerca e l'innovazione industriale," in Simili and Paoloni, *Per una storia del Consiglio Nazionale delle Ricerche*, 224–39. A key example was methanol production.

21. Of total imports, 20.6 percent consisted of coal, coke, mineral oil, and derivatives. Rolf Petri, "Technical Change in the Italian Chemical Industry: Markets, Firms and State intervention," in A. S. Trevis et al., eds., *Determinants in the Evolution of the European Chemical Industry, 1900–1939* (Netherlands: Kluwer, 1998), 280.

22. On Fermi's early career, see Francesco Cordella, Alberto De Gregorio, and Fabio Sebastiani, *Enrico Fermi. Gli anni italiani* (Roma: Editori Riuniti, 2001), chapters 3, 4, and 5.

23. Laura Fermi, *Atoms in the Family*, 35. See also interview with Franco Rasetti by John Kennedy, 1966, in "Fermi Documentary Film Collection: Background Research Materials and Interviews," Harvard Project Physics Papers, Box 1, AIP, 15; Edoardo Amaldi, "Personal Notes on Neutron Work in Rome," in Charles Wiener, ed., *History of Twentieth Century Physics: Proceedings of the International School of Physics "Enrico Fermi," Varenna 1972* (New York: Academic Press, 1977), 317.

24. E. Fermi and Enrico Persico, *Fisica per le scuole medie superiori* (Bologna: Zanichelli, 1938). E. Fermi, *Fisica ad uso dei licei* (Bologna: Zanichelli, 1929).

25. Mafai, *Il lungo freddo*, 43. Emilio Segrè, *A Mind Always in Motion* (Berkeley: University of California Press, 1993), 9.

26. Giovanni Battimelli and Michelangelo De Maria, "La fisica," in Simili and Paoloni, eds., *Per una storia del Consiglio Nazionale delle Ricerche*, 285.

27. Orlando, "Physics in the 1930s," 151.

28. Emilio Segrè, *Enrico Fermi*, 30.

29. Interview with Franco Rasetti by John Kennedy, 1966 in "Fermi Documentary Film Collection, Background Research Materials and Interviews," Harvard Project Physics Papers, Box 1, AIP, 14.

30. Ernest Rutherford, James Chadwick and Charles D. Ellis, *Radiation from Radioactive Substances* (New York: MacMillian, 1930).

31. Battimelli and De Maria, "La fisica," 308. For a broader coverage of non-Italian cases, see also Charles Wiener, "A New Site for the Seminar: The Refugees and American Physics in the Thirties," in D. Fleming, *The Intellectual Emigration, Europe and America, 1930–1960* (London, 1969), 190–228.

32. Jeff Hughes, "Radioactivity and Nuclear Physics," in Mary J. Nye, ed., *The Cambridge History of Science*, Vol. 5 (Cambridge: Cambridge University Press, 2005), 350–74.

33. F. Rasetti, "Sopra un forte preparato di Radio D ottenuto nell'Istituto Fisico di Roma," *La ricerca scientifica* 5 (1934), 3–5.

34. A few years later, Majorana mysteriously disappeared while traveling by ferry from Palermo to Naples; he probably committed suicide. Luisa Bonolis, *Majorana: Il genio scomparso* (Milan: Le Scienze, 2002), 40.

35. E. Fermi, "Radioactivity Induced by Neutron Bombardment," *Nature* 133 (1934), 757. E. Fermi, E. Amaldi, Oscar D'Agostino, F. Rasetti and E. Segrè, "Artificial Radioactivity Produced by Neutron Bombardment," *Proceedings of the Royal Society of London* 146 (1934), 483–500. See also Segrè, *Enrico Fermi*, 77.

36. Graduation certificate, 10 November 1933, in "University Documents," Bruno Pontecorvo Papers, CAC.

37. Interview with Franco Rasetti by John Kennedy, 1966, in "Fermi Documentary Film Collection, Background Research Materials and Interviews," Harvard Project Physics Papers, Box 1, AIP, 14–15.

38. Ibid., 14.

39. Gerald Holton, "Enrico Fermi and the Miracle of the Two Tables," in *Victory and Vexation in Science*, 12.

40. Interview with Enrico Amaldi by Charles Wiener, April 9–10, 1969, AIP. Available at: http://www.aip.org/history/ohilist/4485.html.

41. Laura Fermi, *Atoms in the Family*, 101. See also Amaldi, "Personal Notes on Neutron work in Rome," 317.

42. Letterio Laboccetta to E. Fermi, 26 October 1934, in "Brevetto Neutroni, 1934–35," Carte Nuovo Amaldi, Box 1, Folder 2, AMF. The Italian Ministry of Industry's inventory of Italian patents shows that the taxes for the Italian *privativa* n. 324.458, filed by Labboccetta on Fermi's behalf, were paid for the period 1934–38.

43. E. Amaldi, O. D'Agostino, E. Fermi, B. Pontecorvo, F. Rasetti, and E. Segrè, "Method for Increasing the Efficiency of Nuclear Reactions and Products Thereof," GB 465,045. Filed 26 October 1934. Issued 26 April 1937.

44. The term "slow neutron patent" is used here in a similar manner as for physical processes, e.g., slow neutron detection, or neutron well logging. But of course, both in the "slow neutron patent" and in these processes, neutrons are deployed. The use of the singular rather than the plural is by no means obvious. In Italian, for instance, the patent was known as *brevetto dei neutroni lenti* (slow neutrons patent).

45. On 5 June 1929. Amaldi, "From the Discovery of the Neutron to the Discovery of Nuclear Fission," *Physics Report* 111 (1984): 5–331, on 318.

46. He was the son of Torquato Giannini, vice president of the Dante Alighieri Society, responsible for promoting the study of Italian language and culture abroad. G. Prezzolini, "Il brevetto delle scoperte atomiche italiane comprato dagli SU per trecentomila dollari," *Il Corriere della sera*, 1 August 1953, [copy in "Giannini," Carte Nuovo Amaldi, Box 2, Folder 2, AMF].

47. Gabriello Giannini to E. Fermi, 11 November 1935, in "Brevetto Neutroni, 1934–35," Carte Nuovo Amaldi, Box 1, Folder 2, AMF. E. Fermi, E. Amaldi, B. Pontecorvo, F. Rasetti, and E. Segré, "Process for the Production of Radioactive Substances," US patent n. 2,206,634, filed 2 July 1940.

48. G. Giannini to E. Fermi, 11 October 1935, and G. Giannini to E. Fermi, 25 November 1935, in "Brevetto neutroni, 1934–35," Carte Nuovo Amaldi, Box 1, Folder 2, AMF. See also Lassman, "Industrial Research Transformed," 315–17.

49. Segrè, *A Mind Always in Motion*, 66.

50. Philips to E. Segrè, 20 September 1935, in "Brevetto Neutroni, 1934–35," Carte Nuovo Amaldi, Box 1, Folder 2, AMF. On Philips' industrial research activities see also Kees Boersma, "Tensions within an Industrial Research Laboratory: The Philips Laboratory's X-ray Department Between the Wars," *Enterprise and Society* 4 (2003): 65–98, on 77.

51. G. Giannini to Philips, 18 November 1935 in "Brevetto Neutroni, 1934–35," Carte Nuovo Amaldi, Box 1, Folder 2, AMF. On the Philips' Patents Department see A. Heerding, *TheHistory of N. V. Philips' Gloeilampenfabriken*, Vol. 2 (Cambridge: Cambridge University Press, 1986), 333.

52. Philips to E. Fermi, March 1936 in "Brevetto neutroni, 1934–35," Carte Nuovo Amaldi, Box 1, Folder 2, AMF.

53. E. Fermi, "Recenti risultati della radioattività artificiale," *La ricerca scientifica* 6 (1935), 399–402. Two years later, Segrè, in collaboration with physiologists at the University of Palermo, was able to test the efficiency of radioactive phosphorus as tracer for studying metabolic disturbances. Emilio Segrè to CNR, 18 July 1937, in Fond I, Folder 155, CNR Papers, ACS.

54. Leo Szilard, "Memoradum of Possible Industrial Applications Arising on a New Branch of Physics," 28 June 1934, in Spencer R. Weart and Gertrud W. Szilard, eds., *Leo Szilard: His Version of the Facts: Selected Recollections and Correspondence* (Cambridge, MA: MIT Press, 1978), 39.

55. Orso M. Corbino, "Prospetive e risultati della fisica moderna," *La ricerca scientifica* 5 (1934): 615–20.

56. Items relating to the competition for a CNR scholarship are in Fond I, Folder 155, CNR Papers, ACS.

57. Frèdèric Joliot-Curie to B. Pontecorvo, 3 February 1939, in "University Documents," B. Pontecorvo Papers, CAC.

58. Giorgio Amendola, *Lettere a Milano* (Rome: Editori Riuniti, 1973), 76.

59. Bruno and Marianne married only in 1940, mainly to make it possible for Marianne to leave France following the outbreak of World War II. "Livret de famille," 9 January 1940, in "France, Documents," B. Pontecorvo Papers, CAC.

60. Sereni, *Il gioco dei regni*, 302.

61. On Berti, see Mimmo Franzinelli, *I tentacoli dell'OVRA: Agenti, collaboratori e vittime della Polizia Fascista* (Turin: Bollati Boringhieri, 1999), 351.

62. Clara Sereni, "A proposito di Bruno," *Sapere*, April 2004, 42.
63. Bignardi, *Memorie estorte*, 25.
64. On the X-Crise economic analysis, see Marianne Fischman and Emeric Ledjel, "The Quantitative Approach to Business Cycle in 'X-Crise' Group in the 1930s," in HAL-CNRS (http://hal.archives-ouvertes.fr/halshs-00268373). On Vallon, see also Gérard Brun, *Louis Vallon ou, La politique en liberté. De Jaurès à de Gaulle* (Paris: Economica, 1986).
65. Goldschmidt, *Atomic Rivals*, 17.
66. "I am a socialist as I am a physicist. I aim to use in their respective domains the same rules of logic, the same principles, and above all the same methods of free thinking." Michel Pinault, *Frédéric Joliot-Curie* (Paris: Editions Odile Jacob, 2000), 82.
67. Ibid., 85.
68. Mafai, *Il lungo freddo*, 77.
69. Pinault, *Frédéric Joliot-Curie*, 11. See also J. Laberrigue-Frolow, "Bruno Pontecorvo and Paris," in Bilenky et al., eds., *B. Pontecorvo Selected Scientific Works*, 465.
70. For an analysis of Joliot-Curie's dealings with the foundation, as well as the history of the cyclotron at the French institution, see Heilbron and Seidel, *Lawrence and His Laboratory*, 340–44.
71. On Lazard's work, see Alain Beltran, "La 'fée électricité', reine et servante," *Vingtième siècle: Revue d'histoire* 16 (1987): 90–95.
72. Pinault, *Frédéric Joliot-Curie*, 115. B. Pontecorvo, "Isomeric Forms of Radio Rhodium," *Nature* 141 (1938): 785–86. B. Pontecorvo, "Recent Experimental Results in Nuclear Isomerism," *Nature* 144 (1939): 212–13.
73. Ibid., 212.
74. "Ici, l'on fabrique des atomes!" *L'oeuvre*, 6 April 1939 [Copy in "University Documents," B. Pontecorvo Papers, CAC].
75. CNRS scholarship, 30 August 1939, in "University Documents," B. Pontecorvo Papers, CAC.
76. On Lawrence's work and, more generally, the interaction with Rome's physicists, see Heilbron and Seidel, *Lawrence and his Laboratory*, 195–97.
77. Thomas E. Allibone, "Metropolitan-Vickers Electrical Company and the Cavendish Laboratory," in John Hendry, ed., *Cambridge Physics in the Thirties* (Bristol: Hilger, 1984), 150–73, on 171.
78. G. Giannini to E. Fermi, 17 March 1936, in "Brevetto Neutroni, 1934–35," Carte Nuovo Amaldi, Box 1, Folder 2, AMF.
79. Heilbron and Seidel, *Lawrence and His Laboratory*, 197.
80. Michele Sarfatti, "La persecuzione degli Ebrei in Italia dalle leggi razziali alla deportazione," in G. Luzzatto Voghera, ed., *La Persecuzione degli Ebrei durante il Fascismo. Le leggi del 1938* (Rome: Camera dei Deputati, 1998), 119. See also Giorgio Israel and Pietro Nastasi, *Scienza e razza nell'Italia Fascista* (Bologna: Il Mulino, 1998).
81. Interview with Guido Pontecorvo, 20 November 1979, in "Britain and the Refugee Crisis, 1933–1947," n. 4505, Imperial War Museum, London. "Miss Clayton's Interview," 5 December 1950, in "Bruno Maximovich Pontecorvo: Italian, British, Russian," KV 2/1889, TNA.

82. Bignardi, *Memorie Estorte*, 22. M. Franzinelli, *Delatori: Spie e confidenti anonimi: L'arma segreta del regime Fascista* (Milan: Mondadori, 2001), 135–60. Bignardi, *Memorie estorte*, 22.

83. Swedish consulate's paper, 3 June 1940, in "France, Documents," B. Pontecorvo Papers, CAC.

84. US Vice Consul Taylor G. Gannett's declaration, 26 June 1940, in "United States and Canada, Documents," B. Pontecorvo Papers, CAC. See also Goldschmidt, *Atomic Rivals*, 73.

85. They landed in New York on 20 August 1940. See Mafai, *Il lungo freddo*, 120.

CHAPTER 2

1. D. S. Parasnis, *Principles of Applied Geophysics* (London: Chapman and Hall, 1997 [1962]), 356. Jerry Robertson, *ABC's of Oil* (Evansville, IN: Petroleum Publishers, 1953), 21.

2. Three electrodes were introduced into a borehole to measure the spread of currents through the formations. *Coring* translates the French *carottage*, from the carrot-stick-like samples extracted from a soil to gather information on its composition.

3. Louis A. Allaud and Maurice H. Martin, *Schlumberger: The History of a Technique* (New York: John Wiley & Sons, 1977), 90–103. See also Geoff C. Bowker, *Science on the Run: Information Management and Industrial Geophysics at Schlumberger, 1920–1940* (Cambridge, MA: MIT Press, 1994).

4. Lee C. Lawyer, Charles C. Bates and Robert B. Rice, *Geophysics in the Affairs of Mankind: A Personalized History of Exploration Geophysics* (Tulsa: Society of Exploration Geophysics, 2001 [1982]), 14–22.

5. Ibid., 48.

6. Ibid., 331. See also George E. Sweet, *The History of Geophysical Prospecting* (Suffolk: Neville Spearman, 1978 [1969]), 274–77.

7. Information extracted from the booklet *S. A. Scherbatskoy*, edited by his family and based on an autobiographical note (copy in S. A. Scherbatskoy papers, SI).

8. R. H. Ritchie, "Jacob Neufeld, 15 April 1906–5 April 2000," *Health Physics* 87 (1), 2004: 94–95.

9. Parasnis, *Principles of Applied Geophysics*, 360.

10. Neufeld's laboratory notebook in Box 1, Folder 7, S. A. Scherbatskoy papers, SI. On detectors, see Thaddeus Trenn, "The GM counter of 1928," *Annals of Science* 43 (1986), 111–35. See also Helge Kragh, *Quantum Generations: A History of Physics in the Twentieth Century* (Princeton, NJ: Princeton University Press, 1999); Peter Galison, *How Experiments End* (Chicago: University of Chicago Press, 1987).

11. Well Surveys, "Radioactivity Well Logging–Through Casing," 1 May 1940, in Box 1, Folder 10, S. A. Scherbatskoy papers, SI. The development of radioactivity well logging is documented in papers and patents. See, for instance, William G. Green and Robert E. Fearon,"Well Logging by Radioactivity," *Geophysics* 5:3 (1940): 272–83, and Serge A. Scherbatskoy, "Well

Logging by Measurement of Radioactivity," US 2,219,273, filed 19 June 1939 and issued 22 October 1940.

12. Allaud and Martin, *Schlumberger*, 268.

13. Emilio Segrè, *A Mind Always in Motion*, 160. "Scherbatskoy asked me . . ." suggests that the engineer already knew Pontecorvo, which is not the case. It may be that during his stay in Tulsa, Segrè went on to inform Scherbatskoy about a few other experts in the field who could pioneer the technique, including Pontecorvo.

14. Michel Oristaglio and Alexander Dorozynski, *A Sixth Sense: The Life and Science of Henri-George Doll, Oilfield Pioneer and Inventor* (New York: Overlook Duckworth, 2009), 167–68.

15. Documentation in "Subseries 1, Relationship with Bruno Pontecorvo, 1940–2002," S. A. Scherbatskoy papers, SI.

16. Bruno Pontecorvo, "Possibility of Application of the Curve Neutron Density Curve Plotted against the Source-Detector Distance in Order to Have Information on the Environment, 15.10.1940," Box 1, Folder 6 in S. A. Scherbatskoy papers, SI. On the "AF curve." see Edoardo Amaldi and Enrico Fermi, "Absorption and Diffusion of Neutrons," *Physical Review* 50 (1936): 899–928.

17. Robert E. Fearon, "Well Logging Method and Apparatus," US patent 2,275,748; F. Brons, "Process and Apparatus for Exploring Geological Strata," US patent 2,220,509.

18. R E. Fearon, William L. Russell and B. Pontecorvo, "Preliminary Field Experiment in Scattered Neutron Well Logging, 25.6.1941," Box 1, Folder 6 in S. A. Scherbatskoy papers, SI.

19. B. Pontecorvo and R. Fearon, "Concerning the Securing of Radioactive Sources for Neutron Well Logging, 17.11.1941," Box 1, Folder 6 in S. A. Scherbatskoy papers, SI.

20. On the two methods, see Parasnis, *Principles of Applied Geophysics*, 363–64.

21. Such as the pioneering B. Pontecorvo, "Radioactivity Analyses of Oil Well Samples," *Geophysics* 7:1 (1942): 90–94, which addressed the issue of laboratory sampling through radioactive methods.

22. R. E. Fearon and B. Pontecorvo, "Re: Stevens and Davis' Letter of March 28, 1941, and Paper No. 5 of March 20, 1941 from the Patent Office on Fearon's Application Serial No. 354-294, 8.4.1941," Box 1, Folder 6 in S. A. Scherbatskoy papers, SI.

23. B. Pontecorvo, "Conferences with Dr. Herzog and Mr. Hare of the Texas Company, During a Convention at Huston," Box 1, Folder 6 in S. A. Scherbatskoy papers, SI. In the counter, neutrons strike atoms of boron or lithium, ejecting alpha particles which are detected by counters filled with boron-thriflouride.

24. R. E. Fearon, W. L. Russell, and B. Pontecorvo, "Preliminary Field Experiment in Scattered Neutron Well Logging, 25.6.1941," Box 1, Folder 6 in S. A. Scherbatskoy papers, SI.

25. "In general, sandstone kicks to the right with respect to shale and limestone kicks to the right with respect to sandstone". B. Pontecorvo and

W. L. Russell, "Preliminary Notes on Neutron Log Interpretation: Confidential, 5.1.1942," Box 1, Folder 6 in S. A. Scherbatskoy papers, SI. See also: B. Pontecorvo, "Neutron Well Logging: A New Geological Method Based on Nuclear Physics," *Oil and Gas Journal* 40 (1941): 32–33.

26. Allaud and Martin, *Schlumberger*, 271.

27. B. Pontecorvo, "Trip to Chicago, Swarthmore, and New York, 15.4.1952," Box 1, Folder 6 in S. A. Scherbatskoy papers, SI.

28. B. Pontecorvo and R. E. Fearon, "Concerning the Securing of Radioactive Sources for Neutron Well Logging, 17.11.1941," and S. A. Scherbatskoy, W. L. Russell and B. Pontecorvo, "Notes on Prospecting for Strategic Minerals, 26.2.42," Box 1, Folder 6 in S. A. Scherbatskoy papers, SI.

29. B. Pregel to B. Pontecorvo, 24 November 1942 in "US and Canada, Documents," B. Pontecorvo Papers, CAC. Goldschmidt, *Atomic Rivals*, 139–40.

30. Ibid., 171.

31. On Fermi's work in the United States: Segrè, *Enrico Fermi*, chapter 4. See also Giulio Maltese, *Enrico Fermi in America: Una bibliografia scientifica: 1938–1954* (Bologna: Zanichelli, 2003), and James W. Cronin, ed., *Fermi Remembered* (Chicago: University of Chicago Press, 2004).

32. Segrè, *Enrico Fermi*, 106. Laboratory work on fission had by then taken place at Columbia, in Copenhagen (Otto Frisch), and in Paris (Joliot-Curie, Von Halban, and Kowarski).

33. On the letter to President Roosevelt, see Rhodes, *The Making of the Atom Bomb*, 307–12. On the German project, see Mark Walker, *German National Socialism and the Quest for Nuclear Power, 1939–1949* (Cambridge: Cambridge University Press, 1989). On Szilard, see William Lanouette, *Genius in the Shadow: A Biography of Leo Szilard, the Man Behind the Bomb* (Chicago: University of Chicago Press, 1992).

34. Maltese, *Enrico Fermi in America*, 88–89.

35. On the etymology, see Segrè, *Enrico Fermi*, 116. Fermi told Segrè that the term "pile" had been coined not with Alessandro Volta's *pila* (i.e., the first battery) in mind, but in reference to a heap.

36. Maltese, *Enrico Fermi in America*, chap. 6.

37. Ibid., 113. On the technical details of Fermi's work, see also Hoddeson et al., *Critical Assembly*, chapter 3.

38. Rhodes, *The Making of the Atom Bomb*, 442.

39. On disputed meanings of the acronym MAUD, see Margaret Gowing, *Britain and Atomic Energy, 1939–1945* (London: Macmillian, 1964), 45.

40. Ibid., 50–51. On the rescue of heavy water, see Per Fridtjof Dahl, *Heavy Water and the Wartime Race for Nuclear Energy* (Bristol: Institute of Physics Publishing, 1999). Gowing, *Britain and Atomic Energy*, 109.

41. Ibid., 110.

42. Ibid., 191.

43. George C. Laurence, "Early Years of Nuclear Energy Research in Canada," 4 (published by Atomic Energy of Canada Limited in May 1980, available on the Canadian Nuclear Society website: http://www.cns-snc.ca/history/early_years/earlyyears.html). See also G. C. Laurence, "Canada's Par-

ticipation in Atomic Energy Development," *Bulletin of the Atomic Scientists* 3 (1947): 325–28.

44. Ibid., 5.

45. See Gowing, *Britain and Atomic Energy*, 65–66. George Thomson, "Charles Galton Darwin," *Bibliographical Memoirs of the Fellows of the Royal Society* 9 (1963): 77.

46. Gowing, *Britain and Atomic Energy*, 234–35.

47. A small heavy water reactor was eventually built at Argonne.

48. Halban was "impetuous and vacillating in decisions and unreasonable in his demands of the administrative staff in Ottawa and unfair in criticizing them." Laurence, "Early Years of Nuclear Energy Research in Canada," 4.

49. Gowing, *Britain and Atomic Energy*, 277.

50. B. Pontecorvo, "5000 Kw Hetero P.9 Plant Discussion; Lattice Group n. 2, 25.4.1944," AB 2/644, and B. Pontecorvo, "5000 Kw Hetero P.9 Plant Discussion; Lattice Group n. 3, 2.5.1944," AB 2/645, TNA.

51. Henry Arnold, "Summary of Pontecorvo's Journeys during Wartime," 9 November 1950, KV 2/1888, TNA.

52. B. Pontecorvo, "Some Information on Physical Data Obtained on a Recent Trip to Chicago, 24.4.1944," AB 6/643 and B. Pontecorvo, "Some Physical Data Obtained on Last Chicago Trip, 5.6.1944, Blue Print, Secret," AB 2/652, TNA.

53. Joint Committee on Atomic Espionage, *Soviet Atomic Espionage* (Washington DC: Printing Office, 1953), 2.

54. B. Pontecorvo, "Concerning 23 Production in the Reflector of the Polymer Plant, 1944," AB 2/657, TNA. Alan Munn and B. Pontecorvo, "Spacial Distribution of Neutrons in Media Containing Iron, Lead and Bismuth, 1944," AB 2/305, TNA. B. Pontecorvo, "Some Data Useful in Shielding Problems, 1944," AB 2/655, TNA.

55. Robert Bothwell, *Nucleus. A History of Atomic Energy of Canada Limited* (Toronto: University of Toronto Press, 1988), 159. Donald G. Hurst, "Overview of Nuclear Research and Development," in *Canada Enters the Nuclear Age: A Technical History of Atomic Energy of Canada Limited* (Montreal & Kingston: McGill-Queen's University Press, 1997), 1–32.

56. B. Pontecorvo and D. West, "The Fission Properties of Radium 226 and Protoactinium 233, 1.12.1945," AB 2/318, TNA.

57. NRX Project. Progress Report for August 1944, AB 2/917, TNA.

58. "ZEEP has been used for important research on the behavior of neutrons in reactors and other purposes by B. W. Sargent and others." Laurence, "Early Years of Nuclear Energy Research in Canada," 10.

59. Robert Oppenheimer, "Physics in the Contemporary World," *Bulletin of the Atomic Scientists* 4:3 (1948), 141–81. See also Thorpe, *Oppenheimer*, 188–92.

60. B. Pontecorvo to Glenn Seaborg, 2 February 1946, in "Scientific Correspondence, 1945–1950," Bruno Pontecorvo Papers, CAC. The offer paralleled others from important organizations such as the University of Michigan and the General Electric Laboratory in Schenectady, New York.

61. Bothwell, *Nucleus*, 105.
62. S. A. Scherbatskoy, W. L. Russell and B. Pontecorvo, "Notes on Prospecting for Strategic Minerals, 26.2.42," Box 1, Folder 6 in S. A. Scherbatskoy Papers, SI.
63. As noted in the patent B. Pontecorvo, "Method and Apparatus for Geophysical Exploration," US 2,349,753, filed February 5, 1942, and issued May 3, 1944. See also B. Pontecorvo, "Results of Experiments to Determine the Possibility of Measuring the Variations of Th/U Ratio in Sedimentary Formations, 16.10.1940," Box 1, Folder 6 in S. A. Scherbatskoy Papers, SI.
64. Robert Bothwell, *Eldorado: Canada's National Uranium Company* (Toronto: University of Toronto Press, 1984), 10–25.
65. Ibid., p. 73; Pregel also fell afoul of General Leslie Groves, who did not like his clout on uranium supply. By the end of 1942, Groves sought control of 1,200 tons of high-grade uranium from Congo amassed by the African Metal Corporation in Staten Island, New York. See Jonathan E. Elmerich, *Gathering Rare Ores: The Diplomacy of Uranium Acquisition, 1943–1954* (Princeton, NJ: Princeton University Press, 1986), 7.
66. Roger V. Devlin, "Tulsan Locates Essential Uranium Ore," *Tulsa Tribune*, 16 October 1945 (copy in Box 1, Folder 11 in S. A. Scherbatskoy Papers, SI).
67. B. Pontecorvo, "Report on Trip to Port Radium, September 1944 (secret)," AB 1/648, TNA.
68. Ibid.
69. Robert D. Nininger, *Minerals for Atomic Energy* (Toronto: Van Nostrand, 1954), 26.
70. Devlin, "Tulsan Locates Essential Uranium Ore."
71. Jonathan E. Elmerich, *Gathering Rare Ores*, 10.
72. Bothwell, *Eldorado*, 144. On the importance of Murray Hill Area in US nuclear intelligence, see also Ziegler and Jacobson, *Spying Without Spies*, 21–33.
73. This was a subsidiary of Union Carbide and Carbon Company. Ibid., 43.
74. "Dr. Pontecorvo's notes on a meeting held on 31.10.1944." Copy of this document can be found in F. H. Burstall, H. Carmichael, A. H. Gillieson, and J. Hardwick, "Report on a Technical Conference on Prospecting Problems Held in Washington the 24–25–26 January 1946," AB 2/67, TNA.
75. Segrè, *Enrico Fermi*, 71–73.
76. For example, he thought using isotopes of chlorine or bromine in the following reactions:

$$\nu + {}^{37}Cl \rightarrow {}^{37}Ar + e^-.$$

Evidence of neutrino bombardment would be obtained by extracting ^{37}Ar from the liquid carbon tetrachloride (CCl_4). B. Pontecorvo, "Inverse β-process," Report PD-205, National Research Council, Division of Atomic Energy, Chalk River, 10. Published in Bilenky et al., eds., *B. Pontecorvo Selected Scientific Works*, 21–26.

77. As discussed in Bernice W. Sargent, "Agenda for Meeting to Discuss Nuclear Physics Program, 16.6.1946," AB 2/747, TNA.

78. Laurence, "Early Years of Nuclear Energy Research in Canada," 6.

79. Kragh, *Quantum Generations*, chapter 13. See also Yataro Sekido and Harry Elliot, eds., *Early Days in Cosmic Rays* (Dordrecht: D. Reidel Publishing Co., 1985).

80. Edoardo Amaldi supervised the group. Marcello Conversi, Ettore Pancini, and Oreste Piccioni, "On the Decay Process of Positive and Negative Mesons," *Physical Review* 68 (1945): 232.

81. Cesar Lattes, Giuseppe Occhialini, and Cecil Powell, "Observations on the Tracks of Slow Mesons in Photo-graphic Emulsions," *Nature* 160 (1947): 453–56. C. Lattes, G. Occhialini, and C. Powell, "Observations on the Tracks of Slow Mesons in Photo-graphic Emulsions II," *Nature* 160 (1947): 486–87.

82. Kragh, *Quantum Generations*, 103–4.

83. B. Pontecorvo, "Nuclear Capture of Mesons and the Meson Decay," *Physical Review* 72 (1947): 246. In the 1960s, the Italian physicist Giampiero Puppi provided a better understanding of this analogy in terms of universality of weak interactions. See on this Luisa Bonolis, "Un Genio di Via Panisperna," 29.

84. E. P. Hincks and B. Pontecorvo, "Search for Gamma-Radiation in the 2.2 Microsecond Meson Decay Process," *Physical Review* 73 (1948): 257–58.

85. E. P. Hincks and B. Pontecorvo, "The Absorption of Charged Particles from the 2.2 Microsecond Meson Decay," *Physical Review* 74 (1948): 697–98. E. P. Hincks and B. Pontecorvo, "The Penetration of µ-Meson Decay Electrons and Their Bremsstrahlung Radiation," *Physical Review* 75 (1949): 698–99. E. P. Hincks and B. Pontecorvo, "On the Disintegration Products of the 2.2 µsec. Meson," *Physical Review* 77 (1950): 102–20. See also Galison, *Image and Logic*, chapter 3.

86. B. Pontecorvo to E. Amaldi, 19 June 1949, Edoardo Amaldi Papers, Box 142, AMF.

87. Bruno Maximovich Pontecorvo, "Pages in the Development of Neutrino Physics," *Soviet Physics Uspekhi* 26 (1983): 1087–1108, on 1101. In the 1950s, Raymond Davis Jr. attempted to detect neutrinos emitted by the Savannah River reactor, and then went on to detect solar neutrinos. On this see Frank Close, *Neutrino* (Oxford: Oxford University Press, 2010), chapter 6.

88. Wilfrid B. Lewis, *Electrical Counting: With Special Reference to Alpha and Beta Particles* (Cambridge: Cambridge University Press, 1948).

89. B. Pontecorvo, "Recent Developments in Proportional Counter Technique," *Helvetica Physica Acta* 23 (1950): 97–118.

90. David H. Kirkwood, B. Pontecorvo, and Geoff C. Hanna, "Fluctuations of Ionization and Low Energy Beta-Spectra," *Physical Review* 74 (1948): 497–98.

91. Giovanni Fidecaro, "Bruno Pontecorvo: From Rome to Dubna," in Bilenky et al., eds., *B. Pontecorvo Selected Scientific Works*, 472.

92. Ibid., 478.

93. Allaud and Martin, *Schlumberger*, 269.

94. J. L. Arps, "The Arps Corporation, Proposal for a New Company Prepared for Prospective Investors, 1955," Box 8, Folder 5, S. A. Scherbatskoy Papers, SI.

95. R. H. Ritchie, "Jacob Neufeld," 94–95.

96. S. A. Scherbatskoy to B. Pontecorvo, 14 March 1947, and S. A. Scherbatskoy to B. Pontecorvo, 16 March 1947, in "Scientific Correspondence, 1945–1950," Bruno Pontecorvo Papers, CAC.

97. James V. Dunworth and B. Pontecorvo, "Excitation of Indium 113 by X-rays," *Proceedings of the Cambridge Philosophical Society* 43 (1947): 123–26 (Also report AB 2/315, TNA with the same title.)

98. E. Amaldi to B. Pontecorvo, 16 May 1947, in "Scientific Correspondence, 1945–1950," Bruno Pontecorvo Papers, CAC.

99. Fidecaro, "Bruno Pontecorvo: From Rome to Dubna," 474. B. Goldschmidt to B. Pontecorvo, 23 August 1949, in "Scientific Correspondence, 1945–1950," Bruno Pontecorvo Papers, CAC.

100. S. A. Scherbatskoy to B. Pontecorvo, 27 June 1949, in "Scientific Correspondence, 1945–1950," Bruno Pontecorvo Papers, CAC. See also Allaud and Martin, *Schlumberger*, 269.

101. Bothwell, *Eldorado*, 175.

102. W. G. Russell and S. Scherbatskoy, "The Use of Sensitive Gamma Ray Detectors," *Economic Geology* 46:4 (1951): 427–46.

103. Minutes of meeting of the Strategic Minerals Committee, 29 August 1945, in "Uranium Intelligence, Section 1, 1945–1957," AB 1/507, TNA.

104. J. Cockcroft to B. Pontecorvo, 3 February 1947, in "Scientific Correspondence, 1945–1950," Bruno Pontecorvo Papers, CAC.

105. "Uranium Intelligence, Section 1, 1945–1957," AB 1/507, TNA.

106. B. Pontecorvo, "Equipment Required for Experimental Work, 1948," AB 1/648, TNA.

107. C. F. Davidson to J. Hardwick, 14 April 1948, in "Security, General," AB 6/115, TNA.

108. "Scientific Civil Service," *Nature* 161 (1948), 195.

109. Herbert Spence, "Staff Recruiting: Correspondence from Chalk River, Canada and Harwell, 1945," AB 6/171, TNA.

110. Bothwell, *Eldorado*, 78. Guy Hartcup and Thomas Allibone, *Cockcroft and the Atom* (Bristol: Hilger, 1984), 144.

111. J. Dunworth to Herbert Skinner, 16 June 1946, in "Harwell Pile Discussion Group, AB 12/19. See also M. Gowing (assisted by Lorna Arnold), *Independence and Deterrence: Britain and Atomic Energy, 1945–1952. Vol. 1. Policy Making* (London: MacMillian, 1974).

112. Minutes of meeting, "Power Steering Committee, vol. 1, 1947," AB 12/57 and minutes of meeting, "Power Steering Committee, vol. 2, 1948–49," AB 12/74, TNA.

113. Ibid.

114. B. Pontecorvo to E. Amaldi, 2 February 1949, in Box 142, Edoardo Amaldi Papers, AMF.

CHAPTER 3

1. Copy of extract, "MI5 Special Branch Records," 7 November 1950, in "Bruno Maximovich Pontecorvo: Italian, British, Russian," KV 2/1888, TNA.
2. West, *MI5*, 145.
3. Holligsworth and Fielding, *Defending the Realm*, 14.
4. Simone Turchetti, "Use, Refuse or Lock Them Up: A History of Italian Academic Refugees in Britain, 1930/1950" (PhD Diss., University of Manchester 2003). See also Lucio Sponza, *Divided Loyalties: Italians in Britain during the Second World War* (Bruxelles: Peter Lang, 2000), 102.
5. *The Sunday Dispatch*, 14 April 1940, cit. in Peter and Leni Gillman, *Collar the Lot: How Britain Interned and Expelled Its Wartime Refugees* (London: Quartet, 1980).
6. Ruggero Orlando, "Amore-Odio fra Italiani e Inglesi durante e dopo la Seconda Guerra Mondiale," in *Inghilterra e Italia nel '900: Atti del convegno di Bagni di Lucca* (Florence 1973): 193–200.
7. Interview with Guido Pontecorvo, 20 November 1979, in "Britain and the Refugee Crisis, 1933–1947," n. 4505, Imperial War Museum, London.
8. Bondy, *The Emissary*, 168.
9. Segrè, *A Mind Always in Motion*, 172–73.
10. Athan Theoaris, *Chasing Spies* (Chicago: Ivan Dee, 2000), 120. See also Sanford Ungar, *FBI* (Boston: Little, Brown, 1975).
11. Lawrence Badash, "From Security Blanket to Security Risk: Scientists in the Decade after Hiroshima" *History and Technology* 19 (2003): 241–56.
12. Theoaris, *Chasing Spies*, 100.
13. "Ha la moglie ebrea ed ha fatto ebrea la figlia," in E. Fermi to Luigi Federzoni's secretary, 3 December 1938. In "Fermi Enrico" (Folder 189723), Carte Ordinamenti, Segreteria Particolare del Duce, ACS.
14. "La mia visita in America non è in alcun modo connessa con questioni razzistiche ma è determinata da ragioni scientifiche come le mie cinque visite precedenti. Vi prego di smentire a mio nome ogni altra interpretazione inesatta." Luigi Federzoni to Benito Mussolini, 5 January 1939, in "Fermi Enrico" (Folder 189723), Carte Ordinamenti, Segreteria Particolare del Duce, ACS.
15. Memoradum from the Italian embassy, Washington, DC, 8 September 1940, cit. in Maltese, *Enrico Fermi in America*, 76.
16. See "Army Intelligence Report on Fermi and Szilard, August 13, 1940," [copy in Cronin, ed., *Fermi Remembered*, 220].
17. Gowing, *Britain and the Atomic Energy*, 134. Maltese, *Enrico Fermi in America*, 105.
18. Segrè, *Enrico Fermi*, 122–23.
19. Several appeals have been filed to have the FBI file on Pontecorvo released and available for study. What we know about FBI activities is by and large the result of post-defection inquiries which shed light on existing FBI documentation, in particular the MI5 files on Bruno Pontecorvo.
20. FBI report as summarized by James C. Robertson, 31 October 1950, in "Bruno Maximovich PONTECORVO: Italian, British, Russian," KV 2/1888, TNA.

21. J. S. Rowlinson, "The Wartime Work of Hinshelwood and His Colleagues," *Notes and Records of the Royal Society* 58:2 (2004): 161–66.

22. J. H. Wolfenden, BSCO to Major K. M. Bourne, BSC, 4 November 1942, British Secret—US Confidential in "Employment of Dr. Bruno Pontecorvo," AB 1/361, TNA.

23. Heilbron and Seidel, *Lawrence and His Laboratory*, 342–343.

24. Hans Von Halban and Hugh Paxton, "Doppler Effect of Nuclear Resonance Level," *Nature* 141 (1938): 116.

25. Giuliana Gemelli, "Scholars in Adversity and Science Policies," in G. Gemelli, ed., *The "Unacceptables": American Foundations and Refugee Scholars between the Two Wars and After* (Bruxelles: Peter Lang, 2000), 16.

26. Doris T. Zallen, "Louis Rapkine and the Restoration of French Science after the Second World War," *French Historical Studies* 17:1 (1991): 6–37, on 15–16. John Krige, *American Hegemony and the Post-War Reconstruction of Science in Europe* (Cambridge, MA: MIT Press, 2006), 81–83.

27. One of the BSC agents was Ian Fleming, who later went on to portray the agency activities in his celebrated James Bond series of novels. N. West, *A Matter of Trust*, 27. See also William Stevenson, *A Man called Intrepid: The Secret War, 1939–1945* (London: Book Club, 1976).

28. D. B., "Security Service Aspects of Pontecorvo Case," 24 October 1950, in "Security Service Action in the Case of Bruno Pontecorvo," KV 2/242, TNA.

29. E. K. Balls to J. H. Wolfenden, 30 November 1942 and 3 March 1943, Secret, in "Employment of Dr. Bruno Pontecorvo," AB 1/361, TNA.

30. FBI to Stott, 3 February 1943; Stott to FBI, 10 February 1943, and FBI to Stott, 17 Feburary 1943, in "Bruno Maximovich Pontecorvo," KV2/1888, TNA. See also Tim Gibbs, "British and American Counter-intelligence and the Atom Spies, 1941–1950" (PhD diss., University of Cambridge, 2008), chapter 5.

31. Guy Liddell's report to the prime minister, 23 October 1950, in PREM 8/1273, TNA.

32. Tim Gibbs, "British and American Counter-intelligence and the Atom Spies, 1941–1950," chapter 5.

33. Robert C. Williams, *Klaus Fuchs, Atom Spy* (Cambridge, MA: Harvard University Press, 1987), 49.

34. Goldschmidt, *Atomic Rivals*, 241.

35. Donald Avery, "Allied Scientific Co-operation and Soviet Espionage in Canada, 1941–1945," *Intelligence and National Security* 8:3 (1993), 100–129, on 108. See also D. H. Avery, *The Science of War: Canadian Scientists and Allied Military Technology during the Second World War* (Toronto: University Press, 1998).

36. Henry Arnold to Ronnie T. Reed, 3 May 1952, in "Bruno Maximovich Pontecorvo," KV2/1889, TNA.

37. Williams, *Klaus Fuchs*, 69.

38. Goldschmidt, *Atomic Rivals*, 233.

39. Ibid., 219. James Chadwick to John Cockroft, 16 October 1944, in "Work in North America, Canadian Project, Official and Personal Corre-

spondence with Dr. Cockcroft and other staff at Montreal Lab (1944–1947)," James Chadwick Papers, CAC.

40. Commander Arnold, Summary of Pontecorvo's journeys during wartime, 11 November 1950, in "Bruno Maximovich Pontecorvo," KV2/1888, TNA.

41. See Aldrich, *The Hidden Hand*, 102–4. See also John Baker-White, *The Soviet Spy System* (London: Falcon Press, 1948).

42. Goldschmidt, *Atomic Rivals*, 273.

43. Norman Moss, *Klaus Fuchs. The Man Who Stole the Atom Bomb* (London: Grafton, 1989), 115.

44. On this see Avery, "Allied Scientific Co-operation and Soviet Espionage in Canada, 1941–1945," 108–13.

45. J. Chadwick to Leslie Groves, 24 July 1945, in "Work in North America, Canadian Project, Official and Personal Correspondence with Dr. Cockcroft and other staff at Montreal Lab (1944–1947)," James Chadwick Papers, CAC.

46. J. Chadwick to J. Cockcroft, 14 August 1945, in "Work in North America, Canadian Project, Official and Personal Correspondence with Dr. Cockcroft and other staff at Montreal Lab (1944–1947)," James Chadwick Papers, CAC.

47. B. Pontecorvo to J. Cockcroft, 2 February 1946, in "Scientific Correspondence, 1945–1950," B. Pontecorvo Papers, CAC.

48. Aldrich, *The Hidden Hand*, 108.

49. John Earl Haynes and Harvey Klehr, *Venona: Decoding Soviet Espionage in America* (New Haven: Yale University Press, 2000), 184–85.

50. Copy of RCMP minutes files, 25 September 1946 and 31 December 1946, in "Security Service Action in the Case of Dr. Bruno PONTECORVO," KV 4/242, TNA. See also Williams, *Klaus Fuchs*, 144–47.

51. Copy of Annex, 12 April 1946, in "Security Service Action in the Case of Dr. Bruno PONTECORVO," KV 4/242, TNA.

52. Memorandum by the Director General of Scientific Research (Defence), 12 April 1946, in "Security Service Action in the Case of Dr. Bruno PONTECORVO," KV 4/242, NA.

53. Copy of Secret DCOS memorandum in "Security Service Action in the Case of Dr. Bruno PONTECORVO," KV 4/242, NA.

54. "Pontecorvo shared the same fate as myself—removal from the project." Goldschmidt, *Atomic Rivals*, 283.

55. It is my decision to use the plural in "witch hunts." The term "witch hunt" was first popularized in Arthur Miller's 1953 play *The Crucible*, which was about the Salem witch trials, but also a parody of loyalty inquiries which took place during the McCarthy era.

56. Badash, "From Security Blanket to Security Risk: Scientists in the Decade after Hiroshima," 245.

57. Wang, *American Science in an Age of Anxiety*, 256.

58. Ibid., 59. See also Badash, "From Security Blanket to Security Risk: Scientists in the Decade after Hiroshima," 251–52. Following Wang's pioneering work, the bibliography on cases of interaction between US intelligence and

scientists in the wake of the Red Scare has grown considerably. Here I can only refer to a few inspiring works: Schweber, *In The Shadow of the Bomb*; Shawn Kristian Mullet, "Little Man: Four Junior Physicists and the Red Scare Experience" (PhD diss., Harvard University, 2008); Kaiser, "Atomic Secret in Red Hands? Cold War Fears of Theoretical Physicists."

59. M. J. Heale, *McCarthy's Americans: Red Scare Politics in the State and Nation, 1935–1965* (Basingtoke: MacMillan, 1997), 28–53.

60. Other prominent scientists hit by the oath controversy included Geoffrey Chew, Wolfgang Panofsky, Marvin Goldberger, and Robert Serber. David P. Gardner, *The California Oath Controversy* (Berkeley: The University of California Press, 1967), 183. See also Maltese, *Enrico Fermi in America*, 343.

61. "I now feel younger, as we seem to have gone back to the *Fascio*'s times." E. Segrè, *A Mind Always in Motion*, 106.

62. Salvador Edward Luria, *A Slot Machine, a Broken Test Tube: An Autobiography* (New York: Harper & Row, 1984), 185.

63. Wang, *American Science in the Age of Anxiety*, 59.

64. Top secret FBI report on Bruno Pontecorvo, as summarized by J. C. Robertson, 30 October 1950, in "Bruno Maximovich Pontecorvo: Italian, British, Russian," KV 2/1888, TNA.

65. Luria, *A Slot Machine, a Broken Test Tube: An Autobiography*, 182.

66. Williams, *Klaus Fuchs*, 2. West, *A Matter of Trust*, 32.

67. R. Griffith, *The Politics of Fear: Joseph R. McCarthy and the Senate* (Lexington: University of Kentucky Press, 1970), 47.

68. Rhodes, *The Dark Sun*, 413. Griffith, *The Politics of Fear*, 52.

69. Wang, *American Science in the Age of Anxiety*, 262–63.

70. Mafai, *Il lungo freddo*, 28.

71. Reed to B.2, 10 November 1950, in "Bruno Maximovich Pontecorvo," KV 2/1889, TNA. The letter states that the revelation was made "on the day of Fuchs's arrest"—i.e., on 3 February 1950. However, this is in contradiction with the dates of the conference, which occurred on 9–12 February 1950.

72. Michael Perrin, Ministry of Supply to Roger Makins, Foreign Office, 4 November 1950, Secret, in "Defection to USSR of Dr. Pontecorvo," FO 371/84837, NA.

73. Ibid.

74. Ibid. See also H. Montgomery Hyde, *The Atom Bomb Spies*, 126. Bignardi, *Memorie estorte*, 72.

75. T. R. Reed to B.2, 10 November 1950, in "Bruno Maximovich Pontecorvo," KV 2/1889, TNA.

76. Moss, *The Man Who Stole the Atom Bomb*, 155. Williams, *Klaus Fuchs, Atom Spy*, 3.

77. Sereni, *Il gioco dei regni*, 153.

78. H. Jones, "Herbert Wakefield Banks Skinner," *Biographical Memoirs of Fellows of the Royal Society* 6 (1960): 259–67.

79. As shown in the file "Professor Herbert Wakefield Banks SKINNER (1) / Erna SKINNER (2)," KV 2/2080–2082, TNA.

80. Peter Rowlands, *120 Years of Excellence: The University of Liverpool Physics Department, 1881–2001* (Liverpool: PD Publications, 2001), 27.

81. James Mountford's undated statement, "Appointment for 2nd Chair of Physics at Liverpool University," File D.498/2/5/1, LUL.

82. Ibid.

83. "At the beginning of May 1950 we pressed the Ministry of Supply to hasten the arrangements for Pontecorvo's transfer from Harwell to some post where he would not have access to secret information, such as the post at Liverpool University." R. Makins to M. Perrin, Secret, 9 November 1950, in "Defection to USSR of Dr. Pontecorvo," FO 371/84837, TNA.

84. Philip Dee to J. Mountford, 10 May 1950, in "Second Chair of Physics: Correspondence," File P.2343/17, LUL.

85. Report of the Selection Committee for the Second Chair of Physics, 14 June 1950, in "Second Chair of Physics: Correspondence," P.2343/17, LUL.

86. J. Chadwick to J. Mountford, 30 April 1950, and J. Chadwick to J. Mountford, 14 May 1950, in "Second Chair of Physics: Correspondence," P.2343/17, LUL.

87. J. Mountford to B. Pontecorvo, 6 June 1950, in "Scientific Correspondence, 1945–1950," B. Pontecorvo Papers, CAC.

88. J. Mountford to B. Pontecorvo, 7 July 1950, in "Scientific Correspondence, 1945–1950," B. Pontecorvo Papers, CAC.

89. B. Pontecorvo to J. Mountford, 24 July 1950, in "Scientific Correspondence, 1945–1950," B. Pontecorvo Papers, CAC.

CHAPTER 4

1. The patent was rejected on the grounds that Fermi had published his findings in papers. A claim before the US Patent Office Board of Appeals was overruled on 12 May 1941. The Canadian patent is E. Fermi et al., Radioactive Isotope Production, CA 407,558, issued 22 November 1942.

2. See B. Pontecorvo, "Method of Geophysical Prospecting," US 2,508,772, issued 23 May 1950, filed 31 October 1942; B. Pontecorvo, "Well Surveying," US 2,398,324, issued 9 April 1946, filed 10 August 1943.

3. G. Giannini to E. Segrè (copy to Fermi), 18 April 1939, in "Emilio Segrè," Box 11, Folder 13, Enrico Fermi Papers, UCA.

4. Jacob Neufeld, "Claim of S. A. Scherbatskoy to a Share in the Proceeds From the Sale of the US Patent 2.206.634 to the Government of the US," 14 July 1976, in Box 1, Folder 3, S. A. Scherbatskoy's Papers, SI.

5. "Fubini-Pontecorvo Agreement," 18 November 1949, Exhibit 7 in R. A. Anderson, W. F. Smith, and W. B. Holton, "Memoranda and Documents as to Settlement of Docket n. 2, 6.12.1951" Box 30, Folder 9, AEC Papers (RG326), NARA. On Fubini, see also Charles H. Townes, *How the Laser Happened: Adventures of a Scientist* (Oxford: Oxford University Press, 1999), 129–36, and David G. Fubini, *Let Me Explain: Eugene G. Fubini's Life in Defense of America* (Santa Fe: Sunstone Press, 2009).

6. "Senate Hearing on Atomic Energy, Atomic Bomb Patents," *Bulletin of the Atomic Scientists* 1:7 (1946): 10–11. See also Stephane Groueff, *Manhattan Project: The Untold Story of the Making of the Atom Bomb* (Boston: Little, Brown, 1967), 29–30 and 133; Alex Wellerstein, "Patenting the Bomb:

Nuclear Weapons, Intellectual Property, and Technological Control," *Isis* 99:1 (2008): 57–87.

7. Rhodes, *Making of the Atomic Bomb*, 503.

8. "Senate Hearing on Atomic Energy, Atomic Bomb Patents," 10. See also Wellerstein, "Patenting the Bomb," 59.

9. David Hawkins, "Toward Trinity," in D. Hawkins, ed., *Project Y: The Los Alamos Story* (San Francisco: Tomash, 1983), 1–259, on 34. Legislation on "secrecy orders" was first produced at the end of World War I. Reintroduced during World War II, it was permanently adopted as Public Law 256 in 1952, thereby allowing the US Navy, Air Force, and Army to cover with secrecy letters patent and prohibit the issuing of a patent until further decisions by the US military agencies involved. Special provisions contained in US Code 37 allowed delaying the issuing of patents upon inventions "important to the armament or defense of the United States." Casper W. Ooms, "Atomic Energy and U.S. Patent Policy. Part 2: Patent Provisions of the Atomic Energy Act," *Bulletin of the Atomic Scientists* 2:11 (1946): 30–31.

10. Wellerstein, "Patenting the Bomb," 79. Copies of Fermi's wartime patents are in Box 19, Folders 9 and 10, Fermi Papers, UCA.

11. Gowing, *Britain and Atomic Energy*, 207–12.

12. Ibid., 209.

13. E. Segrè to E. Fermi, 7 December 1943, in "Emilio Segrè," Box 11, Folder 13, E. Fermi Papers, UCA. On the inconclusive meeting: E. Segrè to Robert Lavender, 29 July 1944, Box 9, Folder 2, E. Fermi Papers, UCA.

14. "I could not disclose to him all the developments related to the Patent, but within limits imposed by secrecy I indicated to him that important interests were anxious to secure the Patent rights." Fermi's memorandum, 12 January 1943, Box 19, Folder 2, E. Fermi Papers, UCA.

15. G. Giannini to Vannevar Bush [copy to Fermi], 19 October 1945, Box 19, Folder 2, E. Fermi Papers, UCA.

16. The former US vice president, Henry Wallace, argued that the McMahon's bill could "foster and develop economic, medical and other peaceful uses of atomic fission and its by-products." H. A. Wallace, "Supports the McMahon Bill," *Bulletin of the Atomic Scientists* 1, no. 5 (1946): 6–7. On the JCAE establishment, see Richard G. Hewlett and Oscar E. Anderson, Jr., *The New World: A History of the United States Atomic Energy Commission, Vol. 1* (Berkeley: University of California Press, 1990), 411.

17. Ibid., 513.

18. Captain Lavender, whose JCAE hearing took place on 2 February 1946, stressed that control on patents was established "under police power" rather than "eminent domain." "Senate Hearing on Atomic Energy, Atomic Bomb Patents," 10.

19. "The Revised McMahon Bill," *Bulletin of the Atomic Scientists* 1, no. 9 (1946): 2–5. See also Byron S. Miller, "A Law Is Passed: The Atomic Energy Act of 1946," *University of Chicago Law Review* 15, no. 4 (1948): 799–821, and Hewlett and Anderson, *The New World*, 495–98. Others, such as an AEC adviser on patents, emphasized its merits: "The act provides an elaborate but

flexible procedure designed to prevent the acquisition of patents in the field in which the AEC is given absolute governmental monopoly." Casper W. Ooms, "Atomic Energy and U.S. Patent Policy, Pt. 2: Patent Provisions of the Atomic Energy Act," *Bulletin of the Atomic Scientists* 2, no. 11 (1946): 30–31.

20. "Fermi and to a lesser degree myself have been extremely generous in patent matters with the Govt. as Mr. Lavender knows, and I think he is trying to pull the rope too much." E. Segrè to G. Giannini (copy to Fermi), 28 May 1946, in "Emilio Segrè," Box 11, Folder 13, E. Fermi Papers, UCA.

21. G. Giannini to E. Segrè (copy to Fermi), 2 February 1946, Box 19, Folder 2, E. Fermi Papers, UCA. Bush wanted to wrap up all matters regarding wartime research before devoting more time to the establishment of the National Science Foundation. See Daniel J. Kevles, *The Physicists: The History of a Scientific Community in Modern America* (Cambridge, MA: Harvard University Press, 1971), 344–46, and Jessica Wang, "Liberals, the Progressive Left, and the Political Economy of Postwar American Science: The National Science Foundation Debate Revisited," *Historical Studies in the Physical Sciences* 26 (1995): 139–66, on 146.

22. G. Giannini to E. Segrè (copy to Fermi), 2 February 1946, Box 19, Folder 2, E. Fermi Papers, UCA.

23. L. Bernard to G. Giannini (copy to Fermi), 3 January 1947, Box 19, Folder 2, E. Fermi Papers, UCA.

24. G. M. Giannini and Co., "Application for Just Compensation and the Determination of a Reasonable Royalty Fee under Section 11 of the Atomic Energy Act of 1946," in "Neutron Patent," Box 2, Folder 2, Carte Nuovo Amaldi, AMF.

25. E. Segrè to E. Fermi, 15 November 1948, in "Emilio Segrè," Box 11, Folder 13, E. Fermi Papers, UCA.

26. Franco Rasetti to G. Giannini, 28 April 1948, in "Neutron Patent," Box 2, Folder 2, Carte Nuovo Amaldi, AMF.

27. E. Fubini to B. Pontecorvo, 1948, in "Scientific Correspondence, 1945–1950," B. Pontecorvo Papers, CAC.

28. B. Pontecorvo to E. Amaldi, 23 March 1949, Box 142 Edoardo Amaldi Papers, AMF.

29. He was the owner of a poultry farm in Buckinghamshire. Interview of Giovanni David Pontecorvo by Michael Wade (MI5), 2 November 1950, in "Bruno Maximovich Pontecorvo: Italian, British, Russian," KV2/1888, TNA.

30. AEC Press Release: Appointment of Bennett Boskey as deputy to general counsel, 10 July 1949, in "Office of the Secretary, General Correspondence, 1946–1951," Box 25, AEC Papers (RG 326), NARA. See also B. Boskey, "'39: Not Shy, not Retiring," *Harvard Law Bulletin* 52:1 (2000). Available at http://www.law.harvard.edu/news/bulletin/2000/fall/closing_main.html.

31. A memorandum of understanding, executed on 14 August 1947, ruled out claims against the US government relating to "the use of patents or inventions" by Italians. B. Boskey, "Response to the Application of G. M. Giannini and Company, Inc.," AEC-PCB, Docket No. 2, 6 June 1946, 3–9 (copy in "Neutron Patent," Box 2, Folder 2, Carte Nuovo Amaldi, AMF).

32. Ibid. 10.

33. Ibid., 11–14.

34. Hector Holmes's opinion, 23 September 1952, Appendix A in R. Anderson, W. Smith, and W. Holton, "Memoranda and Documents as to Settlement, 6.12.1951," in "Patents G. M. Giannini & Co., Docket 2," Box 30, Folder 9, AEC Papers (RG 326), NARA.

35. R. A. Anderson, "Representative Cases That Might Come Before the PCB," March 1948, in Office of the Secretary, General Correspondence, 1946–1951, Box 6, AEC Papers (RG 326), NARA.

36. PCB Supplemental Report, 20 April 1948, in Office of the Secretary, General Correspondence, 1946–1951, Box 6, AEC Papers (RG 326), NARA.

37. Restrictions 5.B.i and 5.B.ii in Section 10 "Control of Information" of the Atomic Energy Act allowed the FBI to check and provide the AEC administrators with information on "character, association, and loyalty" of individuals. See Ungar, *FBI*, 90.

38. Wang, *American Science in the Age of Anxiety*, 95 and 219–20. See also Peter J. Westwick, "Secret Science: A Classified Community in the National Laboratories," *Minerva* 38 (2000): 367–77.

39. On Parnell Thomas accusations, see Richard G. Hewlett and Francis Duncan, *Atomic Shield: A History of the United States Atomic Energy Commission, Vol. 2, 1947–1952* (Berkeley: University of California Press, 1990), 89.

40. General Provision 80.4 and Hearing, Provision, 80.42.g in "AEC General Rules of Procedures on Applications for Determination of Reasonable Royalty Fee, Just Compensation, or Grant of Award for Patents, Inventions or Discoveries," *US Federal Register*, 8 May 1948, 2487–88.

41. If the case was not amicably settled, it would end up in the US Court of Claims first and the US Supreme Court afterwards. Bennett Boskey, "Inventions and the Atom," *Columbia Law Review* 50 (1950): 433–47.

42. Ibid., 442. Boskey has claimed that he did not remember whether or not the case of the slow neutron patent informed the writing of his paper: "My best recollection is that the article discussion was dealing with a hypothetical case, perfectly likely to arise, rather than any specific case." B. Boskey, private communication, 5 June 2006.

43. Personnel Security Review Board meeting, 16–18 September 1949, in "Office of General Counsel," Box 32, AEC Papers (RG326), NARA. Individual files for review are at the NARA, but they are classified.

44. E. Segrè to B. Pontecorvo, 31 December 1948, in "Scientific Correspondence, 1945–1950," B. Pontecorvo Papers, CAC.

45. Geoffrey Patterson to MI5 Director General, 13 November 1950, in "Security Action in the Case of Pontecorvo," KV 4/252, TNA.

46. Heilbron and Seidel, *Lawrence and His Laboratory*, 262. See also Thornton's obituary: Owen A. Chamberlain, Carl Helmholtz, David L. Judd, and E. Segrè, "Robert Lyste [sic] Thornton, Physics: Berkeley," Courtesy of Academic Senate, Berkeley Division, 1986 [Available online at http://content.cdlib.org/xtf/view?docId=hb767nb3z6&chunk.id=div00119&brand=calisphere&doc.view=entire_text].

47. Segrè, *A Mind Always in Motion*, 145 and 209.

48. Hewlett and Anderson, *The New World*, 647.
49. Theoaris, *Chasing Spies*, 86–87.
50. Security is the "rock" upon which lawyers "stumbled" while reviewing the atomic patent cases. Prosecutors could be aware of evidence produced in secret that could not be used by the defendants, who could only "work in the dark." Boskey, *Inventions and the Atom*, 443.
51. Reed to B.2, 10 November 1950, in "Bruno Maximovich PONTECORVO: Italian, British, Russian," KV 2/1889, TNA. The seven delegates were Warren C. Johnson, J. M. B. Kellogg, Willard F. Libby, R. L. Thornton, F. de Hoffmann, Cyril Smith, and B. Boskey. See "Minutes of Meeting, Fourth International Declassification Conference (Canada, UK and USA)," 9–12 February 1950, Secret, AB 6/635, TNA. In my Italian work *Il caso Pontecorvo* I formulated the hypothesis that it was Boskey and not Thornton who informed Cockcroft about Pontecorvo's security circumstances. This is because at the time I had no evidence showing that Thornton, aside from Boskey, held private talks with Cockcroft during the classification meeting. The conversation was about coordinating action between US and British atomic agencies in relation to royalty claims on the slow neutron patent. See also S. Turchetti, "A Contentious Business: Industrial Patents and the Production of Isotopes, 1930–1960," *Dynamis* 29 (2009): 191–218.
52. Geoffrey Patterson to MI5 Director General, 13 November 1950, in "Security Action in the case of Pontecorvo," KV 4/252, TNA.
53. G. Giannini to E. Segrè, 20 February 1950, in "Giannini," Box 2, Folder 2, Carte Nuovo Amaldi, AMF. In contrast with Segrè, Fermi, and Pontecorvo, Rasetti had not contributed to wartime nuclear programs. See Valeria Del Gamba, *Il ragazzo di Via Panisperna: L'avventurosa vita del fisico Franco Rasetti* (Turin: Bollati Boringhieri, 2007).
54. E. Fermi to Joseph Volpe, 29 June 1949, and J. Volpe to E. Fermi, 3 August 1949, Box 19, Folder 4, E. Fermi Papers, UCA. Fermi's intention of leaving the GAC is documented in E. Fermi to E. Segrè, 9 January 1950, Box 19, Folder 4, E. Fermi Papers, UCA.
55. 377th AEC meeting. Proposed Amendment of Section 12 (c) of the Atomic Energy Act of 1946, 3 February 1950, in Office of the Secretary, General Correspondence, 1946–1951, Box 6, AEC Papers (RG 326), NARA.
56. E. Fermi to L. Bernard, 15 April 1950, in "Neutron Patent," Box 2, Folder 2, Carte Nuovo Amaldi, AMF.
57. On the "minority report," see Peter Galison and Jeremy Bernstein, "In Any Light: Scientists and the Decision to Build the Superbomb, 1952–1954," *Historical Studies in the Physical Sciences* 19 (1989): 267–347, on 317 and 336. See also Maltese, *Enrico Fermi in America*, 348, and Rhodes, *Dark Sun*, 401–2.
58. L. Bernard to G. Giannini (copy to Segrè and Fermi), 5 April 1950, Box 19, Folder 4, E. Fermi Papers, UCA.
59. AEC, "Isotopes: A Three-Year summary of US Distribution, August 1949," in "Radioisotopi," Box 57, Carte della Direzione Generale Affari Economici, MAE. John Krige and Angela Creager have shown that US isotope exporting activities did not exclusively entail economic aspirations, but also political and diplomatic urgencies. See A. Creager, "Radioisotopes As Political Instruments,

1946–1953," *Dynamis* 29 (2009): 219–40; J. Krige, "The Politics of Phosphorous-32: A Cold War Fable based on Fact," in Doel and Söderqvist, eds., *The Historiography of Contemporary Science, Technology, and Medicine*, 153–71.

60. L. Bernard to G. Giannini (copy to Segrè), 5 May 1950, Box 19, Folder 4, E. Fermi Papers, UCA.

61. E. Segrè to E. Fermi, 10 December 1949, Box 11, Folder 13, E. Fermi Papers, UCA.

62. E. Fermi to G. Giannini, 18 August 1950, in "Neutron Patent," Box 2, Folder 2, Carte Nuovo Amaldi, AMF.

63. G. Giannini to E. Fermi, 24 August 1950, in "Neutron Patent," Box 2, Folder 2, Carte Nuovo Amaldi, AMF.

64. " . . . col governo (Giannini) potrebbe anche rompersi il collo." E. Segrè to E. Fermi, 15 November 1948, in "Emilio Segrè," Box 11, Folder 13, E. Fermi Papers, UCA.

65. "Scientists sues U.S. over Atom Patent," *New York Times*, 22 August 1950. "Il governo americano citato per danni per lo sfruttamento di un brevetto di Enrico Fermi," *Corriere della Sera*, 23 August 1950. "Una scoperta di Fermi che provoca un processo," *Il Messaggero*, 23 August 1950.

66. "Il governo USA condannato per truffa a Enrico Fermi. Lo scienziato italiano ha accusato il governo americano d'aver costruito atomiche senza pagare i diritti all'inventore," *L'Unità*, 23 August 1950.

67. Lawrence J. Bernard to Ross G. Gray, 12 November 1954 in Carte Nuovo Amaldi, AMF.

68. Montgomery-Hyde, *The Atom Spies*, 133.

69. Moorehead, *The Traitors*, 190. The emphasis is mine.

70. Ibid., 190. See also Montgomery-Hyde, *The Atom Spies*, 133–34.

71. H. Skinner, "Unpublished Book Review on Moorehead's The Traitors, 1953," in "Bruno Maximovich Pontecorvo: Italian, British, Russian," KV2/1891, TNA.

72. Fidecaro was one of the few researchers in the Institute of Physics on that day. "Probably Mario [Ageno] and I were the last physicists to talk to Bruno in the Western world." Fidecaro, "From Rome to Dubna," in Bilenky et al., eds., *B. Pontecorvo Scientific Works*, 475.

73. R. T. Reed, "The Case of Bruno Pontecorvo," in "Bruno Maximovich Pontecorvo: Italian, British, Russian," KV 2/1889, TNA.

74. R. T. Reed, "The Case of Bruno Pontecorvo," in "Bruno Maximovich Pontecorvo: Italian, British, Russian," KV 2/1889, TNA, 17.

75. Content of postcard sent by Pontecorvo to Harwell, 31 August 1950, text copied on ibid., 10.

76. Chapman Pincher, "Atom Man Knew Atom Spy," *Daily Express*, 24 October 1950. Montgomery Hyde, *The Atom Bomb Spies*, 133–38.

CHAPTER 5

1. Michael Perrin to Roger Makins, secret and guard, 9 November 1950, in "Defection to USSR of Dr. Pontecorvo," FO 371/84837, TNA.

2. R. T. Reed, "The Case of Bruno Pontecorvo," 13 in "Bruno Maximovich Pontecorvo: Italian, British, Russian," KV 2/1889, TNA.

3. H. Skinner to G. Pontecorvo, 10 October 1950, in "Scientific Correspondence, 1945–1950," B. Pontecorvo Papers, CAC. On Anna's interview: Reed's note, 3 November 1950, in "Bruno Maximovich Pontecorvo: Italian, British, Russian," KV 2/1888, TNA.

4. As reported in the articles "Atomic Expert Missing: Gone to Prague," *Manchester Guardian*, 21 October 1950; *Express* staff reporter, "Atom Man Flies Away," *Daily Express*, 21 October 1950; "Prof. Pontecorvo," *Manchester Guardian*, 22 October 1950; "Missing British Professor: Finnish Enquiries," *Manchester Guardian*, 21 October 1950.

5. Rhodes, *The Making of the Atom Bomb*, 296.

6. Aldrich, *The Hidden Hand*, 380–84.

7. Ibid., 382. On the evolution of the negotiations, see Paul, *Nuclear Rivals*, 103–66.

8. Even if this cooperation "might constitute a technical violation of the Atomic Energy Act." Ziegler and Jacobson, *Spying Without Spies*, 206.

9. Cabinet Office to British Joint Services Mission (BJSM), Emergency Top Secret Cypher Telegram, 20 October 1950, in "Defection to URSS of Dr. Pontecorvo," FO 371/84837, TNA. The BJSM was the overseas agency responsible for representing British interests in the context of the Anglo-American Combined Chiefs of Staff Board; which overviewed military coordination in the wartime alliance.

10. Ibid.

11. BJSM to Cabinet Office, 21 October 1950, and Cabinet Office to BJSM, 23 October 1950, in "Defection to USSR of Dr. Pontecorvo," FO 371/84837.

12. "Professor Pontecorvo Might Have Gained Valuable Information. Mr. Strauss: 'No Means of Holding Him Here.'" *Manchester Guardian*, 24 October 1950.

13. Ibid.

14. "Note on Captain Liddell's interview with the Prime Minister on 23.10.1950 regarding the case of PONTECORVO," top secret, 23 October 1950, "Disappearance of Bruno Pontecorvo from Harwell," PREM 8/1273, TNA.

15. R. Makins to M. Perrin, 9 November 1950, in "Defection to USSR of Dr. Pontecorvo," FO 371/84837.

16. "Scientific Civil Service," *Nature* 161 (1948): 195.

17. Top secret telegram from FO, 24 October 1950, in "Security Service Action in the Case of Bruno Pontecorvo," KV 4/242, TNA.

18. BJSM to Cabinet Office, emergency top secret cypher Telegram, 24 October 1950, in "Defection to USSR of Dr. Pontecorvo," FO 371/84837, TNA.

19. Ibid. See also JCAE, *Soviet Atomic Espionage*, 48.

20. BJSM to Cabinet Office, emergency top secret cypher telegram, 23 October 1950, in "Defection to USSR of Dr. Pontecorvo," FO 371/84837, TNA.

21. See quotation at the beginning of this chapter. Oliver Franks to Roger Makins, Secret, 2 November 1950, in "Security: Disappearance of Professor Bruno Pontecorvo," CAB 126/307, TNA.

22. The session is documented in "Professor Pontecorvo: No Doubt That He is in Russia," *Manchester Guardian*, 7 November 1950. We know that this information was supplied to Strauss from Perrin's remark that "the above paragraphs are extracts from a full brief on the Pontecorvo case which I got together for our Minister."

23. Michael Perrin to Roger Makins, secret and guard, 9 November 1950, in "Defection to USSR of Dr. Pontecorvo," FO 371/84837, TNA.

24. Ibid. This is the reason why the letter was marked as "secret and guard."

25. "Professor Pontecorvo: No Doubt That He Is in Russia," *Manchester Guardian*, 7 November 1950.

26. British Embassy to R. Makins, secret, 13 November 1950, in "Defection to USSR of Dr. Pontecorvo," FO 371/84837, TNA.

27. A. Kellas to Foreign Office, secret, 24 October 1950, in "Security Service Action in the Case of Bruno Pontecorvo," KV 4/242, TNA.

28. Foreign Office to A. Kellas, Helsinki, confidential, 24 October 1950, in "Disappearance of Dr. Bruno Pontecorvo in Finland," FO 371/86439, TNA.

29. E. Segrè to E. Amaldi, 21 October 1950, in "Neutron Patent," Box 2, Folder 2, Carte Nuovo Amaldi, AMF.

30. Ibid.

31. Segrè, *A Mind Always in Motion*, 237.

32. Alvarez's argument was, according to Segrè, that compensation should not be sought because "we would have paid more than a million apiece to avoid being in Italy during the war, and since the United States has taken us and kept us as guests in this country we should give recognition to the US by conferring to them whatever we could." E. Segrè to E. Fermi, 25 October 1950, in "Neutron Patent," Box 2, Folder 2, Carte Nuovo Amaldi, AMF.

33. Giannini press release, 25 October 1950 [copy in Box 19, Folder 7, E. Fermi Papers, UCA].

34. E. Segrè to E. Fermi, 25 October 1950, in "Neutron Patent," Box 2, Folder 2, Carte Nuovo Amaldi, AMF.

35. Segrè, *A Mind Always in Motion*, 237.

36. F. Rasetti to E. Segrè, 3 November 1950, in "Neutron Patent," Box 2, Folder 2, Carte Nuovo Amaldi, AMF.

37. Roger M. Anders, "The Rosenberg Case Revisited: The Greenglass Testimony and the Protection of Atomic Secrets," *The American Historical Review* 83:2 (1978): 388–400, on 396.

38. JCAE, *Soviet Atomic Espionage*, IV-2.

39. E. Fermi, "Statement about Bruno Pontecorvo," 13 March 1951, Box 15, Folder 12, E. Fermi Papers, UCA [copy in JCAE, *Soviet Atomic Espionage*, 39–40].

40. Ibid., 39–40.

41. Ibid., 2–6.

42. L. Bernard to F. Rasetti, 10 April 1951, in "Giannini", Box 2, Folder 1, Carte Nuovo Amaldi, AMF.

43. R. A. Anderson to L. J. Bernard, 14 August 1951, in "Giannini", Box 2, Folder 1, Carte Nuovo Amaldi, AMF.

44. Review of defenses set forth in the response of the OGC in application docket n.2, appendix C in R. Anderson, W. Smith, and W. Holton, "Memoranda and Documents as to Settlement, 6.12.1951," in "Patents G. M. Giannini & Co., Docket 2," Box 30, Folder 9, AEC Papers (RG 326), NARA.

45. Analysis of infringement by the commission and its predecessors, Manhattan District of Fermi et al. patent, appendix B in R. Anderson, W. Smith, and W. Holton, "Memoranda and Documents as to Settlement, 6.12.1951," in "Patents G. M. Giannini & Co., Docket 2," Box 30, Folder 9, AEC Papers (RG 326), NARA.

46. Hector Holmes's opinion, 23 September 1952, appendix A in R. Anderson, W. Smith, and W. Holton, "Memoranda and Documents as to Settlement, 6.12.1951," in "Patents G. M. Giannini & Co., Docket 2," Box 30, Folder 9, AEC Papers (RG 326), NARA.

47. This is the reason why two file series were opened at Scotland Yard. One, "Bruno Maximovich Pontecorvo: Italian, British, Russian," was the personal file containing details of the inquiry into his disappearance. The other, "Security Service Action in the Case of Bruno Pontecorvo, a Scientist at Harwell," contains details on the inquiry into the handling of information on Pontecorvo by the security services.

48. Jim Skardon to MI5, 21 October 1950, in "Security Service Action in the Case of Bruno Pontecorvo," KV 4/242, TNA.

49. M. S. Goodman and C. Pincher, "Clement Attlee, Percy Sillitoe and the Security Aspects of the Fuchs Case," *Contemporary British History* 19 (2005): 67–77.

50. G. T. D. Patterson to P. Sillitoe, MI5 Director General, top secret, 21 October 1950, in "Security Service Action in the Case of Bruno Pontecorvo," KV 4/242, TNA.

51. BJSM to Cabinet Office, emergency top secret cypher telegram, 23 October 1950, in "Defection to USSR of Dr. Pontecorvo," FO 371/84837, TNA.

52. "Note on Captain Liddell's Interview with the Prime Minister on 23.10.1950 Regarding the Case of PONTECORVO," top secret, 23 October 1950, in "Disappearance of Bruno Pontecorvo from Harwell," PREM 8/1273, TNA.

53. D.B., Security Service Aspects of Pontecorvo Case, 24 October 1950, in "Security Service Action in the Case of Bruno Pontecorvo," KV 4/242, TNA. On 1 January 1948 the CIA reported to the MI5 that Pontecorvo's file was NRA.

54. R. Hollis, "Action Taken by the Security Service in Connection with the Pontecorvo Case," 25 October 1950, in "Security Service Action in the Case of Bruno Pontecorvo," KV 4/242, TNA.

55. Cover letter from Guy Liddell, MI5, to D. H. F. Rickett, FO, 25 October 1950, in response to Hollis's report in "Security Service Action in the Case of Bruno Pontecorvo," KV 4/242, TNA.

56. J. Cimperman to R. T. Reed, 21 November 1950, in "Bruno Maximovich Pontecorvo: Italian, British, Russian," KV 2/1889, TNA.

57. G. T. D. Patterson to MI5 Director General, 27 October 1950 in "Security Service Action in the Case of Bruno Pontecorvo," KV 4/242, TNA.

58. DG Note, 2 November 1950, in "Security Service Action in the Case of Bruno Pontecorvo," KV 4/242, TNA.

59. Ibid.

60. R. Hollis to R. T. Reed, secret, 6 November 1950, in "Bruno Maximovich Pontecorvo: Italian, British, Russian," KV 2/1888, TNA.

61. AEC director of security division, J. A. Waters, to Carleton Shrugg, deputy general manager, 10 November 1950, in "AEC General Correspondence," Box 7, AEC Papers (RG 326), NARA.

62. Paul Jacob Pontecorvo to Massimo Pontecorvo, 14 November 1950, in "Bruno Maximovich Pontecorvo: Italian, British, Russian," KV 2/1888, TNA.

63. AEC director of security division, J. A. Waters, to Carleton Shrugg, deputy general manager, 10 November 1950, in "AEC General Correspondence," Box 7, AEC Papers (RG 326), NARA. On Amtorg Trading, see Ted Morgan, *Reds. McCarthyism in Twentieth-Century America* (New York: Random House, 2004), 122–29.

64. Francis Van Dusen (Scherbatskoy's lawyer) to Adrian Fischer, US Department of State, 12 March 1952, Box 1, Folder 5, S. A. Scherbatskoy Papers, SI.

65. S. A. Scherbatskoy to Passport Division, US Department of State, 9 November 1950, and S. A. Scherbatskoy to E. Roosevelt, 12 November 1950, Box 1, Folder 5, S. A. Scherbatskoy Papers, SI.

66. S. A. Scherbatskoy's passport in Box 1, Folder 5, S. A. Scherbatskoy Papers, SI.

67. S. A. Scherbatskoy to Francis L. Van Dusen, 11 March 1952, Box 1, Folder 5, S. A. Scherbatskoy Papers, SI.

68. Ibid.

69. Hartcup and Allibone, *Cockcroft and the Atom*, 144.

70. C. Pincher, "Atom Man Knew Atom Spy," *Daily Express*, 24 October 1950. Other headlines included "Atom Man Flies Away," 21 October 1950; "Atom Family in Russia," 22 October 1950; "Atom House Searched," 25 October 1950.

71. C. Pincher, "Atom Man Not Screened," *Daily Express*, 27 October 1950.

72. David Vincent, *The Culture of Secrecy: Britain, 1832–1998* (Oxford: Oxford University Press, 1998), 194–203. See also Peter Hennessy and Gail Brownfeld, "Britain's Cold War Security Purge: The Origins of Positive Vetting," *Historical Journal* 25 (1982): 965–74.

73. Ibid., 967. See also Aldrich, *The Hidden Hand*, 383.

74. "A-Men 'Lock' Themselves in Britain: Passports Given Up To Prove Loyalty," *Daily Mirror*, 26 October 1950; R. Bedford, "Is the Price of British Citizenship Too Cheap?" *Daily Mirror*, 27 October 1950.

75. C. Pincher, "Perturbed Men," *Sunday Express*, 6 November 1950. On Peierls's action, see Sabine Lee, "The Spy That Never Was," *Intelligence and National Security* 17, no.4 (2002): 77–99, on 95. Original copy of the Peierls statement and correspondence with other foreign-born scientists mobilized against Pincher's allegation is in the Rudolf Peierls Papers, Bodleian Library, Oxford.

76. Aldrich, *The Hidden Hand*, 384. Hennessy and Brownfeld, "Britain's Cold War Security Purge: The Origins of Positive Vetting," 969.

77. C. Pincher, "Pontecorvo: The Full Story," *Daily Express*, 26 February 1951.

78. In 1940 De Courcy played a key role in the British foreign secretary's failed attempt to sign a peace treaty with Nazi Germany. He was eventually discredited when he leaked information about Rudolf Hess's daring flight to Britain to British fascist sympathizers. Scott Newton, *Profits of Peace: The Political Economy of Anglo-American Appeasement* (Oxford: Clarendon Press, 1997), 172–75.

79. "De Courcy Link Missing Scientist to Russian War Plans: Says He Took Scheme for H-Bomb, Guided Missiles, Cosmic Ray Air Defense," *The Commonwealth*, 5 March 1951 [copy in "Bruno Maximovich Pontecorvo: Italian, British, Russian," KV 2/1890, TNA].

80. G. Patterson to P. Sillitoe, 17 April 1951, in "Bruno Maximovich Pontecorvo: Italian, British, Russian," KV 2/1888, TNA.

81. Giuliana Pontecorvo to Guido Pontecorvo, 26 October 1950, in "Bruno Maximovich Pontecorvo: Italian, British, Russian," KV 2/1892, TNA.

82. Reed's note, 8 November 1950, in "Bruno Maximovich Pontecorvo: Italian, British, Russian," KV 2/1888, TNA.

83. Massimo and Maria Pontecorvo to Guido Pontecorvo, 6 November 1950, in "Bruno Maximovich Pontecorvo: Italian, British, Russian," KV 2/1888, TNA.

84. R. T. Reed, "The Case of Bruno Pontecorvo," in "Bruno Maximovich Pontecorvo: Italian, British, Russian," KV 2/1889, TNA, 13. The comments are in P. Sillitoe to Coloner Warren, secret, 7 November 1950, in "Bruno Maximovich Pontecorvo: Italian, British, Russian," KV 2/1888, TNA.

85. R. T. Reed, "The Case of Bruno Pontecorvo," in "Bruno Maximovich Pontecorvo: Italian, British, Russian," KV 2/1889, TNA, 14. On Giovanni David's interview: P. Sillitoe to Coloner Warren, secret, 7 November 1950, in "Bruno Maximovich Pontecorvo: Italian, British, Russian," KV 2/1888, TNA.

86. W. B. Mann's report, 31 October 1950, in "Bruno Maximovich Pontecorvo: Italian, British, Russian," KV 2/1888, TNA. On Mann's activities, see Goodman, *Spying on the Nuclear Bear*, 17.

87. Reed's note on information from MI6, 16 November 1950, in "Bruno Maximovich Pontecorvo: Italian, British, Russian," KV 2/1889, TNA.

88. Reed's note on information from Arnold, 16 November 1950, in "Bruno Maximovich Pontecorvo: Italian, British, Russian," KV 2/1889, TNA.

89. *Evening News*, 16 November 1950.

90. Reed's note, 16 November 1950, in "Bruno Maximovich Pontecorvo: Italian, British, Russian," KV 2/1889, TNA.

91. Moorhead's questionnaire, 27 September 1951, in "Bruno Maximovich Pontecorvo: Italian, British, Russian," KV 2/1890, TNA. We know about Waters's activities from correspondence between Italy's Carabinieri and the Ministry of Foreign Affairs. Carabinieri to Direzione Generale Affari Pubblici, segreto, 8 December 1950, in "Pontecorvo," Folder 480, Fondo Cassaforte, DGAP, MAE.

92. Reed, "The Case of Bruno Pontecorvo," in "Bruno Maximovich Pontecorvo: Italian, British, Russian," KV 2/1889, TNA, 14.

93. Reed's note on Pincher's article, 26 February 1951, in "Bruno Maximovich Pontecorvo: Italian, British, Russian," KV 2/1889, TNA.

94. Reed's note, 18 September 1951, and Moorhead's questionnaire, 27 September 1951, in "Bruno Maximovich Pontecorvo: Italian, British, Russian," KV 2/1889-90, TNA. On Sillitoe's propaganda effort, see N. West, "Fiction, Faction and Intelligence," in L.V. Scott and P. D. Jackson, eds., *Understanding Intelligence in the Twenty-First Century: Journeys in the Shadows* (New York: Routledge, 2004), 123-25. On Moorhead, see Ann Moyal, *Alan Moorehead: A Rediscovery* (Canberra: National Library of Australia, 2005).

95. Moorehead, *The Traitors*, 171. See also A. Moorehead, "Bruno Pontecorvo," *The Sunday Times*, 29 June 1952.

96. Rebecca West, *The Meaning of Treason* (London: Virago, 1965), 240.

97. Gibbs, "British and American Counter-intelligence and the Atom Spies," chapter 5.

98. H. Skinner, "Unpublished Book Review on Moorehead's *The Traitors*, 1953," in "Bruno Maximovich Pontecorvo: Italian, British, Russian," KV2/1891, TNA.

99. H. Arnold to J. C. Robertson, MI5, 8 October 1952, in "Bruno Maximovich Pontecorvo: Italian, British, Russian," KV 2/1891, TNA.

100. "Soviet Seize Pontecorvo as 'British Spy,'" *Sunday Pictorial*, 11 November 1951.

101. C. P. Snow, *The New Men*, (London, 1954). John Halperin, *C.P.Snow: An Oral Biography* (Brighton: St. Martin's Press, 1983), 163.

102. Ian Fleming, *Casino Royale* (London: Jonathan Cape, 1953).

103. See chapter 3.

104. Giuliana Tabet to Guido Pontecorvo, 26 October 1950, in "Bruno Maximovich Pontecorvo: Italian, British, Russian," KV 2/1889, TNA.

105. MI6 report, 16 January 1951, in "Bruno Maximovich Pontecorvo: Italian, British, Russian," KV 2/1889, TNA.

106. MI6 (Broadway) to Reed, 27 November 1950, in "Bruno Maximovich Pontecorvo: Italian, British, Russian," KV 2/1889, TNA.

107. Copy of letter from MI6, 7 December 1954, in "Lew Kowarski: French," KV 2/2591, TNA.

108. W. B. Mann's report, 31 October 1950, in "Bruno Maximovich Pontecorvo: Italian, British, Russian," KV 2/1888, TNA.

109. Reed's note, 16 November 1950, in "Bruno Maximovich Pontecorvo: Italian, British, Russian," KV 2/1889, TNA.

110. H. Skinner, "Unpublished Book Review on Moorehead's *The Traitors*, 1953" in "Bruno Maximovich Pontecorvo: Italian, British, Russian," KV2/1891, TNA.

111. Chancery, British Embassy, Moscow, to permanent under-secretary's department, Foreign Office, 26 February 1951, top secret, in "Bruno Maximovich Pontecorvo: Italian, British, Russian," KV2/1889, TNA.

CHAPTER 6

1. All citations from "Dr. Pontecorvo's Atomic Post in Moscow. 'Given Refuge' in Soviet Union: Work for Industry," *Manchester Guardian*, 1 March

1955, aside from "ashamed [. . .]," from "Pontecorvo Says: 'I Work for Russia,'" *Daily Mirror*, 1 March 1955.

2. Giuseppe Avolio, *Emilio Sereni: Ortodossia politica e genialità scientifica* (Rome: Agra, 1999).

3. For an overview of this political transition, see Paul Ginsborg, *A History of Contemporary Italy: Society and Politics, 1943–1988* (New York: Palgrave, 2003), chaps. 3 and 6. On the PP history, see Ruggero Giacomini, *I partigiani della pace: Il movimento pacifista in Italia e nel mondo negli anni della Guerra Fredda* (Milan: Vangelista, 1984).

4. Ibid., 7–20. See also Mafai, *Il lungo freddo*, 26.

5. Lawrence Wittner, *One World or None: A History of World Nuclear Disarmament through 1953*, Vol. 1 (Stanford, CA: Stanford University Press, 1993), 191–210.

6. Pinault, *Joliot-Curie*, 460–61.

7. Wiener, *One World or None*, 183. Wiener suggests that the appeal branded as war criminal any government "that first used" atomic weapons, thus including the US bombing of Hiroshima and Nagasaki. But an Italian version of the appeal uses the future tense: "any government . . . that *will use*." "Per un plebiscito nazionale di pace contro le armi atomiche," Copy in "PCI Executive Committee, Minutes of Meeting, 1950," Archivio PCI, Box 190, FIG.

8. "Noi sappiamo che la scienza è un fattore dominante del nostro tempo e un elemento fondamentale non solo nella stabilizzazione della pace dei popoli, ma anche per un sempre più progressivo miglioramento del loro tenore di vita. Troppo spesso le applicazioni della scienza sono state poste al servizio della guerra, al servizio di una minoranza che ne ha tratto profitto e ne trae profitti illeciti. Gli uomini di scienza non possono essere complici di quei regimi che sfruttano la scienza per scopi ingiusti." From an article in the PCI newspaper *L'Unità*, 1 November 1949 cit. in Giacomini, *Partigiani della pace*, 82.

9. Patrick M. S. Blackett to B. Pontecorvo, 31 March 1949, in "Scientific Correspondence, 1945–1950," B. Pontecorvo Papers, CAC.

10. Mary Jo Nye, *Blackett: Physics, War and Politics in the Twentieth Century* (Cambridge, MA: Harvard University Press, 2004), 89–93.

11. Albert Einstein to B. Pontecorvo, 5 May 1947, in "Scientific Correspondence, 1945–1950," B. Pontecorvo Papers, CAC.

12. For a coverage of this confrontation see Hewlett and Duncan, *Atomic Shield*, 265–67.

13. Vladislav M. Zubok, "Stalin and the Nuclear Age," in John Gaddis, Philip Gordon, Ernest May, and Jonathan Rosenberg, *Cold War Statesmen Confront the Bomb. Nuclear Diplomacy since 1945* (Oxford: Oxford University Press, 1999), 39–61.

14. Wiener, *One World or None*, 289.

15. Piano di Lavoro per la campagna di pace contro le armi atomiche, 29 April 1950, in "PCI Executive Committee, Minutes of Meeting, 1950," Archivio PCI, Box 190, FIG.

16. Giacomini, *I partigiani della pace*, 102.

17. Wittner, *One World or None*, 193.

18. R. T. Reed, The Case of Bruno Pontecorvo, 14, in "Bruno Maximovich Pontecorvo: Italian, British, Russian," KV 2/1889, TNA.

19. As the PCI newspaper *L'Unità* claimed. See chapter 4.

20. Elena Aga-Rossi and Victor Zaslavski, *Togliatti e Stalin: Il PCI e la politica estera Staliniana negli archivi di Mosca* (Bologna: Il Mulino, 1997), 262. See also Massimo Caprara, *L'Inchiostro verde di Togliatti* (Milan: Simonelli, 1996), 89.

21. As documented in Gianni Donno, *La Gladio rossa del PCI, 1945–1967* (Soveria Mannelli, Catanzaro: Rubbettino, 2001), 67.

22. Anna di Biagio, "The Marshall Plan and the Founding of Cominform, June–September 1947," in F. Gori and S. Pons, eds., *The Soviet Union and Europe in the Cold War, 1943–1953* (Milan: Feltrinelli, 1996), 208–45.

23. The financial details of this operation are in Natalija I. Egorova, "Stalin's Foreign Policy and the Cominform, 1947–1953," in Gori and Pons, eds., *The Soviet Union and Europe in the Cold War*, 200. See also Wittner, *One World or None*, 171.

24. Giovanni Gozzini and Renzo Martinelli, *Storia del PCI: Dall'attentato di Togliatti all'VIII Congresso* (Turin: Einaudi, 1998), 158–69.

25. Sereni's speech, 28–29 September 1950, in "PCI Executive Committee, Minutes of Meeting, 1950," Archivio PCI, Box 190, FIG.

26. Gozzini e Martinelli, *Storia del PCI* , 151.

27. On expatriates' activities in Prague: Rocco Turi, *Gladio rossa. Una catena di complotti e delitti, dal dopoguerra al caso Moro* (Venice, Marsilio: 2004). See also Giuseppe Fiori, *Uomini ex: Lo strano destino di un gruppo di comunisti italiani* (Turin: Einaudi, 1993).

28. West, *The Meaning of Treason*, 243.

29. R. T. Reed, The Case of Bruno Pontecorvo, 17, in "Bruno Maximovich Pontecorvo: Italian, British, Russian," KV 2/1889, TNA.

30. Ibid., 18.

31. Bedrik Geminder (Czech Communist Party) to Rudolf Slánsky (party chairman), 12 August 1949, quoted in Gozzini and Martinelli, *Storia del PCI*, 154. On the "Moscow Gold," see Valerio Riva, *Oro di Mosca: I finanziamenti sovietici al PCI* (Milan: Mondadori, 1999), and Giovanni Cervetti, *L'Oro di Mosca. La verità sui finanziamenti sovietici al PCI raccontata dal diretto protagonista* (Milan, Baldini & Castoldi, 2000). A commentary on these works is in Richard Drake, "The Soviet Dimension of Italian Communism," *Journal of Cold War Studies* 6, no. 3 (2004): 115–19.

32. Massimo Caprara, "Seniga: Quando il PCI nel 1954 provò a impacchettarmi," *Corriere della Sera*, 17 August 1992.

33. Maurizio Caprara, "Una delle operazioni più riuscite è stata la fuga di Bruno Pontecorvo," *Corriere della Sera*, 18 June 1992. On the PCI clandestine operations, see also Maurizio Caprara, *Lavoro riservato: I cassetti segreti del PCI* (Milan, Feltrinelli: 1996).

34. "Atomic Expert Missing. Gone to Prague," *Manchester Guardian*, 21 October 1950.

35. Minute 291, 23 April 1953, in "Bruno Maximovich Pontecorvo: Italian, British, Russian," KV 2/1891, TNA.

36. J. C. Robertson to R. T. Reed, 2 April 1953, in "Bruno Maximovich Pontecorvo: Italian, British, Russian," KV 2/1891, TNA.

37. Ibid.

38. Clara Sereni, private communication, 10 May 2007.

39. William Hood, *Mole: The True Story of the First Russian Intelligence Officer Recruited by the CIA* (London: Weidenfeld and Nicholson, 1982). In the same year the former KGB official Peter Deriabin defected as well. P. Deriabin (with Frank Gibney), *KGB: The Secret World* (New York: Time Inc., 1959).

40. Montgomery Hyde, *The Atom Bomb Spies*, 133; Moorhead, *The Traitors*, 189.

41. Sereni's diary, 1950, in "Carte di Emilio Sereni," FIG.

42. Maurizio Caprara, "Una delle operazioni più riuscite è stata la fuga di Bruno Pontecorvo," *Corriere della Sera*, 18 June 1992.

43. Sereni's Diary, 1950, in "Carte di Emilio Sereni," FIG.

44. R. T. Reed, The Case of Bruno Pontecorvo, 17, in "Bruno Maximovich Pontecorvo: Italian, British, Russian," KV 2/1889, TNA.

45. Maurizio Caprara, "Una delle operazioni più riuscite è stata la fuga di Bruno Pontecorvo," *Corriere della Sera*, 18 June 1992.

46. Sereni's Diary, 1950, in "Carte di Emilio Sereni," FIG.

47. "Le Potenze atlantiche hanno deciso di preparare la guerra atomica. Per loro le armi nucleari sarebbero armi legittime. Nei quattro anni in cui ho vissuto nell'URSS ho potuto convincermi che il popolo sovietico, tutto il popolo sovietico, vuole la pace e che il Governo dell'URSS prende tutte le misure possibili per impedire la guerra. Vorrei rivolgermi a tutti gli uomini onesti, e in particolare agli scienziati, ai fisici che ho conosciuto, con cui ho lavorato, di cui sono amico, che stimo, uomini certo di grande intelligenza, per scongiurarli di prendere posizione. Oggi non si può certo restare in disparte," *La Nave*, undated press cutting in "Carte dei partigiani della pace," FIG.

48. Giacomini, *I partigiani della pace*, 170–74.

49. PCI Executive Committee, Minutes of Meeting, 6 December 1950, PCI Archive, Box 190, FIG.

50. In November 1951 the Vienna meeting of the World Peace Council reiterated these points. Wiener, *One World or None*, 185.

51. Ibid., 188.

52. Gozzini and Martinelli, *Storia del PCI*, 179.

53. Giacomini, *I partigiani della pace*, 276.

54. Egorova, "Stalin's Foreign Policy and the Cominform," 203.

55. Russell's proposal was also known as the Russell-Einstein Manifesto. On the East-West collaboration, see Matthew Evangelista, *Unarmed Forces. The Transnational Movement to End the Cold War* (Ithaca:, NY Cornell University Press, 1999), 25–38.

56. On the implications of the Geneva conference, see John Krige, "Atoms for Peace," in J. Krige and K. Henrik Barth, eds., *Global Power Knowledge: Science, Technology, and International Affairs, Osiris Issue 21* (Chicago: University of Chicago Press, 2006), 174–80.

57. "Pontecorvo on His Work in Russia: Denial of Military Applications," 5, in *The Times*, 5 March 1955.

58. Ibid.
59. The most successful popularizing account of this interpretation is given in West, *The Meaning of Treason*, 238–47.
60. Proposal of Major Lloyd George, 9 March 1955, in "Citizens Deprived of Citizenship, Including Bruno Pontecorvo (1950)," FO 372/3790, TNA.
61. B. Pontecorvo to Comrade Slavin, 20 April 1955, in "Citizens Deprived of Citizenship, Including Bruno Pontecorvo (1950)," FO 372/3790, TNA.
62. Segrè, *A Mind Always in Motion*, 262–63.
63. PCI Executive Committee, Minutes of Meeting, 1950, in PCI Archive, Box 266, FIG.
64. Giacomini, *I partigiani della pace*, 276.
65. "Tempestosa seduta al Senato per una provocazione comunista," *Il Quotidiano*, 6 March 1955.
66. Vittorio Zincone, "Suvarov e Sereni," *Il Tempo*, 8 March 1955.
67. "Lì [Pontecorvo] ebbe un incarico nei gabinetti scientifici di Joliot-Curie . . . [e si incontrò anche con] quel senatore comunista che proprio in questi giorni si è fatto ricordare alla nazione come disertore e come organizzatore di attentati alle spalle dei nostri soldati in guerra, Emilio Sereni, cugino e compagno parigino del traditore Pontecorvo!" "Pontecorvo fuoriuscito a Parigi collaborava con l'attivista Emilio Sereni," *Il Tempo*, 11 March 1955.
68. Sereni's diary, 1955 and 1976, Carte di Emilio Sereni, FIG.

CHAPTER 7

1. R. T. Reed, The Case of Bruno Pontecorvo, 18 in "Bruno Maximovich Pontecorvo: Italian, British, Russian," KV 2/1889, TNA.
2. For more details see V. D. Dzhelepov, "The Genius of Bruno Pontecorvo," in Bilenky et al., eds., *B. Pontecorvo Scientific Works*, 487–93 (available online: http://pontecorvo.jinr.ru/dzhelepov.html). See also Pontecorvo, "Pages in the Development of Neutrino Physics," 1087–1108.
3. Mafai, *Il lungo freddo*, 206.
4. Ibid., 257.
5. Dzhelepov, "The Genius of Bruno Pontecorvo," 488.
6. Zubok, "Stalin and the Nuclear Age," 45–49.
7. On these aspects, see Paul R. Josephson, *Red Atom: Russia's Nuclear Power Program from Stalin to Today* (Pittsburgh: University of Pittsburgh Press, 2005).
8. Ethan Pollock, *Stalin and the Soviet Science Wars* (Princeton, NJ: Princeton University Press, 2006), 91.
9. Mafai, *Il lungo freddo*, 189.
10. Holloway, *Stalin and the Bomb*, 53; Kojevnikov, *Stalin's Great Science*, 132.
11. Kojevnikov, *Stalin's Great Science*, 150–58. See also Rhodes, *Dark Sun*, 40 and 150–53.
12. Rhodes, *Dark Sun*, 306.
13. Holloway, *Stalin and the Bomb*, 304; Kojevnikov, *Stalin's Great Science*, 155.
14. Holloway, *Stalin and the Bomb*, 144.

15. Mafai, *Il lungo freddo*, 194.
16. L. D. Ryabev, ed., *Atomic Project of USSR: Documents and Materials, Vol. 1, 1938–1945, Part II* (Moscow: Nauka, Fizmatlit, 1998), 464. In Russian. I am thankful to Alexei Kojevnikov for pointing me toward this document.
17. Holloway, *Stalin and the Bomb*, 105.
18. Ioffe's recollections mix facts and fiction. It is very unlikely, for instance, that Pontecorvo was asked "how to prepare uranium blocks," because he was an expert in heavy water reactors; it is also unlikely that this is what Russian experts were after, since they had already successfully manufactured graphite reactors. The memoir was originally published in Russian in the magazine *Novy mir* in 1999. The extended version, translated by Forrest Rhoads, Milla Grin, and Mort Hamermesh, was published as Boris L. Ioffe, "A Top Secret Assignment," in M. Shifman, ed., *At the Frontiers of Particle Physics: Handbook of QCD, Boris Ioffe Festschrift*, vol. 1 (Hackensack, NJ: World Scientific, 2001), 18–64. Details on Pontecorvo on 33–35.
19. JCAE, *Soviet Atomic Espionage*, v and 2.
20. G. Hanna and B. Pontecorvo, "The β-spectrum of H_3," *Physical Review* 75 (1949): 985–86. See also Pontecorvo, "Pages in the Development of Neutrino Physics," 1096.
21. Anthony Evans, *Tritium and its Compounds* (London: Butterworths, 1966), 218.
22. Rhodes, *Dark Sun*, 335.
23. Steven J. Zaloga, *The Kremlin's Nuclear Sword: The Rise and Fall of Russia's Strategic Nuclear Forces, 1945–2000* (Washington: Smithsonian Institution Press, 2002), 8. Kojevnikov, Rhodes, and Holloway also stress the existence of this deficiency.
24. Uranium prospecting studies were pioneered in the USSR by Vladimir Vernadski, who was also the founder of Saint Petersburg's Radium Institute. Kojevnikov, *Stalin's Great Science*, 120. Rhodes, *Dark Sun*, 27–28.
25. Ibid., 213. See also Zaloga, *The Kremlin's Nuclear Sword*, 6, and Kojevnikov, *Stalin's Great Science*, 148.
26. Holloway, *Stalin and the Bomb*, 102.
27. Elmerich, *Gathering Rare Ores*, 174. On the intelligence reports, see chapter 2.
28. Ibid., 93.
29. Zaloga, *Kremlin's Nuclear Sword*, 9.
30. Kojevnikov, *Stalin's Great Science*, 148.
31. "New Czech Atom Plants: Uranium Find," *Daily Telegraph*, 21 March 1951.
32. Major General, General Officer Commanding-in-Chief, British Element, Trieste Force to BOX 500, Parliament Street, Secret, 13 April 1951, in "Bruno Maximovich Pontecorvo: Italian, British, Russian," KV 2/1890, TNA.
33. Metamorphic rocks that were originally sediments and volcanic tuffs (ash deposits) intruded in masses of granite before the Cambrian Age. Nininger, *Minerals for Atomic Energy*, 43–48.
34. From 1955 Central Administration of Research and Mining of Radioactive Raw Materials (in Czech ÚSVTRS, with headquarters in Příbram); from

1967 as Czechslovak Uranium Industry (ČSUP); and following the dissolution of Soviet Union, from 1989, as DIAMO. Information from DIAMO website: "History of the Company" at http://www.diamo.cz/en/history.

35. Robert J. Cathro, "Uranium Production from East Germany, Czechoslovakia and Eldorado, Northwest Territories after 1945," *CIM Bulletin* 98 (2005): 67–70. CIM is the Canadian Institute of Metallurgy.

36. Ibid.

37. On these similarities between Ferghana Valley and the Colorado Plateau, see Zaloga, *Kremlin's Nuclear Sword*, 284; Nininger, *Minerals for Atomic Energy*, 67.

38. Reed's minute, 30 January 1952, in "Bruno Maximovich Pontecorvo: Italian, British, Russian," KV 2/1891, TNA. K.M.T. is the Kuomintang, i.e., the Taipei-based government of the Nationalist Republic of China.

39. "Atomic Centre in Sinkiang: Pontecorvo There," *Manchester Guardian*, 10 November 1951. Patrick Maitland, "Central Asian Mystery: Soviet Activity in Chinese Desert Province," *The Scotsman*, 4 June 1951. Maitland was a British journalist, economic geographer, and conservative MP whose area of expertise was energy policy. See Andrew Roth, "The Earl of Lauderdale: Journalist, Tory MP and rightwing peer who was a 'Suez rebel,'" *Guardian*, 8 December 2008.

40. SO (I) to Admiralty, 10 December 1951, Confidential Message (By Secure Means) in "Bruno Maximovich Pontecorvo: Italian, British, Russian," KV 2/1890, TNA.

41. Li Xuetong, personal communication, 11 November 2009. See also Li Xuetong, *The Chronicle of Dr. Weng Wenhao* (Jinan: Shandong Education Press, 2005, in Chinese) and Grace Y. Shen, "Xuetong Li's Weng Wenhao nian pu" (book review), *ISIS* 99 (2008): 874–75.

42. SLO (Pakistan) to MI5 Director-General, Secret, 27 February 1952, in "Bruno Maximovich Pontecorvo: Italian, British, Russian," KV 2/1891, TNA.

43. Reed's memorandum, 16 March 1953, in "Bruno Maximovich Pontecorvo: Italian, British, Russian," KV 2/1891, TNA. Ioffe, "A Top Secret Assignment," 35.

44. F. F. Segesman, "Well Logging Method," *Geophysics* 45 (1980): 1667–84. Donald D. Snyder and David B. Fleming, "Well Logging: A 25-year Perspective," *Geophysics* 50 (1985): 2506–29.

45. Allaud and Martin, *Schlumberger*, 304–7.

46. See chapter 2.

47. As shown in a letter: Henry Arnold to [unknown], 6 April 1951, in "Bruno Maximovich Pontecorvo: Italian, British, Russian," KV 2/1890, TNA, whose recipient is unknown and content partly undisclosed due to Sec. 3 (4) of the UK Public Records Act. On Reed's interest, see chapter 5.

48. The original letters are not available, but the content of Pontecorvo's proposal is clear from the replies he received. J. Neufeld to B. Pontecorvo, 23 January 1950 and 10 April 1950; S. A. Scherbatskoy to B. Pontecorvo, 10 April 1950, in "Scientific Correspondence, 1945–1950," B. Pontecorvo Papers, CAC.

49. S. A. Scherbatskoy, "Nuclear Well Logging," US 2,648,012 filed 5 October 1949, issued 4 August 1953, and its "continuation in part" patent: S. A.

Scherbatskoy, "Nuclear Well Logging," US 3,869,608, filed 1 July 1953, issued 4 March 1975.

50. S. A. Scherbatskoy to J. H. Castel, PGAC, 29 November 1949, copy in "Scientific Correspondence, 1945–1950," B. Pontecorvo Papers, CAC.

51. J. Neufeld to B. Pontecorvo, 23 January 1950, in "Scientific Correspondence, 1945–1950," B. Pontecorvo Papers, CAC.

52. H. Skinner, "Unpublished Book Review on Moorehead's *The Traitors*, 1953" in "Bruno Maximovich Pontecorvo: Italian, British, Russian," KV2/1891, TNA.

53. B. Yerozolimski, personal communication, 5 November 2009. On Flerov's contribution, see A. P. Aleksandrov, Ya. V. Zel'dovich, K. A. Petrzhak, and I. M. Frank, "Georgii Nikolaevich Flerov," *Uspekhi Fiziki Nauk* 109 (1973): 617–19, on 617.

54. B. Yerozolimski, personal communication, 5 November 2009. On the history of the operation model, see A. I. Kedrov et al., "Neutron-Gamma Well Logging Apparatus With More Sensitivity for Chlorine," *Applied. Radiation and Isotopes* 48 (1997): 1647–48.

55. Allaud and Martin, *Schlumberger*, 305–6.

56. In 1973, Pontecorvo and Frank jointly wrote the eulogy for their colleague Fedor L. Shapiro. See B. M. Pontecorvo and I. M. Frank, "Fedor L'Vovich Shapiro," *Uspekhi Fiziki Nauk* 16 (1973): 295–97.

57. Yerozolimski, personal communication, 5 November 2009. See also B. G. Erozolimski, A. S. Shkol'nikov, and A. I. Isakov, "Application of Neutron Pulse Sources for Investigation in Oil Wells," *Atomic Energy* 9 (1961): 144–45. A review on the use of micro-particle accelerators for oil prospecting is in V. M. Zhaporezhetz and E. M. Filippov, "The Use of Accelerators of Charged Particles in Investigating Bore-Holes by the Methods of Radioactive Logging," in N. Rast, ed., *Applied Geophysics. USSR* (London: Pergamon Press, 1962), 397–421.

58. Yerozolimski, personal communication, 10 November 2009.

59. Yerozolimski, personal communication, 13 November 2009.

60. Ioffe, "A Top Secret Assignment," 34.

61. R. E. Fearon and J. M. Thayer, "Well-Logging Radiation Sources," US 3.071.690, issued January 1, 1963. Its importance is stated in V. M. Zhaporezhetz and E. M. Filippov, "The Use of Accelerators of Charged Particles in Investigating Bore-Holes by the Methods of Radioactive Logging," 397.

62. Nicholas Rast, introduction to Rast, ed., *Applied Geophysics. USSR*, 7.

63. S. I. Savosin and V. I. Sinitsyn, "Application of Methods of Nuclear Geophysics in Ore Prospecting, Exploration and Development," *Atomic Energy* 18 (1965): 81–84. This is because in old wells, finding the "oil-water" contact point can be particularly useful. Yerozolimski, personal communication, 13 November 2009.

64. D. L. Spencer, "The Role of Oil in Soviet Foreign Economic Policy," *American Journal of Economics and Sociology* 25, no. 1 (1966), 91–107.

65. Debbie Mitchell, "Fort Worth Scientist Linked to Defector," *Fort Worth Star Telegram*, 1976, copy in in Box 1, Folder 3, S. A. Scherbatskoy's Papers, SI. As for Pontecorvo's royalties, some, like Montgomery Hyde, have argued

that Pontecorvo received his share of royalties (Montgomery Hyde, *Atom Bomb Spies*, 142). Others, like Amaldi, have claimed that he never did. See E. Amaldi (with P. Angela), *Intervista sulla materia dal nucleo alle galassie* (Bari: Laterza, 1980), 44.

66. David J. Dallin, *Soviet Espionage* (New Haven and London: Yale University Press, 1955), 465.

67. Baker White, *Soviet Spy System*, 26–27.

68. Montgomery Hyde, *Atom Bomb Spies*, 130.

69. Williams, *Klaus Fuchs*, 136–50.

70. The "Cambridge five" spy ring also included Anthony Blunt and, allegedly, John Cairncross.

71. Pincher, *Too Secret Too Long*, 151. Pincher, *Their Trade is Treachery*, 49–50. See also Costello, *Masks of Treachery*, 533.

72. Pincher, *Too Secret Too Long*, 152.

73. Costello, *Masks of Treachery*, 533. In a personal communication to the author, Pincher has not confirmed such a claim: "His [Wright] purpose was to help me to write the first inside story of MI5 and to promote his belief that the Russians had penetrated it. [. . .] I cannot remember whether or not Wright and I talked about Pontecorvo." Pincher, personal communication, 19 November 2002.

74. Peter Wright (with Paul Greengrass), *Spycatcher* (Melbourne: Heinmann, 1987). The accusations against Hollis were unfounded.

75. C. Andrew and O. Gordievsky, *KGB*, 312–13.

76. Pavel Sudoplatov, *Special Tasks*, 174–76 and 180–82.

77. Zubok, "Stalin and the Nuclear Age," 44. See also Rhodes, *Dark Sun*, 81.

78. A similar claim appears also in Baggott, *Atomic*, 184. But Pontecorvo and Fermi were not part of any antifascist network operating in Rome in the 1930s, for the reasons explained in chapter 1.

79. Haynes and Klehr, *Venona*, 314. These findings had already become available in 1997, thanks to the work of Joseph Albright and Marcia Kunstel.

80. "Let's admit it: this is only a theory–one that will need a lot of concrete and reinforcing rods before it can stand the test of history." Joseph Albright and Marcia Kunstel, "Did the United States Have *Any* Secrets?" *Bulletin of the Atomic Scientists* 54 (1998): 50–53.

81. Henry Arnold, "Summary of Pontecorvo's Journeys during Wartime," 9 November 1950, in "Bruno Maximovich PONTECORVO: Italian, British, Russian," KV 2/1888, TNA.

CHAPTER 8

1. S. H. Paul, *Nuclear Rivals*, 198

2. Rhodes, *Dark Sun*, 162. See also D. Kaiser, "The Atomic Secret in Red Hands? American Suspicions of Theoretical Physicists during the Early Cold War," 30, and Weart, "Secrecy, Simultaneous Discovery and the Theory of Nuclear Reactors," 1049–60.

3. Goodman and Pincher, "Clement Attlee, Percy Sillitoe and the Security Aspects of the Fuchs Case," 66–77.

4. See, for instance, Wang, *American Science in the Age of Anxiety*.

5. This was the claim put forward by Pontecorvo and reported in books that resulted from interviews with him, such as Mafai, *Il lungo freddo*. The physicist Giorgio Salvini made the claim even more strongly: "The very fact is that he was a convinced believer in communism as inspiring and ruling over the world, as one can be convinced of a religious 'credo.' I use this expression for he himself told me that he was a firm believer in the 'verb' [meaning 'holy word'] as coming from Stalin, Molotov, and the others. This he told me in 1990 while staying in Rome, very often in my studio. He also told me that by this time he was no longer sure and convinced, but that it took time to discover that this 'credo' was inconsistent and even wrong economically and ethically." G. Salvini, "Bruno Pontecorvo, a Great Physicist, a Great Man," in A. Bonetti, I. Guidi, and B. Monteleoni, eds., *Cosmic Ray, Particle and Astrophysics: A Conference in Honor of Giuseppe Occhialini, Bruno Pontecorvo and Bruno Rossi* (Rome: Accademia Nazionale dei Lincei, 1997), 5. As a disclaimer I can only reiterate the point that none of the extensive archival documentation I have managed to look into during the past ten years shows any clear political or religious belief.

6. For an overview of recent cases, which also shows how much we have yet to learn about key episodes in the nuclear age, see J. T. Richelson, *Spying on the Bomb* (New York: W. W. Norton & Co., 2007).

7. An important and more recent case is discussed in G. Hecht, "The Power of Nuclear Things," *Technology and Culture* 51 (2010), 1–30.

Bibliography

Aga-Rossi, Elena, and Victor Zaslavsky. *Togliatti e Stalin: Il PCI e la Politica Estera Staliniana negli Archivi di Mosca.* Bologna: Il Mulino, 1997.

Albright, Joseph, and Marcia Kunstel. "Did the United States Have *Any* Secrets?" *Bulletin of the Atomic Scientists* 54 (1998): 50–53.

Aldrich, Richard. *The Hidden Hand: Britain, America and Cold War Secret Intelligence.* London: John Murray, 2001.

Aleksandrov, A. P., Ya. V. Zel'dovich, K. A. Petrzhak, and I. M. Frank. "Georgii Nikolaevich Flerov." *Uspekhi Fiziki Nauk* 109 (1973): 617–19.

Allaud, Louis A., and Maurice H. Martin. *Schlumberger: The History of a Technique.* New York: John Wiley & Sons, 1977.

Allibone, Thomas E. "Metropolitan-Vickers Electrical Company and the Cavendish Laboratory." In *Cambridge Physics in the Thirties*, edited by John Hendry, 150–73. Bristol: Adam Hilger, 1975.

Amaldi, Edoardo. "Personal Notes on Neutron Work in Rome." In *History of Twentieth Century Physics: Proceedings of the International School of Physics "Enrico Fermi," Varenna 1972*, edited by Charles Wiener, 294–325. New York: Academic Press, 1977.

———. "From the Discovery of the Neutron to the Discovery of Nuclear Fission." *Physics Report* 111 (1984): 5–331.

Amaldi, Edoardo, with Piero Angela. *Intervista sulla materia dal nucleo alle galassie.* Bari: Laterza, 1980.

Amaldi, E., and E. Fermi. "Absorption and Diffusion of Neutrons." *Physical Review* 50 (1936): 899–928.

Amendola, Giorgio. *Lettere a Milano*. Rome: Editori Riuniti, 1973.
Anders, Roger M. "The Rosenberg Case Revisited: The Greenglass Testimony and the Protection of Atomic Secrets." *American Historical Review* 83 (1978): 388–400.
Andrew, Christopher. *The Defence of the Realm: The Authorized History of MI5*. London: Allen Lane, 2009.
Andrew, Christopher, and Oleg Gordievsky. *KGB: The Inside Story of Its Foreign Operations from Lenin to Gorbachev*. London: Hodder and Stoughton, 1990.
Andrew, Christopher, and Vasili Mitrokhin. *The Mitrokhin Archive: The KGB in Europe and the West*. London: Penguin, 1999.
Asor Rosa, Alberto. "Il fascismo alla conquista del potere (1919–1926)." In *Storia d'Italia: Dall'unità ad oggi*, Vol. 4, edited by A.A. Rosa, 1358–1470. Turin: Einaudi, 1975.
Avery, Donald. "Allied Scientific Co-operation and Soviet Espionage in Canada, 1941–1945." *Intelligence and National Security* 8 (1993): 100–129.
———. *The Science of War: Canadian Scientists and Allied Military Technology during the Second World War*. Toronto: Toronto University Press, 1998.
Avolio, Giuseppe. *Emilio Sereni: Ortodossia politica e genialità scientifica*. Rome: Agra, 1999.
Badash, Lawrence. "From Security Blanket to Security Risk: Scientists in the Decade After Hiroshima." *History and Technology* 19 (2003): 241–56.
Badash, L., J. O. Hirshfielder, and H. P. Broida, eds. *Reminiscences of Los Alamos, 1943–1945*. Dordrecht: Reidel, 1980.
Balmer, Brian. "A Secret Formula, A Rogue Patent and Public Knowledge about Nerve Gas: Secrecy as Spatial-Epistemic Tool." *Social Studies of Science* 36 (2006): 691–722.
Baggott, Jim. *Atomic: The First War of Physics and the Secret History of the Atom Bomb: 1939–1949*. London: Icon, 2009.
Baker White, John. *The Soviet Spy System*. London: Falcon Press, 1948.
Battimelli, Giovanni, and Michelangelo De Maria. "La fisica." In *Per una storia del Consiglio Nazionale delle Ricerche*, edited by R. Simili and G. Paoloni, 281–311. Bari: Laterza, 2001.
Beltran, Alan, "La "Fée électricité," reine et servante." *Vingtième Siècle: Revue d'histoire* 16 (1987): 90–95.
Bignardi, Irene. *Memorie estorte ad uno smemorato*. Milan: Feltrinelli, 1999.
Bilenky, S.M., T. D. Blokhintseva, I. G. Pokrovskaya, and M. G. Sapozhnikov, eds. *B. Pontecorvo, Selected Scientific Works: Recollections on B. Pontecorvo*. Bologna: Società Italiana di Fisica, 1997.
Boersma, Kees. "Tensions Within an Industrial Research Laboratory: The Philips Laboratory's X-ray Department Between the Wars." *Enterprise and Society* 4 (2003): 65–98.
Bondy, Ruth. *The Emissary: A Life of Enzo Sereni*. London: Robson, 1977.
Bonolis, Luisa. *Majorana: Il genio scomparso*. Milan: Le Scienze, 2002.
———. "Un Genio di Via Panisperna." *Sapere*, April 2004, 24–33.
Boskey, Bennett. "Inventions and the Atom." *Columbia Law Review* 50 (1950): 433–47.

———. "'39: Not Shy, Not Retiring." *Harvard Law Bulletin* 52 (2000).
Bothwell, Robert. *Eldorado: Canada's National Uranium Company*. Toronto: University of Toronto Press, 1984.
———. *Nucleus: A History of Atomic Energy of Canada Limited*. Toronto: University of Toronto Press, 1988.
Bowker, Geoff C. *Science on the Run: Information Management and Industrial Geophysics at Schlumberger, 1920–1940*. Cambridge, MA: MIT Press, 1994.
Brun, Gérard. *Louis Vallon ou, La politique en liberté: De Jaurès à de Gaulle*. Paris: Economica, 1986.
Caprara, Massimo. *L'inchiostro verde di Togliatti*. Milan: Simonelli, 1996.
Caprara, Maurizio. *Lavoro riservato: I cassetti segreti del PCI*. Milan: Feltrinelli, 1996.
Cathcart, Brian. *The Fly in the Cathedral: How a Small Group of Cambridge Scientists Won the Race to Split the Atom*. London: Penguin, 2004.
Cathro, Robert J. "Uranium Production from East Germany, Czechoslovakia and Eldorado, Northwest Territories after 1945." *CIM Bulletin* 98 (2005): 67–70.
Cervetti, Giovanni. *L'Oro di Mosca: La verità sui finanziamenti sovietici al PCI raccontata dal diretto protagonista*. Milan, Baldini & Castoldi, 2000.
Chamberlain, Owen A., Carl Helmholtz, David L. Judd, and E. Segrè, "Robert Lyste [sic] Thornton, Physics: Berkeley." Courtesy of Academic Senate, Berkeley Division, 1986. Available online at http://content.cdlib.org/xtf/view?docId=hb767nb3z6&chunk.id=div00119&brand=calisphere&doc.view=entire_text].
Close, Frank. *Neutrino*, Oxford: Oxford University Press, 2010.
Cohen, Bernard L. "Guido Pontecorvo 'Ponte,' 1907–1999." *Genetics*, 154 (2000): 497–501.
Collins, Harry M. *Changing Order: Replication and Induction in Scientific Practice*. Beverley Hills and London: Sage, 1985.
Collins H. M., and Trevor Pinch. *The Golem: What Everyone Should Know About Science*. Cambridge: Cambridge University Press, 1998.
Conversi, Marcello, Ettore Pancini, and Oreste Piccioni. "On the Decay Process of Positive and Negative Mesons." *Physical Review* 68 (1945): 232.
———. "On the Disintegration of Negative Mesons." *Physical Review* 71 (1947): 232.
Corbino, Orso Mario. "Prospettive e risultati della fisica moderna." *La Ricerca Scientifica* 5 (1934): 615–20.
Cordella, Francesco, De Gregorio, Alberto, and Fabio Sebastiani. *Enrico Fermi: Gli anni italiani*. Rome: Editori Riuniti, 2001.
Costello, John. *Masks of Treachery*. New York: William Morrow, 1988.
Coyne, G. V., M. Heller, and J. Życiński, eds. *The Galileo Affair: A Meeting of Faith and Science. Proceedings of the Cracow Conference, 24–27 May 1984*. Vatican City: Specola Vaticana, 1985.
Creager, Angela. "Radioisotopes as Political Instruments, 1946–1953." *Dynamis* 29 (2009): 219–40.
Cronin, James W., ed. *Fermi Remembered*. Chicago: University of Chicago Press, 2004.

Dahl, Per Fridtjof. *Heavy Water and the Wartime Race for Nuclear Energy.* Bristol: Institute of Physics Publishing, 1999.
Dallin, David J. *Soviet Espionage.* New Haven: Yale University Press, 1955.
Del Gamba, Valeria. *Il ragazzo di Via Panisperna: L'avventurosa vita del fisico Franco Rasetti.* Turin: Bollati Boringhieri, 2007.
Dennis, Michael Aaron. "Secrecy and Science Revisited: From Politics to Historical Practice and Back." In *The Historiography of Contemporary Science, Technology and Medicine: Writing Recent Science,* edited by R. Doel and T. Söderqvist, 172–84. London and New York: Routledge, 2006.
Deriabin, Peter, with Frank Gibney. *KGB: The Secret World.* New York, 1959.
Di Biagio, Anna. "The Marshall Plan and the Founding of Cominform, June-September 1947." In *The Soviet Union and Europe in the Cold War, 1943–1953* edited by Francesca Gori and Silvio Pons, 208–45. Milan: Feltrinelli/Istituto Gramsci, 1996.
Donno, Gianni. *La Gladio Rossa del PCI, 1945–1967.* Soveria Mannelli: Rubbettino, 2001.
Drake, Richard. "The Soviet Dimension of Italian Communism." *Journal of Cold War Studies* 6 (2004): 115–19.
Dunworth, James V., and B. Pontecorvo. "Excitation of Indium 113 by X-rays." *Proceedings of the Cambridge Philosophical Society* 43 (1947): 123–26.
Dzhelepov, V. D. "The Genius of Bruno Pontecorvo." In *B. Pontecorvo Selected Scientific Works and Recollections,* edited by S. Bilenky et al., 487–93. Bologna: Società Italiana di Fisica, 1997. Available on line at http://pontecorvo.jinr.ru/dzhelepov.html)
Elmerich, Jonathan E. *Gathering Rare Ores: The Diplomacy of Uranium Acquisition, 1943–1954.* Princeton, NJ: Princeton University Press, 1986.
Erozolimski, B. G., A. S. Shkol'nikov, and A. I. Isakov. "Application of Neutron Pulse Sources for Investigation in Oil Wells." *Atomic Energy* 9 (1961): 144–45.
Evangelista, Matthew. *Unarmed Forces: The Transnational Movement to End the Cold War.* Ithaca, NY: Cornell University Press, 1999.
Evans, E. Anthony. *Tritium and Its Compounds.* London: Butterworths, 1966.
Fermi, Enrico. *Fisica ad uso dei licei.* Bologna: Zanichelli, 1929.
———. "Radioattività indotta da bombardamento di neutroni." *La ricerca scientifica,* 1 (1934): 283.
———. "Radioactivity Induced by Neutron Bombardment." *Nature* 133 (1934): 757.
———. "Recenti risultati della radioattività artificiale." *La ricerca scientifica* 6 (1935), 399–402.
Fermi, E., E. Amaldi, O. D'Agostino, F. Rasetti, and E. Segrè. "Artificial Radioactivity Produced by Neutron Bombardment." *Proceedings of the Royal Society of London* 146 (1934): 483–500.
Fermi, E., and Enrico Persico. *Fisica per le scuole medie superiori.* Bologna: Zanichelli, 1938.
Fermi, Laura. *Atoms in the Family: My Life with Enrico Fermi.* Chicago: University of Chicago Press, 1954.

Fidecaro, Giovanni. "Bruno Pontecorvo: From Rome to Dubna." In *B. Pontecorvo Selected Scientific Works and Recollections*, edited by S. Bilenky et al., 472–86. Bologna: Società Italiana di Fisica, 1997.

Finocchiaro, Maurice A. *The Galileo Affair: A Documentary History*. Berkeley: University of California Press, 1989.

Fiori, Giuseppe. *Uomini ex: Lo strano destino di un gruppo di comunisti italiani*. Turin: Einaudi, 1993.

Fischman, Marianne, and Emeric Ledjel. "The Quantitative Approach to Business Cycle in 'X-Crise' Group in the 1930's." In CNRS Hyper Articles en Ligne, HAL-CNRS (http://hal.archives-ouvertes.fr/halshs-00268373).

Fitzpatrick, Anne. "From Behind the Fence: Threading the Labyrinth of Classified Historical Research." In *The Historiography of Contemporary Science, Technology and Medicine: Writing Recent Science*, edited by R. Doel and T. Söderqvist, 67–80. London and New York: Routledge, 2006.

Fleming, Ian. *Casino Royale*. London: Jonathan Cape, 1953.

Franzinelli, Mimmo. *I tentacoli dell'OVRA: Agenti, collaboratori e vittime della polizia Fascista*. Turin: Bollati Boringhieri, 1999.

———. *Delatori. Spie e confidenti anonimi: L'Arma segreta del regime Fascista*. Milan: Mondadori, 2001.

Fubini, David G. *Let Me Explain: Eugene G. Fubini's Life in Defense of America*. Santa Fe: Sunstone Press, 2009.

Galison, Peter. *How Experiments End*. Chicago: University of Chicago Press, 1987.

———. "Removing Knowledge." *Critical Enquiry* 31 (2004): 229–43.

Galison, Peter, and Jeremy Bernstein. "In Any Light: Scientists and the Decision to Build the Superbomb, 1952–1954." *Historical Studies in the Physical Sciences* 19 (1989): 267–347.

Gardner, David P. *The California Oath Controversy*. Berkeley: The University of California Press, 1967.

Gemelli, Giuliana. "Scholars in Adversity and Science Policies." In *The "Unacceptables": American Foundations and Refugee Scholars between the Two Wars and After*, edited by G. Gemelli, 13–34. Bruxelles: Peter Lang, 2000.

Giacomini, Ruggero. *I partigiani della pace: Il movimento Pacifista in Italia e nel mondo negli anni della prima Guerra Fredda*. Milan: Vangelista, 1984.

Giannetti, Renato. *Tecnologia e sviluppo economico italiano, 1870–1990*. Bologna: Il Mulino, 1998.

———. "Il CNR e le politiche per la ricerca e l'innovazione industriale." In *Per una storia del consiglio nazionale delle ricerche*, edited by Simili and Paoloni, 224–39. Bari: Laterza, 2001.

Gibbs, Tim. *British and American Counter-intelligence and the Atom Spies, 1941–1950*. PhD diss., University of Cambridge, 2008.

Gillman, Peter, and Leni Gillman. *Collar the Lot: How Britain Interned and Expelled its Wartime Refugees*. London: Quartet Books, 1980.

Ginsborg, Paul. *A History of Contemporary Italy: Society and Politics, 1943–1988*. New York: Palgrave, 2003.

Goldschmidt, Bertrand. *Atomic Rivals*. Translated by Georges M. Temmer. New York: Rutgers University Press, 1990. Originally published as *Les rivalités atomiques. 1939–1967* (Paris: Fayard, 1967).

Goodman, Michael S. "British Intelligence and the Soviet Atomic Bomb, 1945–1950." *Journal of Strategic Studies* 26 (2003): 120–51.

———. *Spying on the Nuclear Bear*. Stanford, CA: Stanford University Press, 2007.

Goodman, Michael S., and Chapman Pincher. "Clement Attlee, Percy Sillitoe and the Security Aspects of the Fuchs Case." *Contemporary British History* 19 (2005): 66–77.

Gowing, Margaret. *Britain and Atomic Energy, 1939–1945*. London: Macmillan, 1964.

Gowing, M., with Lorna Arnold. *Independence and Deterrence: Britain and Atomic Energy, 1945–1952. Vol. 1: Policy Making*. London: MacMillian, 1974.

Gozzini, Giovanni, and Renzo Martinelli. *Storia del PCI: Dall'attentato di Togliatti all'VIII Congresso*. Turin: Einaudi, 1998.

Green, William G., and Robert E. Fearon. "Well Logging by Radioactivity." *Geophysics* 5 (1940): 272–83.

Griffith, Robert. *The Politics of Fear: Joseph R. McCarthy and the Senate*. Lexington: University of Kentucky Press, 1970.

Groueff, Stephane. *Manhattan Project: The Untold Story of the Making of the Atomic Bomb*. Boston: Little, Brown, 1967.

Guagnini, Anna. "Patent Agents, Legal Advisers and Guglielmo Marconi's Breakthrough in Wireless Telegraphy." *History of Technology* 24 (2002): 171–201.

Halperin, John. *C.P. Snow: An Oral Biography*. Brighton: St. Martin's Press, 1983.

Hanna, G., and B. Pontecorvo. "The ß-spectrum of H_3." *Physical Review* 75 (1949): 985–86.

Hartcup, Guy, and Thomas E. Allibone. *Cockcroft and the Atom*. Bristol: Hilger, 1984.

Hawkins, David. "Toward Trinity." In *Project Y: The Los Alamos Story*, edited by D. Hawkins, 1–259. San Francisco: Tomash, 1984.

Haynes, John Earl, and Harvey Klehr. *Venona: Decoding Soviet Espionage in America*. New Haven: Yale University Press, 2000.

Heale, M. J. *McCarthy's Americans: Red Scare Politics in the State and Nation, 1935–1965*. Basingtoke: Macmillan, 1997.

Hecht, Gabrielle. "The Power of Nuclear Things." *Technology and Culture* 51 (2010): 1–30.

Heerding, A. *The History of N. V. Philips' Gloeilampenfabriken*, 2 Vols. Cambridge: Cambridge University Press, 1986.

Heilbron, John, and Robert Seidel. *Lawrence and His Laboratory: A History of the Lawrence Berkeley Laboratory, Vol. I*. Berkeley: University of California Press, 1989.

Hennessy, Peter, and Gail Brownfeld. "Britain's Cold War Security Purge: The Origins of Positive Vetting." *The Historical Journal* 25 (1982): 965–74.

Hewlett, Richard J., and Oscar E. Anderson. *The New World: A History of the United States Atomic Energy Commission, 1939–1946, Vol. I.* University Park: Pennsylvania State University Press, 1962.

Hewlett, Richard J., and Francis Duncan. *Atomic Shield: A History of the United States Atomic Energy Commission, 1947–1952, Vol. 2.* University Park: Pensylvania State University Press, 1969.

Hincks, E. P. (Ted), and B. Pontecorvo. "Search for Gamma-Radiation in the 2.2 Microsecond Meson Decay Process." *Physical Review* 73 (1948): 257–58.

———. "The Absorption of Charged Particles from the 2.2 Microsecond Meson Decay." *Physical Review* 74 (1948): 697–98.

———. "The Penetration of μ-meson Decay Electrons and their Bremsstrahlung Radiation." *Physical Review* 75 (1949): 698–99.

———. "On the Disintegration Products of the 2.2 μsec. Meson." *Physical Review* 77 (1950): 102–20.

Hoddeson, Lillian, Paul W. Henriksen, Roger A. Meade, and Catherine Westfall. *Critical Assembly: A Technical History of Los Alamos during the Oppenheimer Years, 1943–1945.* Cambridge: Cambridge University Press, 1993.

Holligsworth, Mark, and Nick Fielding. *Defending the Realm: MI5 and the Shayler Affair.* London: André Deutsch, 1999.

Holloway, David. *Stalin and the Bomb: The Soviet Union and Atomic Energy, 1939–1956.* New Haven: Yale University Press, 1994.

Holton, Gerald. "Fermi's Group and the Recapture of Italy's Place in Physics." In *The Scientific Imagination: Case Studies*, 155–98. Cambridge, MA: Harvard University Press, 1978.

———. "Enrico Fermi and the Miracle of Two Tables." In *Victory and Vexation in Science: Einstein, Bohr, Heisenberg and Others*, 48–64. Cambridge, MA: Harvard University Press, 2005.

Hood, William. *Mole: The True Story of the First Russian Intelligence Officer Recruited by the CIA.* London: Weidenfeld & Nicholson, 1982.

Hughes, Jeff. "Plasticine and Valves: Industry, Instrumentation and the Emergence of Nuclear Physics." In *The Invisible Industrialist: Manufacturers and the Construction of Scientific Knowledge*, edited by Jean-Paul Gaudillere and Ilana Lowy, 58–101. London: Palgrave Macmillian, 1998.

———. *The Manhattan Project: Big Science and the Atom Bomb.* London: Icon Books, 2002.

———. "Radioactivity and Nuclear Physics." in *The Cambridge History of Science, Vol. 5*, edited by M. J. Nye, 350–74. Cambridge: Cambridge University Press, 2005.

Hurst, Donald G. "Overview of Nuclear Research and Development." In *Canada Enters the Nuclear Age: A Technical History of Atomic Energy of Canada Limited*, edited by D. G. Hurst, 1–32. Montreal-Kingston: McGill-Queen's University Press, 1997.

Ioffe, Boris L. "A Top Secret Assignment." In *At the Frontiers of Particle Physics: Handbook of QCD, Boris Ioffe Festschrift*, edited by M. Shifman. New Jersey: World Scientific, 2001).

Israel, Giorgio, and Pietro Nastasi. *Scienza e razza nell'Italia Fascista*. Bologna: Il Mulino, 1998.

Johnston, Sean F. "Implanting a Discipline: The Academic Trajectory of Nuclear Engineering in the USA and UK." *Minerva* 47 (2009): 51–73.

Joint Committee on Atomic Energy. *Soviet Atomic Espionage*. Washington: Printing Office, 1951.

Jones, H. "Herbert Wakefield Banks Skinner." *Biographical Memoirs of Fellows of the Royal Society* 6 (1960): 258–68.

Joravsky, David, *The Lysenko Affair*. Chicago: University of Chicago Press, 1970.

———. *Red Atom: Russia's Nuclear Power Program from Stalin to Today*. New York: Freeman & Co., 2000.

Kaiser, David. "The Atomic Secret in Red Hands? American Suspicions of Theoretical Physicists during the Early Cold War." *Representations* 90 (2005): 28–60.

Kedrov, A. I., N. V. Popov, N. S. Khimchenko, and A. S. Tsybin. "Neutron-Gamma Well Logging Apparatus with More Sensitivity for Chlorine." *Applied Radiation and Isotopes* 48 (1997): 1647–48.

Kevles, Daniel J. *The Physicists: The History of a Scientific Community in Modern America*. Cambridge, MA: Harvard University Press, 1971.

Kirkwood, David H., B. Pontecorvo, and Geoff C. Hanna. "Fluctuations of Ionization and Low Energy Beta-Spectra." *Physical Review* 74 (1948): 497–98.

Kojevnikov, Alexei B. *Stalin's Great Science: The Times and Adventures of Soviet Physicists*. London: Imperial College Press, 2004.

Kragh, Helge. *Quantum Generations: A History of Physics in the Twentieth Century*. Princeton, NJ: Princeton University Press, 1999.

Krige, John. "Atoms for Peace." In *Global Power Knowledge: Science, Technology & International Affairs*, edited by John Krige and Kai-Henrik Barth, 174–80. Chicago: University of Chicago Press, 2006 [*Osiris* 21 (2006)].

———. *American Hegemony and the Post-War Reconstruction of Science in Europe*. Cambridge, MA: MIT Press, 2006.

———. "The Politics of Phosphorous-32: A Cold War Fable Based on Fact." In *The Historiography of Contemporary Science, Technology, and Medicine*, edited by R. Doel and T. Söderqvist, 153–71. London and New York: Routledge, 2006.

Lanouette, William. *Genius in the Shadow: A Biography of Leo Szilard, the Man Behind the Bomb*. Chicago: University of Chicago Press, 1992.

Lassman, Thomas. "Industrial Research Transformed: Edward Condon and the Westinghouse Electric and Manufacturing Research Company, 1935–1942." *Technology and Culture* 44 (2003): 309–36.

Lattes Cesar, Occhialini Giuseppe, and Colin Powell. "Observations on the Tracks of Slow Mesons in Photo-graphic Emulsions." *Nature* 160 (1947): 453–56.

———. "Observations on the Tracks of Slow Mesons in Photo-graphic Emulsions II." *Nature* 160 (1947): 486–87.

Laurence, George C. "Canada's Participation in Atomic Energy Development." *Bulletin of the Atomic Scientists* 3 (1947): 325–28.

———. "Early Years of Nuclear Energy Research in Canada." Atomic Energy of Canada Limited, May 1980. Available online at http://www.cns-snc.ca/history/early_years/earlyyears.html.

Lawyer, L.C. (Lee), Charles C. Bates, and Robert B. Rice. *Geophysics in the Affairs of Mankind: A Personalized History of Exploration Geophysics*. Tulsa: Society of Exploration Geophysics, 2001 [1982].

Lee, Sabine. "The Spy That Never Was." *Intelligence and National Security* 17:4 (2002): 77–99.

Lewis, Wilfrid B. *Electrical Counting: With Special Reference to Alpha and Beta Particles*. Cambridge: Cambridge University Press, 1948.

Luria, Salvador Edward. *A Slot Machine, a Broken Test Tube: An Autobiography*. New York: Harper & Row, 1984.

MacKenzie, Donald, and Graham Spinardi. "Tacit Knowledge, Weapons Design and the Uninvention of Nuclear Weapons." *American Journal of Sociology* 101 (1995): 44–99.

Mafai, Miriam. *Il lungo freddo: Storia di Bruno Pontecorvo, lo scienziato che scelse l'URSS*. Milan: Mondadori, 1983.

———. *L'uomo che sognava la lotta armata: La storia di Pietro Secchia*. Milan: Rizzoli, 1984.

Maltese, Giulio. *Enrico Fermi in America: Una bibliografia scientifica: 1938–1954*. Bologna: Zanichelli, 2003.

Masini, Giancarlo. *Marconi*. New York: Marsilio, 1995

Miller, Byron S. "A Law Is Passed: The Atomic Energy Act of 1946." *University of Chicago Law Review* 15 (1948): 799–821.

Montgomery Hyde, Harford, *The Atom Bomb Spies*. London: Sphere, 1980.

Moorehead, Alan. *The Traitors: The Double Life of Fuchs, Pontecorvo, and Nunn May*. London: Hamish Hamilton, 1952.

Morgan, Ted. *Reds: McCarthyism in Twentieth-Century America*. New York: Random House, 2004.

Moss, Norman. *Klaus Fuchs: The Man Who Stole the Bomb*. London: Grafton Books, 1989.

Moyal, Ann. *Alan Moorehead: A Rediscovery*. Canberra: National Library of Australia, 2005.

Mullet, Shawn Kristian. *Little Man: Four Junior Physicists and the Red Scare Experience*. PhD diss., Harvard University, 2008.

Newton, Scott. *Profits of Peace: The Political Economy of Anglo-American Appeasement*. Oxford: Clarendon Press, 1997.

Nininger, Robert D. *Minerals for Atomic Energy. A Guide to Exploration for Uranium, Thorium and Beryllium*. Toronto: Van Nostrand, 1954.

Nye, Mary J. *Blackett: Physics, War and Politics in the Twentieth Century*. Cambridge, MA: Harvard University Press, 2004.

Ooms, Casper W. "Atomic Energy and U.S. Patent Policy. Part 2: Patent Provisions of the Atomic Energy Act." *Bulletin of the Atomic Scientists* 2 (1946): 30–31.

Oppenheimer, Robert. "Physics in the Contemporary World." *Bulletin of the Atomic Scientists*, 4 (1948): 65–68.

Oristaglio, Michael, and Alexander Dorozynski. *A Sixth Sense: The Life and Science of Henri-George Doll. Oilfield Pioneer and Inventor*. New York: Overlook Duckworth, 2009.

Orlando, Lucia. "Physics in the 1930s: Jewish Physicists' Contribution to the Realization of the 'New Tasks' of Physics in Italy." *Historical Studies in the Physical Sciences* 29 (1998): 141–81.

Orlando, Ruggero. "Amore-Odio fra Italiani e Inglesi durante e dopo la Seconda Guerra Mondiale." In *Inghilterra e Italia nel '900: Atti del Convegno di Bagni di Lucca*, 193–200. Florence: La Nuova Italia, 1973.

Parasnis, D. S. *Principles of Applied Geophysics*. London: Chapman & Hall, 1997 [1962].

Paul, Septimus H. *Nuclear Rivals: Anglo-American Atomic Relations, 1941–1952*. Columbus: Ohio State University Press, 2000.

Petri, Rolf. "Technical Change in the Italian Chemical Industry: Markets, Firms and State Intervention." In *Determinants in the Evolution of the European Chemical Industry, 1900–1939*, edited by A. S. Trevis, H. G. Schröter, E. Homburg, and P. J. Morris, 275–300. Dordrecht: Kluwer Academic Publishers, 1998.

Pinault, Michel. *Frédéric Joliot-Curie*. Paris: Odile Jacob, 2000.

Pincher, Chapman. *Their Trade is Treachery*. London: Sidwick & Jackson, 1982.

———. *Too Secret, Too Long: The Great Betrayal of Britain's Crucial Secrets, and the Cover-up*. London: Sidwick & Jackson, 1984.

Pollock, Ethan. *Stalin and the Soviet Science Wars*. Princeton, NJ: Princeton University Press, 2006.

Pontecorvo, Bruno. "Isomeric Forms of Radio Rhodium." *Nature* 141 (1938): 785–86.

———. "Recent Experimental Results in Nuclear Isomerism." *Nature* 144 (1939): 212–13.

———. "Neutron Well Logging: A New Geological Method Based on Nuclear Physics." *Oil and Gas Journal* 40 (1941): 32–33.

———. "Radioactivity Analyses of Oil Well Samples." *Geophysics* 7 (1942): 90–94.

———. "Inverse ß-process." In *B. Pontecorvo Selected Scientific Works and Recollections*, edited by S. Bilenky et al., 21–26. Bologna: Società Italiana di Fisica, 1997.

———. "Nuclear Capture of Mesons and the Meson Decay." *Physical Review* 72 (1947): 246.

———. "Recent Developments in Proportional Counter Technique." *Helvetica Physica Acta* 23 (1950): 97–118.

Pontecorvo, B., D. H. Kirkwood, and G. C. Hanna. "Nuclear Capture of L Electrons." *Physical Review* 75 (1949): 982.

Pontecorvo, B. M. "Pages in the Development of Neutrino Physics." *Uspekhi Fiziki Nauk* 26 (1983): 1087–1108.

Pontecorvo, B.M., and I. M. Frank. "Fedor L'vovich Shapiro," *Uspekhi Fiziki Nauk* 16 (1973): 295–97.
Popovski, Mark, and Mark Aleksandrovich. *The Vavilov Affair*. Hamden, CT: Archon Books, 1984.
Rasetti, Franco. "Sopra un forte preparato di Radio D ottenuto nell'Istituto Fisico di Roma." *La Ricerca Scientifica* 5 (1934): 3–5.
Rast, Nicholas. Introduction to *Applied Geophysics: USSR*, edited by N. Rast. London: Pergamon Press, 1962.
Rhodes, Richard. *The Making of the Atomic Bomb*. London: Penguin, 1986.
———. *Dark Sun: The Making of the Hydrogen Bomb*. New York: Touchstone, 1995.
Richards, Joan L. "Introduction: Fragmented Lives." *ISIS* 97 (2006): 302–6.
Richelson, Jeffrey T. *Spying on the Bomb*. New York: W. W. Norton & Co., 2007.
Ritchie, R. H. "Jacob Neufeld, 15 April 1906–5 April 2000." *Health Physics* 87 (2004): 94–95.
Riva, Valerio. *Oro di Mosca: I finanziamenti sovietici al PCI* (Milan: Mondadori, 1999).
Robertson, Jerry. *ABC's of Oil*. Evansville, IN: Petroleum Publishers, 1953.
Rowlands, Peter. *120 Years of Excellence: The University of Liverpool Physics Department, 1881–2001*. Liverpool: PD Publications, 2001.
Rowlinson, J. S., "The Wartime Work of Hinshelwood and His Colleagues." *Notes and Records of the Royal Society* 58 (2004): 161–66.
Russell, William G., and S. Scherbatskoy. "The Use of Sensitive Gamma Ray Detectors." *Economic Geology* 46 (1951): 427–46.
Rutherford, Ernest, Chadwick James, and Charles D. Ellis, *Radiation from Radioactive Substances*. New York: MacMillian, 1930.
Ryabev, L. D., ed. *Atomic Project of USSR: Documents and Materials, Vol. 1, 1938–1945, Part 1*. Moscow: Nauka, Fizmatlit, 1998 [in Russian].
Salvini, G. "Bruno Pontecorvo, a Great Physicist, a Great Man." In *Cosmic Ray, Particle and Astrophysics. A Conference in Honour of Giuseppe Occhialini, Bruno Pontecorvo and Bruno Rossi*, edited by A. Bonetti, I. Guidi and B. Monteleoni, 3–12. Rome: Accademia Nazionale dei Lincei, 1997.
Sarfatti, Michele. "La persecuzione degli Ebrei in Italia dalle leggi razziali alla deportazione." In *La persecuzione degli Ebrei durante il Fascismo: Le leggi del 1938*, edited by G. Luzzatto Voghera, 81–107. Rome: Camera dei Deputati, 1998.
Savosin, S. I., and V. I. Sinitsyn. "Application of Methods of Nuclear Geophysics in Ore Prospecting, Exploration and Development." *Atomic Energy* 18 (1965): 81–84.
Schweber, Sylvan S. *In the Shadow of the Bomb: Oppenheimer, Bethe, and the Moral Responsibility of the Scientist*. Princeton, NJ: Princeton University Press, 2000.
"Scientific Civil Service." *Nature* 161 (1948): 195.
Segesman, F. F. "Well Logging Method." *Geophysics* 45 (1980): 1667–84.

Segrè, Emilio. *Enrico Fermi, Physicist.* Chicago: University of Chicago Press, 1970.
———. *A Mind Always in Motion.* Berkeley: University of California Press, 1993.
———. Preface to Giancarlo Masini, *Marconi*, 9–11. New York: Marsilio, 1995.
"Senate Hearing on Atomic Energy, Atomic Bomb Patents." *Bulletin of the Atomic Scientists* 1 (1946): 10–11.
Sekido, Yataro, and Harry Elliot, eds., *Early Days in Cosmic Rays.* Dordrecht: D. Reidel Publishing Co., 1985.
Sereni, Clara. *Il gioco dei regni.* Florence: Giunti, 1993.
———. "A proposito di Bruno." *Sapere*, April 2004, 35–45.
Shen, Grace Y. "Xuetong Li's Weng Wenhao Nian Pu" (book review). *ISIS* 99 (2008): 874–75.
Simili, Raffaella. "La presidenza Marconi." In *Per una storia del consiglio nazionale delle ricerche*, edited by R. Simili and G. Paoloni, 72–127. Bari: Laterza, 2001.
Snow, Charles P. *The New Men.* London: Penguin, 1954.
Snyder, Donald D., and David B. Fleming, "Well Logging. A 25-Year Perspective." *Geophysics* 50 (1985): 2506–29.
Sokal, Alan, and Jean Bricmont. *Intellectual Impostures: Postmodern Philosophers' Abuse of Science.* London: Profile Books, 1998.
Spencer, D. L. "The Role of Oil in Soviet Foreign Economic Policy." *American Journal of Economics and Sociology* 25 (1966), 91–107.
Sponza, Lucio. *Divided Loyalties: Italians in Britain during the Second World War.* Brussels: Peter Lang, 2000.
Stevenson, William. *A Man Called Inrtepid: The Secret War, 1939–1945.* London: Book Club, 1976.
Sudoplatov, Pavel, and Anatoli Sudoplatov. *Special Tasks: The Memoirs of an Unwanted Witness. A Soviet Spymaster.* Boston: Little, Brown & Co., 1994.
Sweet, George E. *The History of Geophysical Prospecting.* Suffolk: Neville Spearman, 1978 [1969].
Theoaris, Athan. *Chasing Spies.* Chicago: Ivan Dee, 2000.
Thomson, George. "Charles Galton Darwin." *Bibliographical Memoirs of the Fellows of the Royal Society* 9 (1963): 69–85.
Thorpe, Charles. *Oppenheimer: The Tragic Intellect.* Chicago: University of Chicago Press, 2006.
"The Revised McMahon Bill." *Bulletin of the Atomic Scientists* 1 (1946): 2–5.
Tittle, C.W. "The Theory of Neutron Logging." *Geophysics* 26 (1961): 27–39.
Tomassini, Luigi. "Le Origini." In *Per una storia del consiglio nazionale delle ricerche*, edited by R. Simili and G. Paoloni, 5–71. Bari: Laterza, 2001.
Toniolo, Gianni. *L'economia dell'Italia Fascista.* Bari: Laterza, 1980.
Townes, Charles H. *How the Laser Happened: Adventures of a Scientist.* Oxford: Oxford University Press, 1999.
Trenn, Taddeus. "The GM Counter of 1928." *Annals of Science* 43 (1986): 111–35.

Turchetti, Simone. "Use, Refuse or Lock them Up: A History of Italian Academic Refugees in Britain, 1930/1950." PhD diss., University of Manchester, 2003.

———. "Compenso in ritardo per i neutroni lenti." *Sapere*, June 2005, 54–63.

———. "For Slow Neutrons, Slow Pay." *Isis* 97 (2006): 1–28.

———. "The Invisible Businessman: Nuclear Physics and Patenting Practices in Italy in the 1930s." *Historical Studies in the Physical and Biological Sciences* 37 (2006): 153–72.

———. *Il caso Pontecorvo: Fisica nucleare, politica e servizi di sicurezza nella Guerra Fredda*. Milan: Sironi, 2007.

———. "A Contentious Business: Industrial Patents and the Production of Isotopes, 1930–1960," *Dynamis* 29 (2009): 191–218.

Turi, Rocco. *Gladio Rossa: Una catena di complotti e delitti, dal dopoguerra al caso Moro*. Venice: Marsilio, 2004.

Ungar, Sanford J. *FBI*. Boston: Little, Brown & Co., 1975.

Vincent, David. *The Culture of Secrecy: Britain, 1832–1998*. Oxford: Oxford University Press, 1998.

Von Halban, Hans, and Hugh Paxton. "Doppler Effect of Nuclear Resonance Level." *Nature* 141 (1938): 116.

Walker, Mark. *German National Socialism and the Quest for Nuclear Power, 1939–1949*. Cambridge: Cambridge University Press, 1989.

Wallace, Henry A. "Supports the McMahon Bill." *Bulletin of the Atomic Scientists* 1 (1946): 6–7.

Wang, Jessica. "Liberals, the Progressive Left, and the Political Economy of Postwar American Science: The National Science Foundation Debate Revisited." *Historical Studies in the Physical Sciences* 26 (1995): 139–66.

———. *American Science in an Age of Anxiety: Scientists, Anticommunism and the Cold War*. Chapel Hill: University of North Carolina Press, 1999.

Weart, Spencer. "Secrecy, Simultaneous Discovery and the Theory of Nuclear Reactors." *American Journal of Physics* 45 (1977): 1049–60.

Weart, Spencer, and Gertrude Szilard, eds. *Leo Szilard: His Version of the Facts. Selected Recollections and Correspondence*. Cambridge, MA: MIT Press, 1978.

Welk, William G. *Fascist Economic Policy*. Cambridge, MA: Harvard University Press, 1938.

Wellerstein, Alex. "Patenting the Bomb: Nuclear Weapons, Intellectual Property, and Technological Control." *Isis* 99 (2008): 57–87.

West, Nigel. *MI5: British Security Service Operations, 1909–1945*. London: Bodley Head, 1981.

———. *A Matter of Trust: MI5, 1945–72*. London: Weidenfeld & Nicolson, 1982.

———. "Fiction, Faction and Intelligence." In *Understanding Intelligence in the Twenty-First Century: Journeys in the Shadows*, edited by L.V. Scott and P. D. Jackson, 122–33. New York: Routledge, 2004.

West, Rebecca. *The Meaning of Treason*. London: Virago, 1965.

Westwick, Peter J. "Secret Science: A Classified Community in the National Laboratories." *Minerva* 38 (2000): 363–91.

Wiener, Charles. "A New Site for the Seminar: The Refugees and American Physics in the Thirties." In *The Intellectual Emigration, Europe and America, 1930–1960*, edited by D. Flaming, 190–228. Cambridge, MA: Harvard University Press, 1969.

Williams, Robert Chadwell. *Klaus Fuchs, Atom Spy*. Cambridge, MA: Harvard University Press, 1987.

Wittner, Lawrence S. *One World or None: A History of the World Nuclear Disarmament through 1953, Vol. 1*. Stanford, CA: Stanford University Press, 1993.

Wright, Peter, with Paul Greengrass. *Spycatcher*. Melbourne: Heinemann, 1987.

Xuetong, Li. *The Chronicle of Dr. Weng Wenhao*. Jinan: Shandong Education Press, 2005 [in Chinese].

Zallen, Doris T. "Louis Rapkine and the Restoration of French Science after the Second World War." *French Historical Studies* 17 (1991): 6–37.

Zaloga, Steven J. *Target America: The Soviet Union and the Strategic Arms Race, 1945–1964*. Novato: Presidio, 1993.

———. *The Kremlin's Nuclear Sword: The Rise and Fall of Russia's Strategic Nuclear Forces, 1945–2000*. Washington: Smithsonian Institution Press, 2002.

Zhaporezhetz, V.M., and E. M. Filippov. "The Use of Accelerators of Charged Particles in Investigating Bore-Holes by the Methods of Radioactive Logging." In *Applied Geophysics: USSR*, edited by N. Rast, 397–421. London: Pergamon Press, 1962.

Ziegler, Charles Z., and David Jacobson. *Spying Without Spies: Origins of America's Secret Nuclear Surveillance System*. Westport, CT: Praeger, 1995.

Zubok, Vladislav M. "Stalin and the Nuclear Age." In *Cold War Statesmen Confront the Bomb: Nuclear Diplomacy since 1945*, edited by John Gaddis, Philip Gordon, Ernest May, and Jonathan Rosenberg, 39–61. Oxford: Oxford University Press, 1999.

Patents Consulted

Amaldi, E., D'Agostino, O., Fermi, E., Pontecorvo, B., Rasetti, F., and E. Segrè. Method for increasing the efficiency of nuclear reactions and products thereof. GB patent 465,045, filed 26 October 1934, and issued 26 April 1937 [assigned to Philips Gloeilampfabriken, Eindhoven, Netherlands].

Brons, Folkert. Process and Apparatus for Exploring Geological Strata. US patent 2,220,509, filed 30 January 1940, and issued 5 November 1940 [assigned to Shell Development Corporation, San Francisco, California]

Fearon, Robert E. Well Logging Method and Apparatus. US patent 2,275,748, filed 28 March 1940, and issued 10 March 1942 [assigned to WSI, Tulsa, Oklahoma].

Fearon R. E., and Jean M. Thayer. Well-Logging Radiation Sources. US patent 3,071,690, filed 30 July 1949, and issued 1 January 1963 [assigned to WSI, Tulsa, Oklahoma].

Fermi, E., Rasetti, F., Amaldi E., Segrè E., and B. Pontecorvo. Radio-active Isotope Production. CA patent 407,558, issued 22 November 1942 [assigned to Giannini & Co. Inc., New York, USA].

Fermi, E., Amaldi, E., Pontecorvo, B., Rasetti, F., and E. Segrè. Process for the production of radioactive substances. US patent 2,206,634, filed 3 October 1935, and issued 2 July 1940 [assigned to Giannini & Co. Inc., New York, USA].

Pontecorvo, B. Method and Apparatus for Geophysical Exploration. US patent 2,349,753, filed 5 February 1942, and issued 3 May 1944 [assigned to WSI, Tulsa, Oklahoma].

Pontecorvo, B. Well Surveying. US patent 2,398,324, filed 10 August 1943, and issued 9 April 1946 [assigned to WSI, Tulsa, Oklahoma].

———. Method of Geophysical Prospecting. US patent 2,508,772, filed 31 October 1942, and issued 23 May 1950 [assigned to WSI, Tulsa, Oklahoma].

Scherbatskoy, Serge A. Well Logging by Measurement of Radioactivity. US patent 2,219,273, filed 19 June 1939, and issued 22 October 1940 [assigned to WSI, Tulsa, Oklahoma].

———. Nuclear Well Logging. US patent 2,648,012, filed 5 October 1949, and issued 4 August 1953 [assigned to Perforating Guns Atlas Corporation, Huston, Texas].

———. Nuclear Well Logging. US patent 3,869,608, filed 1 July 1953, and issued 4 March 1975.

Index

Akers, Wallace, 51–52, 77, 99, 121
Aliptekin, Issa Yusuf Bey, 192
Allibone, Thomas, 35, 230n77, 237n110, 251n69
All-Union Research Institute of Nuclear Geophysics and Geochemistry (VNII-YaGG), 195–97
Alvarez, Louis, 128, 249n32
Amaldi, Edoardo, 45, 61, 119, 127, 181, 182f, 232n16, 236n80, 261n65; on discovery of slow neutrons, 24–27; and participation in Fermi's group, 21–23, 22f
Amendola, Giorgio, 31
Anderson, Herbert, 48
Anderson, Roland, 103–8, 131–32
anticommunism, 85, 93, 215. *See also* security purge; witch hunts
argon, 62. *See also* chlorine-argon method
Arnold, Henry: and investigation on Pontecorvo, 87–91; and Anna Pontecorvo's interrogation, 119–20; 134, 142, 146, 149
Arzamas-16 (Los Arzamas), 185
atom bomb, 1, 2, 8, 55, 71, 84, 140, 156, 183, 189. *See also* nuclear weapons
atomic energy: exploitation of, 8, 10, 98, 100, 146, 209; fundamental research on, 66, 83, 124, 185; international control of, 83–85, 159; military uses of, 9, 51, 100, 101, 103, 154, 159, 170, 175, 179–80, 216; peaceful uses of, 155–58, 160, 174, 183–84, 217
Atomic Energy Act: approval, 100–101; security provisions, 104–5, 109, 122, 245n37, 248n8
Atomic Energy Commission (AEC): AEC Argonne National Laboratory, 51, 131, 234n47; AEC General Advisory Com-

Atomic Energy Commission (AEC) (*continued*)
mittee (GAC), 100, 103, 109–10, 129; AEC Oak Ridge National Laboratory, 51, 63, 89f, 131, 136; AEC Office of General Counsel (OGC), 102, 105, 131; AEC Patent Compensation Board (PCB), 101–7, 109, 130–32; AEC Patent Policy Panel, 101; AEC Personnel Security Review Board, 105; AEC Security Division, 136; establishment, 100; negotiations on slow neutron patent, 101–3, 108–13, 129–32, 217; reaction to Pontecorvo's defection, 125, 127; security administration, 104–8

Atomic Energy Research Establishment (AERE), 2, 3, 11, 90, 108, 116, 124, 200; AERE Isotope division, 143; AERE Power Steering Committee (PSC), 66; Pontecorvo's appointment at, 66–67, 124, 215; Pontecorvo's dismissal from, 91–93; Pontecorvo's postcards to, 108, 116, 117, 119; security investigation at, 69, 87–90, 106–8, 112, 159, 169

atomic espionage, 2, 3, 70, 71, 117, 138, 140, 145, 214, 215; and gender, 148–50; history of, 6–7, 213; JCAE report on, 129–33, 139, 144, 187. *See also* espionage

atomic pile, 48–49, 50, 52, 54, 55, 60, 63, 66, 126, 132, 185, 186, 210, 233n35; Chicago Pile (CP-1), 50f, 51, 57, 97, 98, 202; GLEEP, 66; heavy water pile, 53, 126; ZEEP (Zero Energy Experimental Pile), 54, 125. *See also* nuclear reactor; Nuclear Reactor X (NRX); pile physics

atomic secrets: historiographical perspectives on, 6–7, 100, 130, 211, 212

atoms, 8, 23, 48, 59

Atoms for Peace, 160, 173

atom spy cases, 7, 145, 148

Attlee, Clement, 124, 177; and positive vetting, 134–40 (*see also* positive vetting)

Auger, Pierre, 33, 53f, 76, 78, 186

Austria, 165–67

Bakker, Cornelius, 28
Balls, E. K., 76–77
Baruch, Bernard, 159
Bedford, Ronald, 139
Bergman, Harold, 130
Beria, Lavrentiy, 183–84, 188
Bernal, John Desmond, 157, 168
Bernard, Lawrence: and slow neutron patent proceedings, 101–2; and slow neutron patent settlement, 129–32; and suit on slow neutron patent, 106–13
Berti, Giuseppe, 31, 75, 86, 229n61
beryllium, 25, 45
beta decay. *See* beta process
beta process, 59–62
beta spectrum, 187. *See also* beta process
biography: and historiography, 9; Pontecorvo's biography *Il lungo freddo*, 178, 186
Blackett, Patrick, 158
Blum, Léon, 29, 30, 33
Bohr, Niels, 24f, 48
Borden, William L., 129, 130
Boskey, Bennett: and Fermi's conflict of interest, 108–10, 131, 245n42, 246n51; and response to Giannini's claim, 102–4; on security in AEC legal proceedings, 105–7
Bretscher, Egon, 116, 120, 143
Britain, 1, 2, 3, 4, 19, 35, 37, 138, 157, 174, 218; and Anglo-American relations, 121–24, 127, 188, 205, 210; and Anglo-Italian relations, 143; coverage of the Pontecorvo affair in, 120, 145–47, 198; foreign scientists and, 78, 87, 214, 218; Pontecorvo's flight from, 69, 117, 120, 124, 141,

153, 161, 164, 176; Pontecorvo's permanence in, 66, 86, 144; refugees in, 71–72, 77, 141, 143; and the rescue of French scientists, 51–52, 71; security provisions in, 138–40, 206
British Central Scientific Office (BCSO), 53, 75–76, 82
British Joint Services Mission (BJSM): and Pontecorvo's disappearance, 123–25; 248n9
British Parliament, 123, 175
British Security Coordination (BSC): and Pontecorvo's vetting, 76–77, 134; 80, 148, 201, 213
Buffarini Guidi, Guido, 16
Burgess, Guy, 140, 200
Bush, Vannevar, 49, 74; and negotiations on the slow neutron patent, 98–102

Cabinet Committee on Subversive Activities (GEN 183), 138–39
Cabinet Office (British government), 123
Cabinet Sub-Committee on Positive Vetting, 139
Cassin, Lydia, 148
Cavendish Laboratory (University of Cambridge), 23, 35, 53, 91
Central Intelligence Agency (CIA), 78, 134, 167, 250n53
Centre National de la Recherche Scientifique (CNRS), 34
Chadwick, James, 24; and Anglo-American wartime collaboration, 79–82; and Pontecorvo's appointment at the University of Liverpool, 91–92, 108
chain reaction: in a metaphorical sense, 88, 89f, 98; self-sustaining (or divergent), 48–52
Chalk River: cosmic rays work, 60–64, 66, 82, 83, 142, 149, 202, 203, 210; nuclear research laboratory, 53–56, 56f
Chamberlain, Owen, 176

Chaplin, Charlie, 157
Charles E. G. Howard, twentieth Earl of Suffolk, 51, 71
Cheliabinsk-40, 185
China, 127, 189, 211, 218; and prospecting uranium, 191–92
chlorine, 193, 195
chlorine-argon method, 61, 235n76
Christian Democratic Party (Democrazia Cristiana), 155, 161
Churchill, Winston, 52, 78, 79
Cimperman, John, 133, 135
clandestine: organization, 31, 90 (see Partito Comunista d'Italia [Communist Party of Italy, PCd'I]); migrations, 162, 166, 167; operations, 162–65
Cockcroft, John, 35, 62, 108, 119, 138, 213, 246n51; as head of Harwell, 66; as head of Tube Alloys, 53–58; and the Nunn May case, 80–81; and Pontecorvo's dismissal, 88–92; and Pontecorvo's investigation, 124–26
Cold War, 2, 65, 84, 85, 156, 158, 167, 170, 182, 206, 208, 211; escalation of, 181, 185; end of, 199; history of, 3, 4, 5–9, 10, 121, 198, 202, 205, 215, 217
Collège de France, 33, 76
Colorni, Eugenio, 16–17
Columbia University, 49, 97, 233n32; and its Department of Physics, 10, 79
Combined Development Trust (CDT), 53, 59, 64, 188, 189
Combined Policy Committee (CPC), 53, 80, 121
Commissariat à l'Énergie Atomique, 148, 157
communism, 32, 78, 84, 86, 149, 217
Communist Information Bureau (Cominform): decline, 172, 190; establishment and support to Peace Partisans, 162–65
Communist Party: in Czechoslovakia, 163; in Germany, 133; in the So-

Communist Party (*continued*)
    viet Union, 163, 164, 181; in the
    US, 87. *See also* Partito Comunista Italiano (Italian Communist Party, PCI)
Como, Lake, 3, 61, 90, 95, 143
Compton, Arthur H., 98, 202; and the Uranium Project, 49–51
Conant, James B., 49, 51, 202
Condon, Edward U., 85, 215
Congress (US), 3, 100, 126, 140
Consiglio Nazionale delle Ricerche (CNR), 29, 35, 36, 229n56; CNR Centro di Studio per la Fisica Nucleare, 64, 67; establishment and policy, 19–21; *La ricerca scientifica* (CNR journal), 23, 26–27
Coplon, Judith, 107
Corbino, Orso Mario, 24f, 35; and the discovery of slow neutrons, 26–29; and Fermi's group, 20–21
coring (carottage), 41, 43, 44, 231n2
Corriere della Sera, 112
cosmic rays, 31, 67, 124; research at Chalk River, 60–63
Costello, John, 200
counters: design of, 43, 45, 46, 231n10; scintillation, 63, 194, 195; use in basic research, 61; use in nuclear reactors, 63; use in prospecting activities, 57, 63–66, 194–95
Courcy, Kenneth De, 140, 145, 252n78
Court of Claims (US), 110, 111, 128, 245n41
Curie, Marie, 23, 33
Curie, Pierre, 23
cyclotron, 34, 35, 76, 91, 107, 230n70. *See also* particle accelerator
Czech Communist Party, 163
Czechoslovakia, 49; as focus of clandestine operations, 164–69; and uranium prospecting, 190–92, 211

D'Agostino, Oscar, 22f, 25, 27, 35, 103
Daily Express, 138, 143, 145, 200
Daily Telegraph, 190
Dallin, David, 199, 203
Darwin, Charles Galton, 53
Dean, Gordon, 125, 129
De Benedetti, Sergio, 31, 37, 86
De Gasperi, Alcide, 155
de Gaulle, Charles, 76, 157
Department of Atomic Energy (D.At.En.), 121–24
Department of State (US). *See* State Department (US)
Deputy Chiefs of Staff (DCOS), 82, 83
deuterium, 49, 185, 196
Devlin, Roger, 57
dialectical materialism, 10, 183
Directorate of Naval Intelligence, 192
Directorate of Tube Alloys (DTA), 51, 99, 212
Doll, Henry-George, 44
Dolomites, 15, 95, 165, 168
Donini, Ambrogio, 31, 75, 86
Donovan, William, 78
Dreyfus, Alfred, 10
Dubna, 174, 175, 179, 196, 198. *See also* Joint Institute of Nuclear Research (JINR)
Dunning, John Ray, 76
Dunworth, James, 64, 66, 82
Dwyer, Peter, 80

Edison company, 21
Einstein, Albert, 48, 145; and peace campaigning, 157, 159, 172, 174, 256n55
Eisenhower, Dwight, 160, 173
Eldorado mining company, 57–59
Endicott, James G., 192
enemy aliens, 71–74, 97. *See also* refugees
Erroll, Frederick J., 124
espionage, 2, 39, 67, 87, 135, 206; in Canada, 81–84, 93; counterespionage, 70, 80, 86, 87, 134,

146, 214; FBI espionage operations, 136–37; Pontecorvo's alleged espionage activities, 92, 132, 181–82, 198–203. *See also* atomic espionage
Evening News, 143

Fascism, 30, 74, 85, 155; Fascist regime in Italy, 14, 20, 27, 72, 73; and science policy, 16–21
Fearon, Robert: work for Well Surveys, 42–46, 197
Federal Bureau of Investigation (FBI), 2, 3, 5, 82, 87, 88, 90; and AEC security provisions, 104–8; Italian refugees and, 72–74; and MI5, 123, 125–26, 133–36, 140, 144, 145, 199, 200, 207, 211–14; search in Pontecorvo's house in Tulsa, 75–78, 92, 200–201; second FBI report on Pontecorvo, 83–86, 93. *See also* intelligence service
Federation of Atomic Scientists (FAS), 85, 214
Federzoni, Luigi, 73, 238n13, 238n14
Fergana Valley, Uzbekistan, 189, 191
Fermi, Enrico, 3, 8, 12, 20, 31, 32, 45, 47, 137; and administration of the slow neutron patent, 96–104, 110–12, 128–29, 131–32; conflict of interests, 108–10; exchanges with Pontecorvo in Chicago, 52–54; and his group in Rome, 14, 18, 20–29, 22f, 35–36, 38, 60; and JCAE inquiry, 130; on Pontecorvo's neutrino research, 61–62; and security operations, 73–76; and speculations on atomic espionage, 202; wartime contribution, 48–51, 185, 210
Fermi, Laura, 27
Filippov, Eugenii M., 197
Finland, 2, 127; and Pontecorvo's flight, 144, 150, 169–70
fissile (fissionable) materials, 3, 54, 66, 67, 100, 101, 126, 159, 210
Flerov, Georgy Nicolaevich: and oil prospecting in Russia, 195–96, 260n53
Foreign Office (UK), 77, 91, 176, 208; and management of the Pontecorvo affair, 121–27, 134, 207
Frank, Il'ja, 196, 260n56
Franks, Oliver: and strategy to downplay the Pontecorvo affair, 121–26, 206, 248n21
Front Populaire (Popular Front), 30, 32
Fubini, Eugenio, 25, 97, 102, 106, 132, 242n5
Fuchs, Klaus, 2, 39, 40, 67, 129, 130, 139, 147, 148, 185; consequences of Fuchs' arrest for Pontecorvo, 86–90; and his father, 90; speculations about involvement in espionage, 199–211. *See also* Fuchs, Klaus

Galilei, Galileo, 10, 15
Gardner, Meredith, 87
GEN 183. *See* Cabinet Committee on Subversive Activities (GEN 183)
gender bias, 148–49
General Electric, 21, 28, 234n60
Geological Engineering Company, 41
Geological Survey (UK), 59, 65
Geophysical Measurements Corporation (GMCO), 63, 64
Geophysical Research Corporation, 41
geophysicists: and oil prospecting in the Soviet Union, 193–97; and oil prospecting in the United States, 41
Giannini, Gabriello M.: and AEC settlement, 131–32; as assignor of the slow neutron patent, 27, 28, 35, 101, 102; and disinvestment from patent, 127–28; and petition against the AEC, 108–13, 114, 117, 144, 162, 169, 216
Ginzburg, Vitaly, 185, 188
Glavnoye Razvedyvatel'noye Upravleniye (GRU), 81, 167

Gold, Harry, 2, 87, 129
Goldschmidt, Bertrand, 33, 37, 48, 53f, 64, 80
Gorbachev, Mikhail, 181
Gordievsky, Oleg, 201, 203
Gouzenko, Igor: and Canadian spyring, 80, 82, 199, 200
graphite: and CP-1 experiment, 49–50, 54, 97, 186
Great Bear Lake, 57, 58f, 65, 190, 207
Green, William G.: and the establishment of Well Surveys, 41–44
Greenglass, David, 2, 87, 129
Gromyko, Andrei, 159
Group for Academic Freedom, 85
Groves, Leslie, 51, 59, 99, 137, 235n65; and wartime security provisions, 79–83
Guarin, Paul, 59
Guéron, Jules, 53f, 80, 81
Guttuso, Renato, 157

Hahn, Otto, 34
Halban, Hans Von, 33, 53f, 71, 99, 125, 187; recruiting Pontecorvo in Tube Alloys, 47–48, 51–53, 76, 79; and security restrictions, 78–82
Hanna, Geoff C., 62, 187
Harwell. *See* Atomic Energy Research Establishment (AERE)
Helsinki, 3, 116, 120, 126. *See also* Finland
high-energy physics, 175, 176, 180
Hincks, E. P. "Ted," 60, 61
Hollis, Roger, 80; and Pontecorvo's investigation, 125–26; and Pontecorvo's vetting, 135–36, 139, 200, 206, 213
Holloway, David, 187
Home Office (UK), 171, 175
Home Office, 171, 175
Hoover, J. Edgar, 73, 79, 87, 107, 201, 207; and Pontecorvo's vetting, 133–35; and security inquiries following Pontecorvo's defection, 136–37

Hope, John, 124
Horní Slavkov, 191. *See also* Czechoslovakia; prospecting: uranium
House of Commons, 3, 69, 123, 125. *See also* British Parliament
HUAC (House of Representatives Committee on Un-American Activities): enquiries on subversive activities, 73, 84–85, 104, 129
hydrogen bomb, 130, 185, 187. *See also* atom bomb; nuclear weapons: thermonuclear

Imme, Salvatore, 120
indium, 34, 64
Institut du radium (Institute of Radium, Paris), 23, 24
Institute of Physics (Rome), 14, 20, 24
Intelligence Digest, 140
intelligence service, 6, 7, 11, 133, 137, 167, 201, 209, 214. *See also* Federal Bureau of Investigation (FBI); MI5
inverse beta process. *See* beta process
Ioffe, Boris: revelations on Pontecorvo, 187, 192, 196, 211, 258n18
ionization chamber, 30f, 43, 45, 46, 57
Iron Curtain, 84, 162, 178, 194; and Pontecorvo's flight across, 2, 3, 69, 117, 130, 178, 179–80
Italy, 3, 8; transition to Fascism, 15–21, 27, 28, 29, 31; Jews in, 36–38, 72, 73, 82, 96, 106, 108, 112, 115, 120, 129, 143, 144, 155, 157; and peace campaign, 158–64, 165, 166, 167, 177, 181, 216, 217
Izvestia, 175

Jáchymovské doly (DIAMO), 191, 259n34
Jews, 14, 15, 17, 36, 37
Joe-1, 123, 160, 185
Johnson, Edwin C., 100

Joint Committee on Atomic Energy (JCAE): establishment and activities, 100, 104, 109, 243n16, 243n18; and investigation on atomic espionage, 54, 129–33, 139, 144, 187, 207
Joint Institute of Nuclear Research (JINR), 180. *See also* Dubna
Joliot-Curie, Frédéric, 24, 25, 29, 51, 64, 71, 99, 169, 172, 177, 215; and peace campaign, 157–58; and research in Paris, 32–34; and security provisions, 76–80, 86
Joliot-Curie, Irène, 24, 32, 33, 34

Kamenice. *See* Příbram
Khariton, Yulii, 185
Khrushchev, Nikita, 174
Kirkwood, David H., 62, 187
Kojevnikov, Alexei, 189, 258n16, 258n23
Komitet Gosudarstvennoy Bezopasnosti (KGB), 135, 199, 201
Kowarski, Lew, 33, 51, 54, 71, 78, 99; and surveillance operations, 80, 148–49
Kramish, Arnold, 202, 203
Kremlin, 3, 174, 187
Kuomintang (KMT), 191, 259n38
Kurchatov, Igor Valisyevich: and F-I reactor, 185–86; and prospecting needs, 188–89, 195

L'Unità, 112, 114, 225n19
Labine, Gilbert, 57. *See also* Eldorado mining company
Lamphere, Robert, 87
Langevin, Paul, 33, 37, 86
Lattice Group, 53–54, 66. *See also* Chalk River; Tube Alloys
Lavender, Robert A.: management of atomic patents, 98–101, 243n18, 244n20
Lawrence, Ernest Orlando, 35, 76, 107, 230n76
Lazard, Andrè, 34, 230n71
League of Nations, 28

Lewis, Wilfrid B., 62
Liddell, Guy: briefing with prime minister, 127, 134–35; investigation on Pontecorvo's vetting, 77–78
Lilienthal, David, 105, 125
Lindemann, Frederick, 79
lithium deuteride, 188
Lloyd George, Gwilym, 175
log: combined, 193; Compensated Neutron Log (CNL), 193; Epithermal Neutron Log (EPL), 193–94; gamma, 44, 46; neutron, 45, 46, 63, 64. *See also* well logging
Los Alamos: and administration of patents, 98–100; establishment and activities, 51, 55, 76, 111; and security investigations, 87, 129–30, 202
loyalty, 32, 74, 106, 111, 113, 127, 129, 131, 213, 216, 217, 240n55; federal loyalty program, 84–85, 89, 138; loyalty of scientists in Britain, 126, 139. *See also* witch hunts
Luria, Salvatore, 31, 37, 85, 86
Lysenko, Trofim, 10

Mackenzie King, William Lyon, 52
MacLean, Donald, 140, 200
Mafai, Miriam, 16, 178, 186
Majorana, Ettore, 21, 24, 27, 228n34
Makins, Roger: and strategy to downplay the Pontecorvo affair, 121–26, 122f, 206
Malenkov, Georgy, 173
Manchester Guardian, 138
Manhattan Project, 5, 51, 58, 185; and administration of patents, 97–98; and security enquiries, 76–79, 130, 199, 203, 211. *See also* Los Alamos; wartime nuclear projects
Mann, Wilfrid Basil, 142, 149
Marconi, Guglielmo: and patent legislation, 19–20, 22f, 35

Marshak, Robert, 60
Marshall Plan, 83
Matusel, Milan, 166
MAUD committee, 51, 233n39
McCarthy, Joseph R., 2; consequences of his campaign, 104, 117, 125, 128, 214; McCarthyism, 111; and the witch hunts campaign, 85–87. See also witch hunts
McMahon Act. See Atomic Energy Act
McMahon, Brien, 100, 109, 129. See also Atomic Energy Act
meson, 60–61, 175
Messaggero, Il, 112
Messe, Giovanni, 177
Metropolitan-Vickers, 35, 66
MI5, 11, 69, 77, 80, 88, 92, 124, 126, 139, 164, 192, 201, 208, 211, 214, 215; counterespionage section, 75, 86, 87, 90; and enemy aliens, 71–72, 54; and FBI, 133–36, 137, 140, 199, 200; and investigation in Pontecorvo's disappearance, 142–48. See also intelligence service; Scotland Yard
MI6, 77, 80, 134, 142, 143, 148, 201
Mikhailov, Sergei, 166
Ministry of Supply, 65, 91, 92, 121, 144, 210. See also Department of Atomic Energy (D.At.En.)
Modus Vivendi, 122
Montgomery Hyde, Harford: on Pontecorvo's espionage activities, 199–201; on Pontecorvo's holiday, 113–14; work of: *The Atom Bomb Spies,* 114, 199
Montreal (Canada): Pontecorvo's move to, 47–48; and security investigations, 76–81, 199, 200; and Tube Alloys, 52–54, 187; and uranium prospecting, 57–58
Moorehead, Alan: on Pontecorvo's holiday, 114; *The Traitors,* 146, 199, 209, 253n94; work for MI5, 145–46

Moscow (Russia): and clandestine operations, 31, 163–67; Pontecorvo's flight to (and permanence in), 170–74, 176, 187; research facilities in, 78, 150, 186, 195, 202, 208
Moscow gold (*Oro di Mosca*), 165, 170
Mountford, James : and Pontecorvo's move to Liverpool, 91–92
muon. See meson
Murray Hill Area, 59, 235n72
Mussolini, Benito, 16, 19, 31

Narodnyy Kommissariat Vnutrennikh Del (NKVD), 183; espionage operations, 201–2
National Research Defense Committee (NRDC), 49
Natoli, Antonio, 167
naturalization, 134, 139, 176
Nature, 25, 26, 34, 60, 66, 124
Nehru, Jawaharlal, 174
Nenni, Pietro, 156, 161, 177
Neruda, Pablo, 157
Neufeld, Jacob: work for Well Surveys, 42–44, 63, 136, 194
neutrinos, 59–61, 67, 180, 209, 236n87
neutrons, 47, 209; in nuclear reactions, 24–25, 48, 54; in oil prospecting, 44, 49, 52, 97, 193, 196, 197; slowing down of, 25–28, 49, 97, 193
New York, 27, 48, 75, 76, 77, 135, 201, 202; Pontecorvo's visits to, 37, 47, 74, 209
New York Times, 112
Niepce, Henriette, 148
Nobel Prize, 36, 48, 73, 108, 132, 176, 196
Nordblum, Marianne: flight to Soviet Union, 96, 115–16, 142, 170; marriage with Bruno Pontecorvo, 31, 37, 229n59; planned move to Liverpool, 161, 168; suspicion about her, 147, 149

INDEX

North Atlantic Treaty Organization (NATO), 84, 143, 157, 171, 177, 197
Northwest Territories (Canada), 57, 189
nuclear diplomacy, 121, 159, 171
nuclear disarmament, 112, 172–73
nuclear geophysics: Pontecorvo's expertise in, 64, 67, 97, 181, 210, 211; advances in the Soviet Union, 194–98. See also nuclear physics
nuclear physics, 3, 8, 11, 34, 67, 90, 96, 119, 124, 174, 183–84, 197, 206, 209; advances in Italy, 13, 14, 20–29, 33, 38, 108; application to prospecting problems, 40–43; history of, 209; military applications, 48, 51, 55. See also nuclear geophysics
nuclear proliferation: campaigns against, 84, 111–12, 154–59, 171, 172, 173, 208, 215, 216
nuclear reactor, 40, 48, 51, 53–55, 66, 67, 97, 110, 125, 130, 131, 181; heavy water reactor, 185–88, 210; and nuclear physics research, 60–63; Soviet reactor F-I, 185, 189. See also atomic pile; Nuclear Reactor X (NRX)
Nuclear Reactor X (NRX), 53–55, 59, 60, 63, 66, 125, 126, 130, 186. See also nuclear reactor
nuclear test, 2, 160, 173. See also Trinity; Joe-I
nuclear warfare, 159, 174
nuclear weapons, 1, 79, 112, 158, 159, 162, 172, 184, 216, 218; campaign against, 154, 157, 160, 171, 177; thermonuclear, 109, 173. See also atom bomb; peace campaign
Nunn May, Alan, 2, 81–83, 138, 147, 187, 215

Operation Borodino, 183, 184
Oppenheimer, J. Robert, 13, 55, 130, 215

OSRD (Office of Scientific Research and Development), 49, 51, 57, 74, 98
OSS (Office of Strategic Services), 78
OVRA (Fascist secret police), 72

Pangh de Co Co, 144–46, 166, 191
Paris, 11, 15, 40, 47, 51, 71, 76, 86, 90, 149; Pontecorvo's permanence in, 29–37, 155, 177, 209, 216; Scherbatskoy's visit to Paris, 136–37; suspects about the permanence, 88, 106, 142, 177; World Peace Congress, 157–58
Parlamento (Italian Parliament), 3, 165, 177
Parravano, Nicola, 19
particle accelerator, 91, 196, 215. See also cyclotron
particles, 8, 23, 24, 33, 34, 35, 181; in the context of Pontecorvo's research program, 56–61
Partigiani della Pace (Peace Partisans, PP), 142, 154, 156, 157, 159, 161, 163, 169, 171, 172, 173, 177, 254n3
Partito Comunista d'Italia (Communist Party of Italy, PCd'I), 16, 17, 31, 32, 90
Partito Comunista Italiano (Italian Communist Party, PCI): clandestine operations, 90, 160–72; executive committee, 126, 154, 163, 172, 173; and peace campaign, 142, 160–63, 177; Sereni's role in, 155–56.
Partito Nazionale Fascista (National Fascist party, PNF), 16, 19, 20
Partito Socialista Italiano (Italian Socialist party), 155, 156, 161, 177
patents, 8, 11, 14, 26, 41; French patents on fission, 80, 99; military provisions in the United States, 98–99; Marconi's patent policy, 19; military provisions in the United States, 98–99; patenting activities, 8, 29, 35, 97; on

patents (continued)
  prospecting methods, 45–46, 64, 194, 209; provisions in the Atomic Energy Act, 100–101; and security, 104–8. See also slow neutron patent
Patterson, Geoffrey T. D.: on consequences of Pontecorvo's defection, 133–35; on De Courcy's article, 140–41
Pauli, Wolfgang, 59
Paxton, Hugh C.: acting as referee for Pontecorvo's employment, 76, 79, 107, 137
peace appeal: Pontecorvo's appeal, 215; Stockholm appeal, 157–61, 215, 254n7; Warsaw appeal, 172.
peace campaign, 157–63, 168, 169, 170; lacking luster, 172–76
Pegram, George B., 48
Peierls, Rudolf, 51, 139
Perrin, Francis, 33
Perrin, Michael, 77, 99; and strategy to downplay the Pontecorvo affair, 121–26, 130, 135, 206, 210
persecution, 17, 85, 213, 217
Persico, Enrico, 21
Petroleum Institute (Moscow, Russia), 195–96
Philby, Kim: deceiving MI5, 200–201
Philips Gloeilampenfabriken: and the slow neutron patent, 20–21, 35, 229n50
Picasso, Pablo, 90, 157
pile physics, 40, 48, 52, 66, 67, 126, 187, 206. See also atomic pile; nuclear reactor
Pincher, Chapman: allegations on Philby's role, 200–201, 203; coverage of Pontecorvo's defection, 138–40, 145, 208
pion. See meson
Pisa (Italy), 15, 16, 17, 20, 21, 37
Placzek, George, 47, 78, 80, 186
plutonium: intellectual rights on plutonium production, 97, 103, 109–10, 132; plutonium device, 55; production with nuclear reactors, 54, 55, 130, 185; synthesis of, 50, 81.
Poincaré, Henry, 17, 21, 216
Pontecorvo, Anna, 37, 95, 164; interrogation of, 119, 141, 142
Pontecorvo, Antonio, 114, 115
Pontecorvo, Bruno, 2, 3, 6, 9, 13, 18f, 62f, 102f; and advent of Fascism, 17–18; the alleged spy, 199–203; and employment in Tulsa, 44–47; flight to Russia, 120, 123, 127, 144, 168–70, 215–18; holiday in Italy, 95–96, 114–17, 142, 144; and leaving Europe, 36–37; and oil prospecting in Russia, 184, 186–88, 190–92; participation in Fermi's group activities, 20–26; and permanence in Paris, 30–33; permanence in Russia, 180–83; and planned move to Liverpool, 91–93; reappearance in Moscow, 174–76; religious identity, 15–16; research in Harwell, 66–67; security operations concerning, 69, 74–79, 82, 105–7, 85, 89–90; and the slow neutron patent, 29, 97, 101–2, 112–13, 128; and Soviet nuclear program, 184, 186–88, 190–92; and Tube Alloys, 51–65, 53f
Pontecorvo, Gil, 31
Pontecorvo, Gilberto (Gillo): and Bruno's disappearance, 88–90; and Bruno's holidays, 96, 113, 114; on permanence in Paris, 32; on religious identity, 15
Pontecorvo, Giovanni David, 37, 102, 142
Pontecorvo, Guido, 17, 22, 36, 72, 120, 141, 142
Pontecorvo, Laura, 141, 143, 148
Pontecorvo, Massimo, 16, 37, 119, 141
Pontecorvo, Paolo (Paul Jacob), 17, 36, 136
Pontecorvo, Pellegrino, 15, 37

Pontecorvo affair, 3–5, 11; coverage in newspapers, 137–39, 146–50; Fuchs affair, 39, 86–92, 121, 122, 133, 149; Gouzenko affair, 82, 199; historiographical perspectives, 9–10; management by British diplomats, 120–21, 206–8; Pontecorvo affair, 70, 75, 77, 89f, 113, 205–6, 217–19; security implications, 125–29, 134–36, 212–15; Sereni and the Pontecorvo affair, 165–71
Pontecorvo family, 116, 141, 149
Popov, Piotr, 167
Port Radium. *See* Great Bear Lake
positive vetting, 138–40, 213. *See also* security purge
Pravda, 153, 154, 160, 175, 189, 208
Pregel, Boris, 47, 57
Příbram, 190–91, 259n34
Proceedings of the Royal Society, 25
Progressive Party, 86
proportional counting technique, 62, 187, 188
prospecting, 40, 63, 126, 184, 188, 206, 207, 209; uranium, 4, 55–59, 64–67, 188–93, 210, 211; oil, 40–45, 46, 47, 49, 56, 193–98
Pu-239. *See* plutonium
Pugwash Conference on Science and World Affairs, 91, 174
purge procedures. *See* positive vetting
Pyke, Sumner T., 109

Quercia, Italo Federico, 64

Rabi, Isidor Isaac, 109, 215
Rabinowitch, Eugene, 85
radioactivity, 23, 28, 29, 30, 36, 43, 44, 63, 210
radioisomers, 33–34
radioisotopes, 14, 23, 28, 29, 34, 35, 48, 50, 54, 60, 62, 97, 103, 110, 185–86, 196, 209, 210
radium: re-use of sources in Rome, 23–25; in prospecting research, 45–47, 57

Rapkine, Louise, 76, 137
Rasetti, Franco: research in Rome, 21–27, 22f, 35; and administration of slow neutron patent, 101, 108; and JCAE investigation, 129–32.
Raytheon Corporation, 36, 136
Reed, Ronald ("Ronnie") T.: and investigations on Pontecorvo's disappearance, 141–46, 151, 154; and revelations on Pontecorvo, 164–66, 180, 191, 194, 203
refugees, 40, 70, 77, 212; academic, 36, 48, 97, 139; in Britain, 71–72; political, 31, 72; in the United States, 72–75. *See also* enemy aliens
Rentschler, Harvey, 28
Rhodes, Richard, 202
Robertson, James C., 75, 86, 146, 166, 167
Rome (Italy): Fermi's group in, 20, 21, 28, 29, 35, 74, 130; Imme's press conference in, 120, 123; 1932 conference on nuclear physics, 23, 24f; Pontecorvo's holidays before defection, 95, 114–15, 165–72; Pontecorvo's permanence in, 14, 15, 17, 31, 33, 34, 97, 209; World Peace meeting, 157–61
Roosevelt, Eleanor, 137
Roosevelt, Franklin Delano, 48, 52, 74
Rosenberg, Ethel, 2, 87
Rosenberg, Julius, 2, 87
Rotblat, Joseph, 91
Royal Canadian Mounted Police (RCMP), 82
royalties, 28, 97, 98, 132, 198, 260n65
Russell, Bertrand, 157, 164, 191, 256n55
Russell, William, 42, 64
Rutherford, Ernest, 23, 25, 26

Sakharov, Andrei, 181, 185
Sawbridge, Eric, 144, 147
Scandinavian Airlines (SAS), 115, 116, 165, 170

INDEX 290

Scherbakov, Dimitri, 189
Scherbatskoy, Serge Alexander: FBI security investigation, 136–37; prospecting at Great Bear Lake, 57; and slow neutron patent, 97, 102, 197–98; work for GMCO, 63–65, 141, 191, 194; work for WSI, 42–44, 43f
Schlumberger, Inc., 41, 43, 44, 194
Scotland Yard, 3, 200, 208, 250n47; establishment and activities, 125; investigation on Pontecorvo's defection, 140–49, 167, 169, 190; security action in the Pontecorvo case, 133–37
Seaborg, Glenn, 50, 55, 109
Secchia, Matteo, 165–69
Secchia, Pietro, 165–69, 172
secrecy, 13, 183, 205–6, 211, 215, 218; and management of the Pontecorvo's affair, 67, 70, 122, 135, 207–9; and scientific research, 5, 6; secrecy orders (on patents), 98
security hysteria, 104, 152
security purge, 85, 138, 162, 213. *See also* positive vetting
Segrè, Emilio, 3, 5, 11, 17, 22f, 35, 36, 44, 49, 61, 74, 85, 99f, 109, 110, 132, 182f, 216–17; meeting with Pontecorvo in Moscow, 176; and negotiations on slow neutron patent, 97–102; participation in Fermi's group, 21–28; and reaction to Pontecorvo's disappearance, 127–29; and security inquiry on Pontecorvo, 106–8, 113, 131, 141, 142, 199, 208, 213, 215
Seismic Services Corporation (SSC), 42, 44
Seligman, Henry, 53f, 143, 149
Semyonov, Semyon, 202
Seniga, Giulio: and clandestine PCI operations, 165–70
sensationalist account, 138, 140, 146
Sereni, Emilio, 17, 31, 126; and the decline of peace campaigning, 171–74; downfall, 176–78; role in PCI covert operations, 162–66; role in Pontecorvo's defection, 154–56, 156f, 166–70; Segrè, Pontecorvo and, 216–17
Sereni, Enzo, 17, 72
Sereni, Lea, 167
Shapley, Harlow, 85
Shawcross, Hartley, 81, 87
Shell, 45, 59
Sillitoe, Percy: and the management of the Pontecorvo affair, 133–35, 192, 206, 305; on MI5 need to inform the public opinion, 140, 145
Silverberg, Xenia, 31
Sinkiang province (China), 192
Skardon, William James, 87, 133
Skinner, Erna, 91, 148
Skinner, Herbert Wakefield Banks: on Pontecorvo's planned move to Liverpool, 91–92, 108; writing on the Pontecorvo affair, 145–49, 194, 120
slow neutron patent, 8, 11, 144, 176, 197, 207, 217, 228n44; AEC proceedings and settlement, 101–4, 121, 127–33, 198; early negotiations, 26–29; petition before US Court of Claims, 111–13, 115, 117; wartime negotiations, 96–100. *See also* patents
Smith, Ralph Carlisle, 98, 99
Snow, Charles Pierce (C. P.), 146, 147
socialism, 14; and science, 32–33, 38, 177, 180–81
Society for Protection of Science and Learning, 36
Soddy, Frederick, 34
Sokal, Alan, 10
Soviet Academy of Sciences, 154, 176, 180, 181; and fission research, 185, 188; and Institute of Economics, 197; and VNIIYaGG, 195
Soviet Union, 2, 3, 4, 38, 40, 41, 64, 73, 78, 81, 84, 87, 96, 130, 140,

145, 177, 201, 207, 208, 210, 211, 214, 217; and clandestine operations, 153–55; and oil prospecting, 195–98; and peace campaign, 159–60; Pontecorvo's defection to, 170–74; Pontecorvo's permanence in, 179–88; and uranium prospecting, 189–91
spiv, 146, 147f
spy, 81, 93, 107; networks, 80, 87, 212; Pontecorvo as an (alleged) spy, 2, 4, 9, 121, 137–39, 144, 145, 146, 147, 150, 180, 182, 198–202, 205, 214. *See also* atom spy cases
Stalin prize, 180
Stalin, Joseph, 10, 78, 140, 145, 202; and Cominform activities, 165, 172–74; on the inevitability of war, 159, 160; on peace campaigning, 163; and Soviet nuclear program, 183–85, 188
Standard Oil, 41, 44
State Department (US), 87, 103, 122–23, 137
Stephenson, William, 77, 80
Stockholm (Sweden), 73; and Pontecorvo's flight to Russia, 115–16, 164, 170
Stott Mr., 77; and security documents about Pontecorvo (Stott papers), 133–34, 200, 201
Strauss, George: and parliamentary briefings on Pontecorvo, 39, 69–70, 123–26, 135
Strauss, Lewis, 109
subversive activities, 84
Sudoplatov, Pavel, 201–3
surveillance: and enemy aliens, 71, 73–74; and French scientists, 80, 84, 136, 149, 217
Suslov, Mikhail, 163, 173
Sweden, 64; and Pontecorvo's flight, 115–17, 142, 169, 170
Swift, Gilbert, 42
synchrocyclotron. *See* cyclotron
Szilard, Leo, 13, 29; on patenting activities, 34–35, 98–99; and Uranium Project, 48–52

Taylor, Dennis, 65
Telecommunications Research Establishment (TRE), 65
Teller, Edward, 202
Tempo, Il, 177
Texaco, 41
Thatcher, Margaret, 199, 201
Thomas, John Parnell, 84; on security flaws, 104–5
Thompson, Lesslie, 57
thorium, 43, 54, 57, 65
Thornton, Robert: and revelations about Pontecorvo, 106–8, 128, 135, 199, 213
Times, The, 114, 138, 145
Tizard, Henry, 52, 78
Togliatti, Palmiro, 165, 169, 173
Trabacchi, Giulio Cesare, 24, 27, 103
transmutation, 23, 29, 35
Trinity, 79
Tripartite Security Conference, 139, 140
tritium, 62, 196; and hydrogen bomb, 130, 185–88
Truman, Harry S., 83, 84, 86, 138
Tube Alloys, 11, 52–53, 57, 121, 138, 200, 203; Directorate of (DTA), 48; security administration in, 75–79, 82, 137. *See also* wartime nuclear projects
Tulsa (Oklahoma), 3, 63, 136, 203, 232n13; FBI search in Pontecorvo's house in, 75–77, 93, 200; prospecting research in, 40–44, 57
Tyrrhenian Sea, 15

U-233. *See* uranium
U-235. *See* uranium
United Kingdom Atomic Energy Authority (UKAEA), 122. *See also* Department of Atomic Energy (D.At.En.)
United States, 1, 3, 4, 6, 27, 35, 36, 37, 42, 48, 52, 59, 80, 82, 96,

United States (*continued*)
98, 100, 103, 117, 141, 157, 160, 162, 183, 188, 193, 194, 197, 199, 206, 207, 210, 215, 216, 217, 218; and reactions to Pontecorvo's disappearance, 120, 127, 129, 136, 140; and refugees, 72–74, 75, 76, 78, 97; and security purges, 84–90, 106
University of California, 23, 85
University of Chicago, 48, 50, 97, 203
University of Liverpool, 3, 90, 108, 133
Uranium Committee, 48, 49. *See also* Uranium Project
Uranium Project, 49, 51, 74, 97, 98
uranium, 40, 43, 122, 218; and its fission, 48–52; and Soviet nuclear project, 184–86; search for, 56–60, 64, 65, 67, 188–92, 207, 210 (*see also* prospecting: uranium); and Tube Alloys, 52–56

Vallon, Louis, 32, 230n64
Van De Graaff generator, 33, 34
Veall, Norman, 82
Venona code, 107, 202
Vishinski, Andrei, 160
Volkoff, George M., 54
Volpe, Joseph, 109

Walton, Ernest, 35
wartime nuclear projects, 3, 6, 13, 40, 47, 185. *See also* Manhattan Project; Tube Alloys

Washington Post, 112
Waters, Roger, 143, 208
well logging, 43, 63, 110; neutron well logging, 44–47, 45f, 52, 56, 63, 64, 67, 97, 192, 193–97, 210. *See also* log
Well Surveys Incorporated (WSI): oil prospecting research, 44–47; uranium prospecting research, 56–57, 63, 197
Wenhao, Weng, 192
West, Rebecca, 145, 164, 208
Westinghouse, 28
Wheeler, John, 48
Wick, Gian Carlo, 21, 85
Winnifrith, John, 139
witch hunts, 2, 83, 84, 112, 213, 240n55. *See also* anticommunism; loyalty; persecution; security purge
Wolfenden, John H., 75–77
World Peace Congress, 157–59, 163, 168
World Peace Council, 172, 176, 256n50
Wright, Peter, 200, 261n73

Xuetong, Li, 192

Yerozolimski, Boris, 195, 196
Yukawa, Hideki, 60, 61

Zabotkin, Nikolai, 81
Zel'Dovich, Yakov, 185
Zinn, Walter, 48
Zubok, Vladislav, 202